American Funeral Cars & Ambulances Since 1900

By Thomas A. McPherson

Editing and Design
By George H. Dammann

Box 48, Glen Ellyn, Illinois 60137

OTHER CRESTLINE AUTO BOOKS:

ILLUSTRATED HISTORY OF FORD
(1,400 Illustrations)

50 YEARS OF LINCOLN-MERCURY
(1,600 Illustrations)

60 YEARS OF CHEVROLET
(1,650 Illustrations)

70 YEARS OF BUICK
(1,900 Illustrations

AMERICAN FUNERAL CARS & AMBULANCES Since 1900

By Thomas McPherson

Copyright © By Crestline Publishing Co., 1973

Library of Congress Catalog Card Number 73-80009

ISBN Number 0-912612-05-3

All rights to this book are reserved. No part of this book may be reproduced in any manner whatsoever without express written permission of the author. For further information, contact Crestline Publishing Company, Box 48, Glen Ellyn, Illinois.

Typesetting by George Munson, Naperville, Ill.

Proof and Copy Reading by Linda Hallstrom, Addison, Ill.

Printed in U.S.A. by Wallace Press, Hillside, Ill.

Binding by The Engdahl Co., Elmhurst, Ill.

Cover Design by Bob Williams, Arlington Heights, Ill.

Published by Crestline Publishing Company
Box 48 Glen Ellyn
Illinois U.S.A. 60137

Many Thanks

When this book was first contemplated in the late 1960's it was thought to be an easy task. The author already had what was considered to be the largest collection of pictures and information on this subject known to exist.

Unfortunately, there were some gaps in this collection and its coverage of all of the manufacturers. Several years were spent searching for the information and other material needed to fill these gaps. During this search period it was found that there was not what could be called an abundance of material around pertaining to these vehicles. Further, it was found that even some of the surviving manfacturers did not have complete files. In the end it was necessary to make several trips to various parts of the country to assemble a complete, documented history of these vehicles.

The fascinating material that you are about to view could not have been presented without the co-operation and assistance of many people who contributed freely of their time, knowledge, and in some cases opened their own personal collections and archives for us. Their co-operation greatly assisted us in our effort to make this the foremost authoritative text ever compiled on the subject.

Special thanks are in order to the following people..........Walter M.P. McCall of Windsor, Ontario, a long-time friend and partner in the research and collection of funeral coach and ambulance material and who assisted in compiling this text.

Charles A. Eisenhardt Jr., who wrote the foreward for this book and is the Chairman of the Board, Hess & Eisenhardt Company, Cincinnati, Ohio.

Robert Rossiter, Sales Manager, Miller-Meteor Division, Wayne Corporation, Piqua, Ohio

Dudley Starr, Public Relations Manager, Wayne Corporation, Richmond, Indiana.

Darrel Matzger, General Sales Manager, Funeral Coach and Ambulance, Superior Coach Corporation, Lima, Ohio.

William R. Rousch, Manager, Advertising and Sales Promotion, Superior Coach Corporation, Lima, Ohio.

Joseph A. Diehl, Manager, General Sales Office, Flxible Company, Loudonville. Ohio.

James Kloss, Photographic Department, The Flxible Company, Loudonville, Ohio.

William Knight, Director, Public Relations, Cadillac Division, General Motors Corporation, Detroit, Michigan.

John Conde, American Motors Corporation, Detroit, Michigan

James J. Bradley, Director, The Automotive History Collection, Detroit Public Library, Detroit, Michigan.

Dr. Charles H. Nichols, Director, National Foundation of Funeral Service, Evanston, Illinois.

Howard C. Raether, Executive Director, National Funeral Directors Association, Milwaukee, Wisconsin.

Messers. Habenstein and Lamers, authors of *The History of American Funeral Directing*, published by the National Funeral Directors Association, 1962.

Howard Barnard, Editor, *Casket and Sunnyside Magazine,* New York, New York.

Frank Bradicich, Advertising Manager, *Casket and Sunnyside Magazine,* New York, New York.

Thelma Workman, Executive Secretary (retired), Miller-Meteor Division, Piqua, Ohio.

Others deserving of thanks in varying degrees are ....Jim Hiltz, National Casket Company, Boston; Crane & Breed Company, Cincinnati; Dwain Cotner, Cotner Bevington Division, Wayne Corporation, Blythsville, Ark.; Dominion Manufacturers Limited, Toronto; J. Barnett, General Manager, Woodall Nicholson Limited, Halifax, England; D.H. Hackett, Company Secretary, Coleman-Milne Limited, Bolton, England; G.R. Freeman, Fleet Sales Manager, Stratstone Limited, London, England; Mr. Francois M. Chatillion, Roblot S.A. Paris, France; Morddeutsche Karosserlifabrik Kinrad Pollmann, Bremen, Germany; H.F. Gruhn, Director, The Gruhn Funeral Home, Kitchener, Ontario; Francis Janisse, the Janisse-Bisnett Funeral Home, Windsor, Ontario; Richard Scott, Conestoga College, Waterloo, Ontario, and Ronald E. Schaaf, Shelby, Ohio. Finally I would like to extend many thanks to all of the others, too many to mention, who assisted in any ways and to all of my friends, business associates, and family for their help and confidence. Most of all thanks to my wife Cindi for tolerating this rather unique interest in these vehicles and the inevitable mess that any book entails.

Photo Credits

The major portion of the photographs and illustrations used in this book were taken from the author's private collection while some were obtained from the following outside sources:

The Public Relations Department, The Flxible Company, Loudonville, Ohio.
The Public Relations Department, Superior Coach Corporation, Lima, Ohio.
The Public Relations Department, Miller-Meteor Division, Piqua, Ohio.
The Hess and Eisenhardt Company, Cincinnati, Ohio.
The Public Relations Department, Cadillac Motor Car Division, Detroit, Michican.
The Public Relations Department, American Motors Corporation, Detroit, Michigan.
The Automotive Collection, Detroit Public Library, Detroit, Michigan.
The National Foundation of Funeral Service, Evanston, Illinois.
The private collection of Walter M.P. McCall, Windsor, Ontario, Canada.
The private collection of John A. Conde, Detroit, Michigan.
The private collection of Robert Rossiter, Piqua, Ohio.
The private collection of Crestline Publishing, Glen Ellyn, Illinois.
The Public Relations Department, Norddeutsche Darosserieeabridk Konrad Pollmann, Bremen, Germany.
(A portion of the photo processing for this book was done by John Faul Jr., New Hamburg, Ontario, Canada.)

1911

The very first motorized funeral coach to be used in the nation's motor city, Detroit, was this unit owned by A. Van Leberghe of that city. The car was built by the Alden-Sampson Motor Truck Company that was a division of the United States Motors Company. Note the large glass window on the rear body side features a rayed or sunburst type drape in it, that the driver sits in the open, and the large coach lamps.

Preface

For obvious reasons of public sensitivity and our natural inclination to avoid any thought of our own mortality, the hearse and ambulance field has been the most neglected chapter in the entire history of transportation. The solemn, slow-moving funeral procession is among man's oldest rites, and the fascinating story of the evolution of the sleek funeral coach and ambulance of today deserve to be properly chronicled. This book, the only work of its kind ever attempted, is intended to accomplish this task.

At the turn of the century, when the automotive age was still in its infancy, nearly every buggy and carriage shop in business offered in its product line custom-built hearses, ambulances, and casket wagons. Today many of our leading hearse and ambulance manufacturers can trace their heritage back to one of these early buggy shops. The products of these establishments were magnificent horse-drawn funeral coaches with ornate carvings or plumes. Some of these coaches were even gilded. They were available in a vast array of sizes, from the small child's hearse to mammoth cathedrals on wheels; in shapes, from rounded ends to square boxy designs and in various styles. But, just as today's cars, these vehicles were subject to style changes. Styles for these vehicles maintained a longer life cycle — some ran for as long as ten years in any one form, more than has ever been the case with the automobile.

By 1905 the automobile was much less of an oddity, and a few progressive funeral directors began to mount their former horse-drawn bodies on motorized chassis to keep abreast of the times. This was done in the funeral director's livery or by a local carriage shop. But these transformations were crude at best and generally inadequate. More innovative funeral directors and ambulance operators sought professionally designed motor funeral service vehicles and ambulances designed from the ground up as hearses, ambulances, and casket wagons. They then went through the tiring process of designing and constructing their own cars. The auto hearse was already in use in many parts of Europe, where the automobile was in a much more advanced stage than that of its North American counterparts, and where auto funerals were common.

So, one by one, a few motorized hearses began to take to the streets to begin one of the most interesting chapters in transporation's history. Some of the major hearse manufacturers took notice of these early motorized professional vehicles, and began to realize that the motor car was more than just a passing fad. With this realization they began to turn out commercially-built "auto hearses" in the early summer of 1909. After the initial introduction of the first factory "undertaker's auto" by an Ohio firm, many other hearse makers got into the act with "auto hearses" of their own. Many of these firms simply transferred their horse-drawn coach bodies to motorized chassis much as individuals had done earlier. But a factory conversion is preferable to one done in a home garage, and funeral directors quickly realized this. Orders poured in to the makers as fast as local conditions would permit. Many times this simply meant as soon as a funeral director's local cemetery would allow these new-fangled machines within its gates. Many cemeteries had forbidden entrance to the noisy, dirty auto hearse in an effort to curtail its use. But the cemeteries were not alone in their protests over the new vehicles. Some people were of the opinion that the new auto hearse lacked the dignity of the horse-drawn funeral, and that speeds of up to 15 miles-per-hour were much too fast for any funeral procession and demonstrated an acute lack of respect for the departed.

Needless to say, as the years went by more and more cemeteries were forced to open their gates to these new machines and the market for them expanded enormously. Seeing a very lucrative area for a quick profit, many small companies then took up building funeral coaches and ambulances to a point where almost every other page in the funeral trade journals contained an ad for one coach company or another.

But building a funeral coach or an ambulance in these early days was anything but a simple chore. Before a company could even begin to construct bodies, they had to gather all of the components needed to assemble a car. Engines, clutches, transmissions, rear ends and bearings were all collected from one source or another until all the necessary parts had been obtained. Mounted within the maker's own frame and wearing his own body these coaches were called assembled cars and usually bore the nameplate of the assembler. These assembled cars dominated the scene until the early twenties, when the coach makers began using manufactured chassis of heavy passenger cars or light trucks. At this point Packard, Buick, Pierce Arrow, Studebaker and many other names began to appear more frequently on the pages of the trade journals as custom-built funeral coaches and ambulances.

Through the years the progress of the funeral coach and ambulance has paralleled that of the passenger car — they have developed together. Styling of the coaches has closely followed that of the passenger car with minor, but interesting, deviations. Within this context this is as much the story of the automobile as of the funeral coach and ambulance.

FOREWORD

"... Funeral customs develop as part of the social process; and, although their variety is legend, their growth is the consequence of such impinging factors as physical environment and social, economic, and religious conditions. From within man alone, as an aspect of human nature, comes the desire to care for the dead in the best of all possible ways. Funeralization in its optimum form permits the expression of such desire through a variety of socially approved patterns of behavior. The context of funeralization is dramatic, and its function, basically, is therapeutic. In brief, for all people everywhere, funerals and funeral ceremonies satisfy basic needs, allay suffering, and help rescue death from the horror of meaninglessness."

These sentences conclude the book *Funeral Customs the World Over* by Habenstein and Lamers (Bulfin 1963).

The evidence for this conclusion is found in the practice of most cultures which, since the beginning of recorded history, have publicly viewed their dead and buried them with ceremony.

The advent of television and telstar has permitted the peoples of many nations of the world to see the funeral service ceremony, including the processions for many great leaders — Kennedy, Churchill, DeGaulle, Eisenhower, Nasser, Nehru, Pearson, Pope John XXIII, Truman and Johnson. Sometimes horses and caisson were used. Once there was an army field piece. Once a boat. Another time a train. But in most instances a modern motor hearse such as those illustrated in this book was used at some point during the period of the funeral. Like the other forms of transportation utilized, it gave dignity to the person whose funeral was being held — it gave dignity to man.

The British statesman Gladstone once said,

"Show me the manner in which a nation or a community cares for its dead and I will measure with mathematical exactness the character of its people, their respect for laws of the land, and their loyalty to high ideals."

This "loyalty to high ideals" comes to us here in America as a part of our Judeo-Christian heritage from ages past. It came across the seas with the early settlers in their sailing ships, and with it came the many attitudes, traditions and customs that were a part of their original culture.

Prominent among these was the tradition of the formal religious funeral service and a procession to the church yard burial ground for those whose life had ended. Burial with ceremony and procession has been a practice of man from the beginning of recorded history. In fact the word funeral is derived from the Latin "funeralis" meaning "torchlight procession". It is only natural that this practice should remain as a strong tradition and custom in our country today.

Because in the early days in America artisans and craftsmen were primarily occupied with building the vehicles that were essential to the growing general transportation needs of a new developing country, the first vehicles used for funeral processions were necessarily quite simple. There just wasn't time to create the elaborately decorated horse-drawn vehicles similar to those used for funeral processions in Europe.

As the years passed and more artisans came to the New World bringing their artistic skills, more highly decorated vehicles came into being. The latter part of the 19th century saw the creation of hearses, magnificently designed and built, using the highest degree of coach crafting skills and decorative arts. Hand carved moldings, columns and draperies were incorporated into these designs and the vehicles were finished with the most meticulous care. More than three months were consumed in the painting process alone for finishes were slow drying and it took many coats and much hand rubbing to build a beautiful and durable finish.

The teams of horses that pulled these vehicles were also very carefully matched and selected and they were given exceptional training and care.

As our cities and towns grew and cemeteries were opened farther and farther from the center of town, the funeral processions of those years could consume the greater part of a day and many a carriage driver brought his team and carriage back to the stable after dark.

Shortly after the turn of the century when the first automobiles came into use, automobile hearses were created by taking the elaborate hand carved bodies of the latest horse drawn vehicles and attaching them to an automobile chassis. A short canopy type roof was attached to the body for the protection of the driver but it was not attached to the small windshield and there were no doors to the driver's area. Some of the owners of these converted automobile hearses went so far as to have more than one body fitted for attachment to the same chassis so that a different color body could be used for different funeral services. If the funeral was that of an older person, the black body could be used. If the service was for a middle-aged person, the gray body would be used and for a young person, a white body would be attached to the chassis. The different color bodies were stored on hoists and lowered and bolted to the chassis as required.

However, it didn't take long for the builders to design hearse bodies specifically for an automobile chassis. They were, of course, larger than the horse drawn vehicles and they included such comforts for the driver as a fully enclosed driver's compartment with doors and windows and a roof attached to the windshield.

Our company introduced such a model in 1913 and to the entreprenuers of that era selling an auto hearse presented quite a challenge. Many of the old timers believed that the public would not take to funeral processions moving "at such a rapid pace as 15 miles per hour, and without teams" — it just wasn't dignified to move the procession so fast. But they whittled away at these objections with a brand new list of sales points.

Here's what an automobile hearse catalog said about the benefits of this new method of funeral transportation in 1913-

"THE AUTO HEARSE IS THE MODERN, RAPID AND ECONOMICAL METHOD. IT HAS MANY ADVANTAGES OVER HORSE DRAWN VEHICLES.

It is more economical, it is rapid, which enables one car and operator to do the work of several teams. It has a big advantage for urgent calls, and your customers will readily realize your superior status in this respect. This will prove a big advantage.

It will make long trips to distant places and depots a convenience, not a burden or a worry. With an auto hearse, these trips will not cut into the best part of the day as they do with a horse drawn vehicle. It consumes fuel only while it is in motion. Does not eat while it is idle or standing in the stable, as with horses, therefore, the highest value is obtained for every cent of fuel cost — no waste.

It will run every day, winter and summer, as many hours as you require, not necessary to have days of rest in between as with horses. The auto hearse will perform its work as well in the last hours of the day as well as in the first, and it does not get tired.

Will work equally as fast in the summer as in the winter. No danger of overheated horses in hot weather, or no risk of horses breaking legs from icy street in winter, or delays in having shoes sharpened.

Will not shy at anything, or run away, or bite, or kick — will stand without hitching.

Can be stopped quicker and in a much shorter space than a horse drawn vehicle, which is important in an emergency to avoid collision, etc.

No risk of throwing horses and injuring them by sudden stops. With the use of auto hearses, the streets will be sanitary and clean, and will not wear out as fast as if traveled by horse drawn vehicles exclusively.

As it can be manipulated more readily and in less space than a horse drawn vehicle, it facilitates traffic on congested streets. It requires less stable room, and eliminates the feed loft entirely — not necessary to put in a big quantity of feed at certain periods of the year. No rats or vermin around the premises.

It is safe from diseases common to horses, and will not die or require any attention during the night.

It is the modern and progressive method, and in addition to other advantages, this is good advertisement."

The entrepenuers were successful — and the auto hearse was accepted in funeral service.

The chassis for those early engine powered funeral vehicles were specially designed for they required a longer than passenger car wheelbase. They were assembled using components manufactured by standard component makers as were many makes of passenger cars in those early years. Prominent among the engines used were such names as Continental, Lycoming, Hercules and others. Those early automobiles were certainly simple machines when we compare them to the highly sophisticated mechanisms turned out by Detroit today.

In 1920 a new style of funeral car, the limousine hearse, made its appearance. This style followed the lines of the passenger limousines of the time and by 1925 it replaced the hand carved column type. The carved funeral car reappeared again about 1931 and was built in various versions, some using wood carvings and others cast aluminum, through 1941 when all domestic automotive production was halted by World War II. It did not appear again when production was resumed after the war.

In the post war era the formal limousine style, sometimes called "Victoria" or "Landau", has become by far the most desired style. This seems rather fitting, for a funeral procession to the cemetery is still a most dignified and formal event. It is not only a recognition of a death, it is also testimony to a life, whether it was one of modest achievements or of great renown. It pays a final tribute to this life that has been lived and gives a special dignity and stature to the most modest among us.

One of the great strengths that has come to us from our Judeo-Christian philosophic background is our respect for the individual human being and the formal funeral service ceremony and procession are but a public confirmation of this respect.

In recent years psychologists are strongly recognizing the theraputic benefits to the bereaved pointed out by St. Augustine many centuries ago when he said,

"The care of the funeral, the manner of burial, the pomp of obsequies are rather for the consolation of the living than of any service to the dead."

It has been a very interesting experience to have been associated with this industry, actively for over 40 years and through family activity for many prior years. The progress and professional improvement that have occurred during that span predict, I believe, an even more important future.

Charles A. Eisenhardt, Jr.
Chairman of the Board
The Hess & Eisenhardt Co.

FOREWARD EXPLANATION

Charles A. Eisenhardt, chairman of the board of the Hess & Eisenhardt Company, has often been described as "the grand old man of the carriage trade." He grew up with the business and is often considered the expert spokesman for the field.

1900 – 1908

The most elaborate and outstanding coach of the 19th century was displayed at the Chicago World's Fair of 1893 by the Crane & Breed Company of Cincinnati. Originally designed with West Indian or South American trade in mind, this coach featured immense size, and church-like design, with massive carvings in bold relief, gildings, heavy gold draperies with gold fringes and tassels and gold coach lamps. The total weight of this car was 2,400 pounds. It was used to carry the body of Chicago's assassinated mayor, Carter F. Harrison, and was later sold to M. Raoul Bonnot of New Orleans for $5,000.

At the turn of the century every well-equipped funeral director owned a massive rubber-tired hearse, a utilitarian casket wagon, and often a small, white hearse for children's funerals. The horse-drawn hearse of the early 1900's had ornately carved draperies framed by from six to twelve carved columns, a slightly swelled double "mosque deck" style roof, and a driver's seat connected to the vehicle's body by a gooseneck frame. And it could not be considered complete unless ornamented with a pair of large ornate carriage lamps.

Although the first commercially-produced motorized funeral coaches and ambulances were offered to the trade in volume in 1909, some companies had ventured into the motorized hearse and ambulance field in the early 1900's. F.R. Wood and Sons, well-known carriage builders of New York City and mainly noted for the Wood Electric, built America's first motorized ambulance in 1900. It was put into service with St. Vincent's Hospital of New York City. This vehicle weighed 4,000 pounds and was geared to a speed of 9 miles-per-hour. The car ran on solid rubber tires while being powered by 44 cells grouped in four sets of batteries. By 1901 there were several of these vehicles in use by the larger New York hospitals, but their main drawbacks were that they could not get through horse-clogged city streets and the drivers of other carriages could not easily recognize them as emergency vehicles.

By 1905 some progressive-minded funeral directors with a mechanical bent fashioned their own motorized versions of the new "auto hearse" using the body of a former horse-drawn vehicle. This body was crudely transferred to a passenger car or light truck chassis. But because of their high initial cost, undignified noise levels and, at the time, questionable reliability, the new motorized funeral vehicles and ambulances were slow in gaining professional acceptance.

Built in the late 1890's or the very early 1900's, this Crane & Breed hearse features the then popular mosque deck style roof with eight carved pillars and carved drapery beside the large plate glass windows. The driver's seat is attached to the body by a gooseneck. Note the large coach lamps.

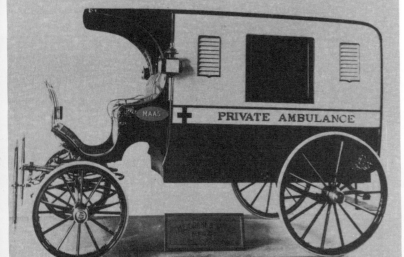

This Crane & Breed ambulance is typical of ambulances of this era and was most likely built in the mid-1800's. Note the graceful body curves and the roof that extends well over the driver's seat. Ventilators on the body sides assured patients of circulated fresh air while the rubber shrouded wheels assured him of a comfortable ride.

Stephens & Bean Undertakers of Fresno, Cal., were among the first morticians to use a motorized funeral coach. The vehicle was built by the firm in their own livery. Mounted on a 1905 Rambler Surrey Type 1 chassis, the body featured a special receptacle for the casket on the left side of the car, with two bucket seats, one in front of the other, on the right side. The firm reported that it was operated at moderate cost and was easily maintained.

One of the first motorized vehicles ever put into service in North America was this ambulance. It was the vehicle that transported the martyred President McKinley to the Buffalo Temple of Music Hospital where he died eight days later. The vehicle itself was most likely built by F.R. Wood, makers of the famous Wood Electric, and ran on solid rubber tires while being powered by four sets of batteries that gave the coach a top speed of nine miles-per-hour.

General Vehicle Company of New York City built this undertaker's wagon in the early Spring of 1908. Sold to Leonard Rouff, the wagon was capable of going 35 miles on one charge of the electric batteries at an average speed of 10 miles-per-hour. The coach featured the latest 1908 type of construction and was arranged with double opening rear doors and a removable shelf for carrying chairs or flowers or other funeral sundries.

This beautiful hearse is an excellent example of the higher class, higher priced style that was dominant through this era. Built by Sayers & Scovill of Cincinnati, this coach features elaborately detailed carvings, eight pillars, a mosque deck roof and a pair of large ornate coach lamps.

Sayers & Scovill built this six-window, eight-pillar, mosque deck hearse in the early 1900's. Note the slender gooseneck attaching the driver's seat to the body. This type of coach usually weighed around 1,600 pounds and sold for an average of $1,500.

The mosque deck style hearse was also the dominant type in use in Canada in this era, as is seen here with this Dominion horse-drawn funeral coach. Built by Dominion Manufacturers of Toronto, Ontario, the car features the eight pillar, six-window style with extremely large coach lamps.

1900 – 1908

This stylish casket wagon was used by T. P. Sampson Undertaking in the late 1900's and featured styling that was prevalent on horse-drawn casket wagons and solid rubber tires. The body was obviously transformed from a horse-drawn coach but the job was done with a degree of skill. Note the carved pillars, large coach lamps and the rayed texture rear compartment window.

This little hearse was mounted on a 1908 Cadillac chassis and although crude, carried styling that was reminiscent of the passenger cars of the era.

The passenger car was in a much more advanced state of development in Europe, and while the North American funeral director and ambulance operator were still using horse-drawn coaches for the most part, the funeral directors in Europe had begun the switch to motorized vehicles. This 1905 English hearse was owned by Reuben Thompson of Sheffeld and was powered by a 12 horsepower Wolseley engine and chassis. The body would accommodate a casket 7 feet long by 2'6" wide. This body was also interchangeable with that of a landaulette and a double tonneau. The coach was reported to have cost its owner only one pound six shillings per mile (one way).

This rather strange hearse was used in Germany in the very early part of this decade and was electrically powered. The highly ornamented and carved casket compartment has no glass or draperies. This car has hard rubber tires.

Another European offering was this French Borniol long distance motor hearse used in Paris in 1908. The coach not only accommodated the casket but carried the pallbearers as well in a closed compartment behind the driver's seat.

1909

America's first automobile funeral (with an auto hearse and a full procession of motor cars) took place in Chicago in March, 1909. This funeral procession drove 11 miles to a cemetery on the outskirts of the city and, despite muddy roads encountered three miles from the place of burial, the entire trip took only an hour. The newspapers made much of this fact and not even they could grasp the significance of the event. As this new vehicle ambled down the streets at the head of the somber procession, indignant onlookers recognized the clattering machine as a hearse and a totally irreverent means of transporting the dead. Little could they appreciate the fact that they were witnesses to an important historical event. This auto funeral was the forerunner of revolutionary changes in the burial customs of the world.

Three months later, on June 15, 1909, the eminent firm of Crane & Breed of Cincinnati, introduced the first commercially-built auto hearses to the trade. The chassis of this new vehicle was strong and featured chain drive. The engine was a four-cylinder water-cooled unit that supposedly started with one turn of the crank and was coupled to a three speed transmission. It developed 30 horsepower and could do 30 miles-per-hour. "Fifteen miles-per-hour faster than any hearse should have to go." said Crane & Breed. Two oil lamps in front and one in the rear, two electric headlamps, a complete set of tools and a set of wire flower baskets to go on the steps were standard equipment. One month later the company introduced a more ornate version of this original coach.

Crane & Breed, an old and highly respected manufacturer of caskets and horse-drawn hearses, ambulances and casket wagons, has always been both a leader and a trend setter in the funeral vehicle field. With the introduction of their new commercially-built coach, the company told the funeral director, "People who continually ride in automobiles object to the long and (to them) uncomfortably close and slow carriage ride. They want speed — a smooth glide to the cemetery, same as downtown or anywhere else — and especially in the larger cities." To further reassure the funeral director that he, too, could have automobile funerals if he were to purchase this new machine they continued, "Taxicabs and touring cars are in garages awaiting your hire, though most attendants at your funerals will use their own vehicles. You will furnish the hearse only."

Other manufacturers, for competitive reasons, quickly followed suit and announced auto hearses of their own. James Cunningham Sons of Rochester, N.Y., was already marketing a commercially-built motorized ambulance and notified the trade that they, too, would soon be manufacturing a motor hearse. The commercially-manufactured funeral coach was here and with it the motor age officially dawned for the funeral business. But, with the majority of the morticians in this era still using the horse-drawn vehicles, there was to be a long evolutionary road ahead for the motor hearse. The horse-drawn coaches were to maintain their strong hold on the field for several years yet even though the handwriting was on the wall and their fate had been previewed in 1909.

H.D. Ludlow was the first funeral director in Chicago to own and operate his own motor hearse. The car was put into operation on January 15, 1909. It was home built and used for several months.

This White Steamer was used by F.F. Roberts and was one of the very first motorized hearses in Chicago. The company had mounted an older horse-drawn coach body on the White Steamer chassis and used the coach in several funerals.

Introduced to the trade on June 15, 1909, this Crane & Breed was the first commercially-built motor hearse to be marketed. The coach was painted black and had very little decoration except for the rather grotesque replica of the tomb of Scipio carved in wood atop the otherwise flat rectangular casket compartment roof. The car was powered by a four-cylinder water-cooled engine started with the crank and had a three-speed transmission. While developing 30 horsepower the coach was able to speed up to 30 miles-an-hour.

Later in the year Crane & Breed introduced their own version of the auto ambulance. This ambulance, with the exception of the extended roof over the driver and the windshield, closely resembled the horse-drawn Crane & Breed ambulance shown in the last chapter.

Cunningham was marketing an auto ambulance when Crane & Breed announced its motor hearse in June. The large grey coach featured automotive styling with a roof over the driver's compartment and a full windshield. It was equipped with a 32 horsepower engine, rubber tires, a heater and a gong. The interior consisted of one suspended cot, two seats for attendants, a system of electric lighting, and an interior of stained mahogany.

Crane & Breed introduced a more distinguished mate to the original auto hearse of July 15. Called the Auto Hearse Model 220, this coach resembled the earlier model in styling but was much more ornate with carved drapery and pillars on the casket compartment sides. Both of these cars rode on a wheelbase of 130 inches with 36-inch wheels.

The tires were 4 inches in size and could be ordered either solid rubber or pneumatic. These cars came complete with two oil lamps in front and one in the rear, two electric headlamps, a complete set of tools and, if desired, a set of wire flower baskets to go on the running boards.

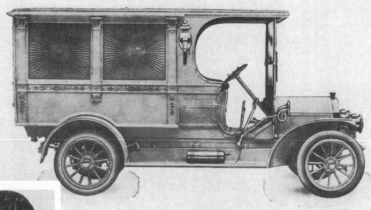

By early Fall, Cunningham introduced the production version of the auto hearse Model 807. Using the chassis and running gear from their auto ambulance, this coach carried a body that was designed to express dignity and tradition. Carved pillars and rayed windows graced the body sides along with an oil fueled coach lamp.

The famed mortuary of Frank E. Campbell, New York City, was the first firm in that city to own and operate a motor hearse. Cunningham built the coach, the second model of auto hearse offered to the trade. The coach rode on a White Steamer chassis and wore styling that Cunningham had previewed to the trade months earlier by means of a drawing. Combining leaded glass windows, eight-pillar design, ornate carvings, and huge carriage lamps, the car bumped the number of commercially built auto hearses offered to the trade, by all makers, to four. Two each from Crane & Breed and Cunningham.

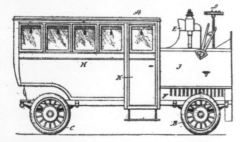

Fred Hulberg of New York City designed and patented this vehicle, called a combination hearse and passenger vehicle. The strange design contained a compartment for accommodating passengers and a separate compartment for the casket. The casket compartment was arranged at one end of the passenger area and opened into the same. Thus, the passengers were riding with the casket. Hulberg began to build this coach in his own repair shop this year.

1909

Shown in the trade magazine The Sunnyside is this auto ambulance owned by J.C. Dodge and Son of Glen Cove, New York. Although no details of this 1909 car were known, it is obvious that it was a home conversion and the chassis is that of a regular passenger car.

Looking like a passenger limousine, this auto-ambulance was owned and operated by the A. M. Ragsdale Company of Indianapolis. The vehicle was converted from a 7-passenger limousine into a hearse or an ambulance. While the chassis is unidentified, it appears to be a Stevens-Duryea 4-cylinder model. The body work was by a local shop.

This little home-built vehicle was the first auto hearse in Baltimore. It was owned by E. Madison Mitchell. Styling was similar to that of commercially built cars of this year while the make of chassis is unknown.

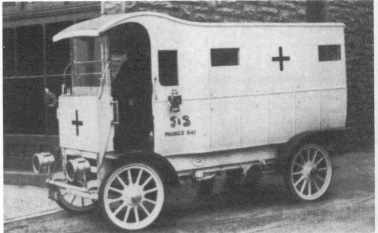

Another interesting auto ambulance shown in the trade journal is this unique vehicle that belonged to the Flanner & Buchanan Undertaking Co. of Indianapolis. It is gasoline powered, with chain drive.

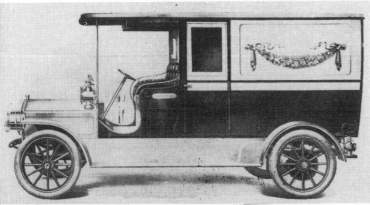

Crane & Breed was now motivated to expand its offerings in the motorized field and late in the year introduced two new models. This combination casket wagon and ambulance featured the same chassis used on the Model 220 auto hearse, with a more modern body design. This particular model was designated the number 1112 and was sold to F.S. Simmons of Camden, N.J.

The other Crane & Breed entry was the model number 1116 combination hearse and casket delivery wagon. Crane & Breed introduced new models in a wide variety in order to retain the lead that it had gained with the introduction of the first auto hearse. This pleasingly styled coach featured two large windows with drapes and four smaller ones beside the main ones. Eight pillars graced the body sides and the roof over the driver had a dip in its styling line. This particular car was sold to J.P. Finley & Son of Portland, Ore.

1910

The cars used in the previous year's premier American automobile funeral were, in the case of the hearse, crude home-built cars. But, after the initial introduction of commercially-built models, the market began to fill with professional makers and a larger variety of styles. Following the growing auto hearse trend, New York City had its first auto funeral in November, 1910. This procession traveled a lengthy route to Cedar Lawn Cemetery in Paterson, N.J. The following month Providence, R.I. had its premier auto funeral with burial taking place in New Bedford, Mass.

Inherited from the passenger car, several new features began to appear on hearses and ambulances in 1910. Among them were windshields, electric-acetylene headlamps, electric-oil carriage lamps, and in some cases electric tail lamps. Prices for these new cars were high, $4,000 to $6,000 apiece for vehicles with an unknown span of reliability. This compared with an average of $1,500 for a horse-drawn coach with a known longevity. But even considering the high initial cash outlay required, the new auto hearse was beginning to find an eager and growing market.

As, one by one, they began entering service with funeral directors, who also operated the local ambulance service, the establishment fortunate enough to have one enjoyed a larger clientele, due largely to its modern equipment. Many times curious people would stop to inspect the new and unique equipage at the funeral director's livery.

Fred Hulberg of New York City, who had designed and patented a unique combination passenger and funeral coach in 1909, finished building his coach in January of this year. An undertaker, Mr. Hulberg built the car in his own repair shop and livery stable. The construction of this vehicle took one year and three months and the cost exceeded $6,000. With a maximum speed of 20 miles-per-hour the coach was powered by a 45-horsepower Continental engine driving through a two speed gearbox. Riding on a wheelbase of 138 inches, this unique vehicle was equipped with four brakes; two on the transom shaft and two on the rear wheels.

A name that was destined to become famous in the field, Rock Falls Manufacturing of Sterling, Ill., entered the funeral coach and ambulance arena with this stylish combination ambulance-hearse. Like most of the other makers, Rock Falls also assembled the cars from the ground up. Note the heavy beveled glass in the rear side windows and the graceful curves of the rear body belt line. The car was finished in white with gold colored highlights and pin striping.

Crane & Breed's Model 222, Plato, was the first really distinct departure from the horse-drawn coach. For inclement weather the car could be equipped with curtains that buttoned on the sides while the driver was further protected by a full roof over his compartment. Advertised as lower by three inches than the company's horse drawn models, the Model 222 Plato was powered by a 45-horsepower motor with four cylinders. This particular coach was delivered to George W. Loudermilk of Dallas.

Crane & Breed continued its combination casket delivery wagon and ambulance, Model 1112. Here it is being driven by its owner, Fithian Simmons of Camden, N.J. This same vehicle was displayed at the Philadelphia Automobile Show in January of 1910. Note the chains on the rear wheels and the chain guard mounted in front of the rear fender.

1910

This White Steamer was equipped with a custom built ambulance body and was operated by the Kings County (N.Y.) Hospital's Department of Public Charities. The body features rather plain styling and screened windows. The car has no windshield or headlights but is equipped with oil lamps and could be closed up in inclement weather by means of snap curtains.

Fred F. Groff of Lancaster, Pa., built his first funeral coach, the Groff combination limousine, hearse and ambulance, mounted on an unidentified chassis.

Offered again for 1910 was the Cunningham Model 807 seen here in a Cunningham factory photograph. Note the intricate carvings in the area between the carved pillars and the ray textured effect behind the rear windows. This coach was delivered to the L.P. Robertson Undertaking Co. It was finished in light grey with gold trim and pin striping. Cunningham completely assembled their own chassis using Continental Red Seal engines and other components supplied by independent makers.

Scully Walton operated this Cunningham private ambulance styled with eight side windows on each side. Full draperies were featured in each window and the coach's rear compartment featured the most modern emergency equipment.

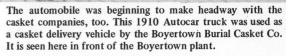

The automobile was beginning to make headway with the casket companies, too. This 1910 Autocar truck was used as a casket delivery vehicle by the Boyertown Burial Casket Co. It is seen here in front of the Boyertown plant.

Grabowsky, a Detroit based truck maker offered the trade this casket wagon on its Power Wagon chassis. It featured a powerplant that was removable for service. Clearly visible in this picture are the chain driven rear wheels and chains and the starting crank. This particular casket wagon was in the service of the W.R. Hamilton Undertaking Co. of Detroit.

1911

To the swelling ranks of manufacturers came several new makers of professional funeral coaches and ambulances. One of the old leaders in the horse-drawn hearse and ambulance field this year began to experiment with the motorized style cars. Sayers and Scovill, established in Cincinnati in 1876 to build high-quality coaches, built several prototype motorized cars in 1911 but did not advertise or otherwise promote the fact that they were thinking of entering the motorized era at this point.

Other makers to join the motor age in 1911 included such names as The Dorris Motor Car Co. with some new ideas of their own and the U.S. Carriage Co., long-time builders of high quality horse-drawn vehicles.

Styles in this year retained that overall transformed horse-drawn vehicle look due mainly to the fact that most of the makers at this point were still marketing the horse-drawn vehicles and their styles predominated. Yet, there was a hint of a coming trend toward a fully enclosed style coach. A few makers introduced models with fully enclosed driver's compartments, while with others the driver was still exposed to the elements. This was also a year when the hazards of starting a car by hand cranking were first tackled. Several types of starters were available to car owners, one of which was the compressed air starter. The Geiszler Starting Device was made for almost every type of car and was supposed to start the car with the aid of an electrical charge. Without a doubt some of these new devices found their way to the fledgling funeral coach and ambulance of the era.

The United States Carriage Co. began producing motorized coaches this year. This is an example of the Model 1241 motor ambulance, which features body side ventilators and an electric gong. All U.S. Carriage Co. vehicles featured the hood straps that were found on performance cars of the era.

The most elaborate example of U.S. Carriage Company's coach work, built in Columbus, Ohio, this coach featured intricate carvings below the rear body side windows and six carved pillars.

This handsome Cunningham funeral coach was one of that company's most attractive offerings of the year. Cunningham built its own hearse bodies and began to build complete engines this year. But it would also mount bodies on any other make of chassis.

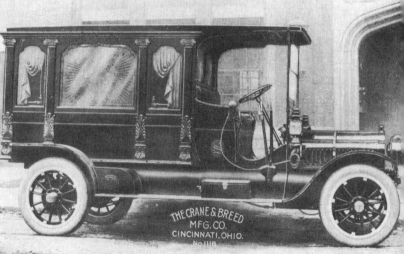

This Crane & Breed combination hearse, casket-delivery and flower wagon was the Model 1116, furnished to H. Samson of Pittsburg. Powered by a 45-50-horsepower four-cylinder engine and using chain drive, this coach had demountable rim tires measuring 36x5 inches and included one extra case, tube and rim complete with cover and side fastenings. The rayed cloth behind the center rear compartment window was removable. The window glass was beveled and etched with artistic patterns.

Cunningham also offered this Model 867 that featured removable ray textured drapery panels behind the side windows. Detailed carvings were used below the rear side windows. This coach was built to order for J.S. Waterman and Sons of Boston.

1911

Another company to join the auto hearse age was Blue Ribbon Auto & Carriage Co. of Bridgeport, Conn. The company built both automotive and horsedrawn coaches, with this being the Model 26 Private Ambulance.

Featuring unitized construction of the engine, gearcase and radiator, this 30-horsepower casket coach was built by the Dorris Motor Car Co. on the standard touring car chassis. This was the company's first entry into the funeral coach field.

Established in 1861, the S.C. Pease & Sons Co. of Hyde Park, N.Y., entered the motor hearse field in this year. The firm was owned by James F. and Frank E. Pease and prior to the introduction of motorized coaches they built horse-drawn coaches, hearses, undertaking wagons, embalming buggies, and ambulances. The line of funeral service vehicles was transferred to motorized chassis, such as this example, sold to F.C. Graham Undertaking of Hyde Park.

This interesting custom funeral coach was a one-off vehicle built by a local carriage shop for James T. Chandler of Wilmington, Del., at a cost of $2,000. Powered by a two-cylinder opposed engine the vehicle could go up to 40 miles-per-hour. The chassis is a standard Autocar unit built at Ardmore, Pa.

The Livingston Funeral Home of Hastings, Neb., constructed this little hearse on a Model T chassis. The ornately decorated, fully enclosed body was interchangeable with that of a regular passenger car. Note the drapery in the driver's compartment as well as the ray-type drapes in the casket compartment.

Built by the New York Transportation Co., this electric casket wagon was owned and operated by Fairchild & Sons Funeral Directors of New York City and was one of four operated by this large firm. These cars had a total running capacity of from 40 to 70 miles in range. The bodies were interchangeable with that of a brougham passenger car, so that these cars could be used as part of the livery stable.

1912

Called the greatest funeral car achievement of the age, this funeral omnibus was constructed by the L. Glesenkamp, Sons & Company of Pittsburg. This was the first funeral car built by the Glesenkamp firm and was sold to J.J. Flannery Brothers of Pittsburg at a cost of $10,000. The company's advertisements called this the largest and finest funeral car in the world and with some merit to the claim. The vehicle, mounted on a Packard truck chassis, could carry both the casket and a full entourage of mourners all in the same quarters.

The interior of the Glesenkamp funeral omnibus was magnificently upholstered in dark blue leather and equipped with electric lights, electric fans, and telephones to communicate with each compartment. Over five hundred feet of wire was required to make these connections. Also included in the coach's interior equipment were water coolers, washstands and other conveniences equal to the equipment of a Pullman railway coach. Woodwork inside was solid mahogany and all the glass was beveled. The windshield hung on brass hinges and was arranged to swing in any angle. The coach made an exhibition trip through Pittsburg from the Flannery Funeral Home to the Allegheny Cemetery six miles away with not a bump, jar or sound to disturb the passengers. The coach could accommodate 32 persons and the casket.

Another of the evolutionary years, 1912, heralded the entry of still more manufacturers into the funeral coach and ambulance field. Some notable truck makers joined the carriage makers and mounted ambulance and hearse bodies on their own chassis and sold these vehicles through their dealer outlets. Autocar and Lippard Stewart were among the truck makers to join the field in this year. Meanwhile, some smaller companies without the resources to assemble their own vehicles from the ground up began to build their bodies to fit manufactured chassis of heavy passenger cars or light trucks. Some of these chassis were even extended to lengthen the wheelbase. At this point, some of the smaller makers began to mount their bodies on the popular Ford Model T chassis using the chassis as delivered or extending it.

Charles F. Kettering of Dayton Engineering Laboratories Co. (Delco) developed the electric self-starter which was adopted by Cadillac. This system, accompanied by a generator battery lighting system, was to have enormous impact on the development of electrical systems in the funeral coach and ambulance.

The advent of the fully enclosed style funeral coach and ambulance was gaining favor, especially in the northern parts of the country. In this year several more makers offered this style of coach to their customers. These vehicles began to set their own style while being sold beside the older horse-drawn coaches.

Electric vehicles were still popular in major cities and were having hearse and ambulance bodies mounted on them. This year a fleet of three electric ambulances was delivered and put into service with the U.S. Steel Co. Rock Falls Manufacturing Co. of Sterling, Ill., built these little electric ambulances which were used in the U.S. Steel Company's steel mills in Gary, Ind., South Chicago, Ill., and Milwaukee. They were of the fully enclosed, private ambulance style and were, at the time, the epitome in the ambulance field.

A cleanly styled casket wagon was Cunningham's Model 867. It featured four large windows in the rear casket compartment with ray type draperies and mildly carved pillars on the body sides.

A massive bulk of carvings, this Cunningham features beautiful styling. It was painted white. Note the elaborate driver's quarter window and the carriage lamps. This coach has a mosque deck type roof similar to that on horse-drawn coaches shown in earlier chapters.

1912

Cunningham built this ambulance Model 774 and delivered it to J.H. Wann & Son of Chattanooga in the early spring of this year. The signs in the rear windows, showing that this is a private ambulance, were quite common in this era as were ambulances operated by funeral homes or small hospitals.

Delivered to Lindsey & Son Undertaking of Kansas City, Mo. this elaborately carved hearse was built by Rock Falls of Sterling, Ill. This was the Model 31 hearse/casket wagon and featured ray texture drapes in the rear side windows and intricate carvings on the body sides.

Rock Falls built these electric ambulances for the United States Steel Company's immense mills at Gary, Ind., South Chicago, Ill., and Milwaukee. The steel company maintained private hospitals for the care of its workmen injured in these factories.

Another Rock Falls, this Model 56 funeral car shows even more elaborate draped carvings on the rear body sides and has a fully enclosed driver's compartment. Note the large carriage lamps and the decorative metal window shield in the rear compartment. The coach is built on a White light truck chassis.

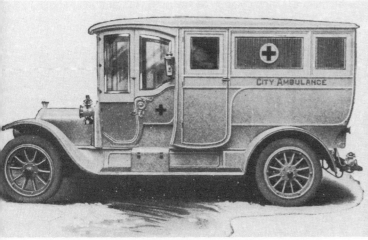

This fully enclosed ambulance was built by Rock Falls for the city of Madison, Wis, this year. Rock Falls built its own chassis with components from the larger suppliers like Borg & Beck, Timken, Continental and Eaton.

Another Rock Falls ambulance, the Model 55, was sold to the F.B. King & Co. of Youngstown, Ohio. The open driver's compartment had etched initials in the small quarter window. Heavy beveled glass was used in the rear compartment windows and the body work was devoid of any carving or ornamentation.

1912

Crane & Breed made a big switch in this year from their own assembled chassis to a Winton chassis. The Winton chassis used a six-cylinder engine and was built to exceedingly high standards. The Crane & Breed Winton Special Six funeral coach was Model 1117-½, featuring a highly carved body. Selling for $4,600, this coach went to the W.R. Hamilton & Co. Undertakers of Detroit, where it was photographed for an advertisement in front of the Masonic Hall.

This fully enclosed funeral coach was also built by Crane & Breed on the 48 horsepower Winton Special chassis. It had a wheelbase of 152 inches and sold for $4,800. This particular model was sold to the George L. Thomas Co. of Milwaukee. It featured heavy beveled glass in all of the windows and small carving details.

Peter Kief was the proprietor of the G. Dessecker Co. that was founded in 1869 in New York City. This uniquely styled hearse was delivered to the Charles Stumpf Funeral Home and resembled a box mounted on the rear of a coupe style passenger car. This coach also shows some carving on the body sides and an extremely large window in the casket compartment with ray type drapery.

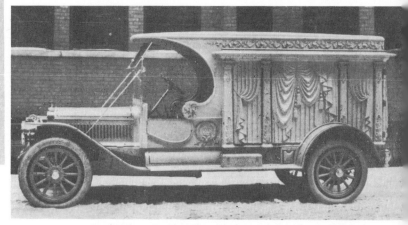

The Winton Special Six chassis used by Crane & Breed produced 48 horsepower and was self-cranking. Tires were 37x5 and the wheelbase was 152 inches. This model, the 222 Plato, was delivered to the George W. Loudermilk firm of Dallas, Texas. Loudermilk was one of the Crane & Breed company's best customers, always using Crane & Breed vehicles. Note the massive drapery carvings on the rear body sides and the stylish curvature of the body around the driver's compartment, the detailed ivy carved above the belt line, and the eight pillar design.

This White ambulance features slightly different body work and was used as a private ambulance by Turner & Stevens Co. of Pasadena, Cal. During its use as an ambulance by this firm, the car clocked over 100,000 trouble-free miles.

Bentley Undertakers of Chicago had this electric hearse built especially for them. The large glass area was complemented by heavy draperies and a small amount of carvings between the pillars.

1912

Another ambulance of the era was built by the White Motor Car Co. of Cleveland. This car follows the style and ambulance specifications seen since the late horse-drawn models dominated the scene.

Lippard-Stewart of Buffalo, N.Y., was well known for trucks. This year it began offering a line of funeral service vehicles. Built on their own chassis, the Lippard-Stewart Undertakers Car sold for $2,300. The chassis design closely follows that of the French Renault, a prestige car in that era.

Cunningham, like Crane & Breed and several other makers, by now offered a wide choice of models and styles in the line of motor hearses. This Model 822 casket wagon features double ray textured windows on each side and a minimal amount of carving with an open driver's compartment.

This pair of Cunningham ambulances was owned and operated by the Rochester General Hospital. Notice the gongs, acetylene lamps and the ambulance signs over the windshields. Taking it for granted that the two chaps standing beside the cars are drivers, note the overall size of the vehicles in relation to these men.

Great Eagle (U.S. Carriage Co.) produced this attractive undertaker's limousine, Model 1240. A 10-passenger car, it was mounted on their own chassis, with a 72-horsepower six-cylinder engine built by Rutenber. It had a wheelbase of 141 inches.

Designated the Model 1260, this Great Eagle funeral coach was built by the U.S. Carriage Co. of Columbus, Ohio, for Brown & Willett of Los Angeles. With carved draperies and a fully enclosed driver's compartment this was a very attractive vehicle.

1912

Another ambulance of yet another design was this Knox, built for the city of Holyoke, Mass. The high, narrow body, devoid of any ornamentation, is quite reminiscent of a casket wagon. Note the oil lamp hanging from the driver's seat.

This combination limousine-ambulance was a unique vehicle when it was owned by J.A. Costa Funeral Home of Cambridge, Mass. Custom-built for the firm, it was considered the last word in utility and general excellence of design. It should be noted that within a few years this type of vehicle would become quite common.

Frank E. Campbell & Son of New York City had always operated a fleet of small electric casket wagons for use around the city and as service cars. This car ran on four batteries and had a top speed of 37 miles-per-hour with a 70-mile radius.

Autocar, now famous for trucks, began to build some funeral service vehicles this year and offered them in three series. This Model S-4 looked similar to the earlier horse-drawn vehicles with carved pillars and very large coach lamps. The driver's compartment on this model is open, as the fully enclosed car was still somewhat rare. This car was owned by William Necker of New York City.

One of the few Sayers and Scovill hearses to be motorized and marketed was sold to Joseph Grawler, Sons Undertaking of Washington, D.C. It was the first motor hearse in that city. A rather stark looking coach, all S&S cars were built on a chassis completely assembled by S&S and built at the Cincinnati plant.

In the late Spring of the year Peter Kief succeeded the G. Dessecker Co. and produced this unique motor hearse on an Autocar truck chassis. A special department of the plant was devoted to creating drapes of any material, color or design to fit the hearses. The company, located in New York City, built a wide variety of coach types and styles and also reupholstered or refinished coaches of older vintages to customer specifications.

1913

This was the year that the new auto hearse began to prove its reliability to the skeptics. In August, 1913, a new Great Eagle funeral coach began the first cross country funeral trip. Beginning in New York City, the coach traveled the paths and muddy lanes that served as roads to San Francisco in the course of the month. Weighing 6,560 pounds during the trip, the coach made the journey without any difficulty or mechanical problems, thus reinforcing the fact that the motor age had truly dawned for the funeral serrvice.

The year also saw a good many larger and more prominent carriage makers joining the ranks moving toward motorized vehicles. One of these was Sayers and Scovill of Cincinnati Although they had experimented with the idea of a motor hearse in 1911, they did not introduce their car until this year. S & S was established in 1876 by William A. Sayers, a carriage maker, who had teamed up with A.R. Scovill, who acted as the banker, and formed a company to produce high quality buggies, carriages, ambulances and hearses.

Other makers to join the motorized hearse and ambulance field in this year were Peerless, White, and Kunkle. Styles were again more evolutionary in scope than anything else. The fully enclosed car seen last year was again seen on more and more funeral coaches and ambulances and was becoming almost as common as the sedan-type passenger car.

Shown at the start of the trip, the Great Eagle funeral coach sits in front of the Lanyon Funeral Home in San Francisco. The coach was equipped with a 72-horsepower Rutenber six-cylinder engine and Hoover springs that sustained a load of 6560 pounds during the trip. The heavy drapery in the rear windows was fringed and the driver's compartment was fully enclosed.

This Great Eagle funeral coach was photographed in the western United States on its coast to coast funeral trip. Proving the reliability of the "new-fangled" motor hearse, this car, fully loaded, traveled from San Francisco to New York City. The extra equipment hanging on the side of the car and on the running boards was needed for the trip across the virtually roadless prairies.

The Model 1252 Great Eagle funeral coach featured four massive plate glass windows in the casket compartment. This was another model in the Great Eagle line, using the same chassis as the coach that made the cross-country funeral trip. This coach was delivered to the firm of George J. Schoedinger & Brothers of Columbus, Ohio.

Great Eagle had been producing motor cars for six years when it introduced this invalid carriage (ambulance), claimed to be the most progressive vehicle available to the trade. This interesting coach was owned and operated by the E.T. Shaw Undertaking Parlours of Birmingham, Ala.

Designated the Great Eagle Model 1268, this ambulance was operated by Brown Myers of Columbus, Ohio. This ambulance reflects a completely different style from the other one presented on this page. Great Eagle coaches were built by the United States Carriage Company of Columbus, Ohio, a firm founded in 1907 to build passenger cars, specialty cars, and custom bodies for any chassis.

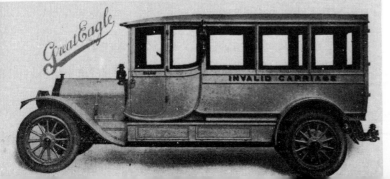

This Cunningham, owned by Wann & Son, shows more carved surfaces and an eight column design. The fully enclosed style car was becoming increasingly popular. The large, ornate carriage lamps were still an integral part of the design.

Cunningham also built this attractive open-front limousine-type ambulance. Designated the Model 938, this coach was sold to G.S. Butler of Cumberland, Maryland.

Lewis & Maycock of New Haven, Conn., owned this beautiful 1913 Cunningham funeral coach. The drapery panels in the rear compartment were pieces of hand-carved wood painted a dull color to give the appearance of being real draperies.

This coach is quite interesting, as it is mounted on an International Harvester High Wheel chassis. This vehicle is the only high wheel funeral coach that the author has ever seen. It was built by Emerson Potter for use in his funeral home in Louistown, Penn.

Cunningham made a major contribution to the field this year with a fully enclosed coach. Designated the Model 896 Limousine Funeral coach, this was a preview of the wave of the future, the limousine body style.

The Peerless Motor Co. of Cleveland, announced the fact that they were entering the funeral coach and ambulance field this year. With substantial construction, this was the 1913 Peerless ambulance with open type driver's compartment.

Sayers & Scovill began the full production and marketing of motorized funeral coaches this year. This was the Model 1100, Twelve-Column funeral coach. This particular coach was sold to Stack-Falconer Co. of Omaha. It was designed from the ground up as a motorized hearse, using a chassis especially assembled by S&S for funeral coach work. The company continued to market a line of horse-drawn equipment in addition to the motorized vehicles.

The H.H. Williams Undertaking Co. owned the first motorized funeral coach in the Hawiian Islands. The Honolulu based funeral home operated this Rambler chassied hearse with wicker filled oval rear window and an open driver's compartment. The coach also had an electric starter.

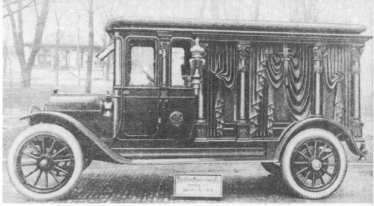

Rock Falls built this massively carved hearse. This was the Rock Falls Model 58 and was delivered to George Wagoner of St. Louis. It featured the increasingly popular fully enclosed style.

This unusual vehicle was called a Dorris Draped Motor Catafalque and was used for the funeral of Adolph Bush, millionaire brewer. The vehicle was purchased by the Wagoner Undertaking Company of St. Louis, who conducted the funeral. Dorris was also active in the passenger car business and made several funeral service vehicles.

An interesting little vehicle, this Rock Falls electric ambulance was the Model 66. Featuring beveled plate glass windows and a fully enclosed body, this coach was built for Hileman & Gindt of Waterloo, Iowa.

The Michigan Hearse and Carriage Company was established at Grand Rapids, and entered the motorized funeral coach and ambulance field this year. It built both horse-drawn and motorized funeral cars, ambulances and casket wagons. This is the the auto hearse, Model 1002. Mounted on a white light truck chassis and with their own uniquely styled body, this coach still had an open driver's compartment.

Crane & Breed continued to use the Winton Special Six chassis. This was Model 1116 FW Portland combination casket delivery wagon and hearse. It had four doors and removable scroll windows at the sides of the driver's compartment. Finished in grey it sold for $4,350. This version was sold to Stephens & Bean of Fresno, Cal., with an identical model finished in black going to the Whitmer Brothers of Dayton.

Another funeral omnibus mounted on a Packard chassis. This model would seat 22 people, including the minister, funeral director and mourners. The casket was transported in the forward compartment just behind the driver and in front of the other passengers. Several omnibusses of this type were built and tried by funeral directors to transport the complete funeral assembly in one vehicle and many more would be built in the succeeding years but they never really caught on.

1913

White built this interesting ambulance that was delivered to Weisz & Samuals Undertaking Co. of Sioux City, Iowa. This vehicle appears to have had an unusual two-tone paint finish. Notice that there are several horse-drawn carriages and buggies in the background.

The body on this White chassis appears to have been all glass and the driver's compartment was fully enclosed. The coach belonged to Honeywell & Painter of Newark, N.J.

Lippard-Stewart continued to offer a line of funeral service vehicles this year, with this being the carved-sided funeral coach. They built on their own chassis. This cleanly styled coach is an excellent example of the style of the era. Notice the very large coach lamp and its ornate styling.

This White ambulance, operated by the Flynn Froelk & Company, covered over 100,000 miles of service with no major trouble and was, years later, still giving dependable and low cost service. This particular model had an open driver's compartment and was equipped with thermos bottles of both hot and cold water, one cot, and full draperies.

Another Lippard-Stewart model, called the Economy Casket Wagon, sold for $2,300. It was powered by a 30-horsepower Continental engine. The driver on this less expensive model was more exposed to the elements than on the other model. The car had a rather stark exterior style, with frontal styling similar to the Renaults, Franklins, and Keetons of the day.

W.I. Winegarner of Columbus, Ohio, owned this Kunkel Model 150 carved hearse. The body was fully enclosed and featured an eight column design. An interesting feature of this car is the "W" etched into the quarter window of the driver's compartment. There was no glass in the casket compartment. Kunkel built its own chassis.

1914

The building of funeral coaches and ambulances had become big business and many of the larger manufacturers were now selling their chassis to customers who would in turn be able to have any make of body mounted thereon. Styles had progressed within their evolution to a point where fully enclosed type bodies were now offered by all of the manufacturers within their regular lines. Cars of this era were high and narrow looking and appeared as if a good breeze would topple them. The transportation of the pallbearers to the burial site had also been tackled by the coach manufacturers with the introduction of the bearers' coach. These were large sedan/wagon type cars with a full width rear seat and folding occasional seats along the lines of a modern limousine. They would seat six in the rear compartment in comfort and luxury.

The motor hearse was by now easily displacing the older horse drawn vehicles and the market for the horse drawn coaches was rapidly dwindling. One livery in New York City could not sell its second-hand horse vehicles and was thus shipping these to Cuba where the market was better for this type of coach and where they would fetch better prices. While new makers continued to enter the market in this year, it was but a preview of what was to come. Within the next ten years there would be more makers involved in the construction of these vehicles than ever again in their history.

One of the Mitchell & Company's exhibits at the Canadian National Exposition in Toronto, was this carved hearse on the Chalmers chassis. Claimed to have been one of the major attractions in the Transportation Building, the carved coach featured ornate drapery carvings, detailed roof surround carvings, eight pillars, a fully enclosed driver's compartment and large coach lamps. Mitchell & Company was located in Ingersoll, Ontario.

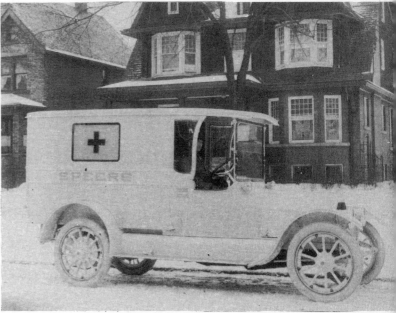

This Cadillac wears a body built by Seaman and was delivered to the Speers Funeral Home. Notice the different styling given to the driver's quarter window. In this era most ambulances were painted white or a light grey.

Another Seaman Cadillac, this time a hearse. Carving was kept to a minimum with carved draperies at the sides of the smaller rear casket compartment windows. The driver's compartment on this model is fully enclosed.

The Seaman Body Company of Milwaukee, built a few funeral coaches and ambulances to order on manufactured chassis. This ambulance is mounted on a Cadillac V-8 chassis and was delivered to the Department of Public Health, Emergency Hospitals, in San Francisco.

1914

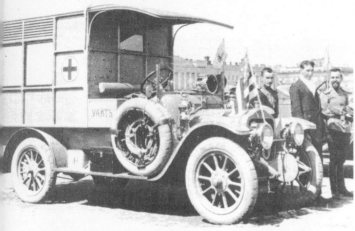

A war had broken out in Europe and some manufacturers began to build and sell ambulances to foreign powers. This White ambulance was delivered to Czarist Russia where it served on the front. It is shown here wearing the American flag as it was delivered to Russian troops.

The Premier Motor Manufacturing Co. of Indianapolis built passenger cars from 1903 until 1925, and built at least one funeral coach this year. Utilizing the Premier six-cylinder engine and the 148-inch wheelbase chassis, this coach was sold to Flanner & Buchanan. It featured eight columns and drapery carvings on the rear of the fully enclosed body.

Crane & Breed, one of the leaders in this field, offered this attractive hearse on the Winton Six chassis. Finished in grey overall, this coach featured eight carved columns with six oval plaques set into the rear body sides on a field of intricate carvings. The Crane & Breed trademark in these years was "Quality First."

Advertised as "fine autos for the fine undertaking trade" these Crane & Breed vehicles were built to the highest standards in the industry and mounted on the excellent Winton Six chassis. This Model 110½ FW, had an open driver's compartment for the southern and middle tier states and came with a fully enclosed driver's compartment for northern states. All of the carved drapery and other carvings were done by hand, right down to the drape tassels.

This Crane & Breed Model 1116½ featured a fully enclosed body that had swelled rear doors and a ray textured drape in the center rear windows, while the areas beside the center windows had draped carvings within. Mounted on the Winton Six chassis, this coach was delivered to Stewart & Fix Undertakers of Shelbyville, Ind.

Because the operating costs were lower, 90% of the taxicabs, busses, and trucks of this era operated on a four-cylinder engine. But Crane & Breed felt that a six was better for smoothness and overall comfort. Winton was among the first passenger car makers to adopt the six-cylinder engine in 1908 and Crane & Breed were the first to adopt the Winton chassis to funeral coach and ambulance work. This ambulance was the Model 1127 FW Colonial limousine style ambulance. The rear compartment windows had beveled glass in each of the little panes and a piece of stained glass in the center of the large rear window with the cross insignia.

1914

Sayers & Scovill had a full line of models now and were giving some of the other makers a real run for the money. This Model 1325 has carved drapery panels, eight column design, and an ornately carved initial plaque behind the front doors. Note the slightly curved window at the extreme front corner of the fully enclosed driver's compartment.

The S&S Model 1025 was an interesting combination ambulance and pallbearer's coach sold to the Shirley Brothers Funeral Home of Indianapolis. The coach had removable seats for pallbearers or could be used as an ambulance. The car was finished in white with gold striping and highlights.

This full ambulance was S&S Model 1400. It had an open driver's compartment and a body with only four carved columns. The carriage lamps on some of the coaches were now becoming smaller.

Horace Link & Company of Harris, Ill., owned this elaborately carved S&S Model 1125 hearse. Sayers & Scovill built their own chassis and used either a Continental or a Lycoming engine.

Mounted on a Ford chassis, this Millspaugh & Irish six-in-one combination coach was an example of the unusual type of vehicles that emanated from this Indianapolis based company. This car could be used as a passenger car, a pallbearer's limousine, a casket wagon, a hearse, or a sedan ambulance. The interior was convertible to any of these uses with only a few minutes work. All windows were fitted with blinds.

The Michigan Hearse and Carriage Co. offered this Model 1012 on a White chassis. This car was equipped with hard rubber tires.

1914

Nance & Pool of Puducah, Ky., were the proud owners of this Cunningham funeral coach. The fully enclosed body featured four full-sized carved columns and four shorter carved wooden columns framing six plate glass windows in the casket compartment. Heavy drapes were hung inside these windows.

Pallbearer's coaches were becoming more and more popular throughout the industry. This limousine style model was built by Cunningham and had leaded glass windows above the regular windows. It is of the fully enclosed body style. This coach belonged to Indianapolis funeral directors Hisey & Titus.

H.M. Patterson & Son of Atlanta, owns this huge Cunningham carved funeral coach. Note the size of the coach in relation to Mr. Patterson. This model has an open driver's compartment and the casket compartment is highly ornamented with carved draperies.

This elaborately carved Cunningham was designated the Model R-981 and was classed as a combination funeral coach and casket wagon. Cunningham offerings were now almost totally closed cars.

No two coaches are ever entirely alike, as is shown here with another Cunningham hearse style R-981. This model is finished in black and has more ornate carriage lamps and a glass window within the carved rear compartment.

Cunningham also produced a line of high quality passenger cars. This is the Cunningham funeral limousine that would seat seven passengers and was used primarily by hearse liveries and funeral directors. This car employed the same type of running gear and chassis as was used on the ambulances and funeral coaches built by Cunningham.

1914

The Rock Falls Manufacturing Co. of Sterling, Ill., built this combination hearse-casket wagon for Skeeles Brothers of Chicago. The single beveled plate glass window on each side featured ray textured drapes and the driver's compartment of this Model 72 was open. The small amount of carving on the body sides added a nice effect to the overall appearance.

This Rock Falls hearse featured considerably more draped carvings on the rear body sides and beveled plate glass in the driver's compartment quarter windows. The carriage lamp also has a draped overlay to add to the dignified effect.

Another Rock Falls is this Model 76 funeral coach. This was a very popular model and was built on the standard Rock Falls chassis with a 54-horsepower four-cylinder engine.

The Riddle Coach and Hearse Co. was established in 1831 in Ravenna, Ohio, and entered the field of motor funeral coaches and ambulances this year. The products were widely known throughout Ohio for quality and longevity. The first coaches were built on White chassis, with this silver-gray ambulance going to the Koch Brothers of Bradford, Penn.

With heavy carved drapery effect this Rock Falls Model 58 was one of the most expensive cars in the line in this year. The carved front windows were interesting.

This Riddle, on a White chassis, was delivered to Spiker & Kline of Canton, Ohio. It featured intricate carvings in the four-large-column, four-small-column design. The tassels and the heavy drapery seen in the casket compartment window were made of heavy broadcloth or velvet.

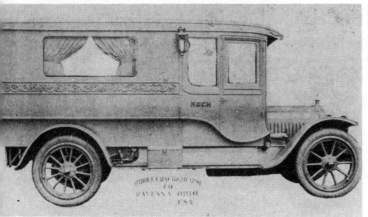

1914

The Lafayette, Ind., firm of C.H. Bradshaw owned this silver-grey ornately carved hearse built by Riddle on a White chassis. All of the carvings were done by hand and carved in huge pieces of wood.

L.O. Ream of Ada, Ohio, was the proud owner of this Kunkel combination ambulance and casket car. A fully enclosed coach, this car was built on a Kunkel chassis.

Kunkel built this ambulance for H.C. Hieber of Cleveland. It was designated the Model 162 ambulance. Kunkel of Galion, Ohio, built bodies for either motor or horse-drawn undertaker's vehicles or complete coaches. The large bell on this ambulance was typical of this type of vehicle during this era. It functioned as a warning just as the siren and the red lights do today. The car uses a Winton chassis.

A familiar face to antique car buffs, this Model T wears a body built by Kunkel and designated the Model 155 casket coach. The plain, high, square design blended with the Ford chassis with ease.

G.A. Schnabel & Sons of Pittsburg began the construction of motor ambulances and funeral cars this year. This one, constructed on a 1912 Packard chassis, was the Model 25, which sold for $2,000. This company specialized in building modern style coaches on used chassis.

Another Schnabel creation, designated the ambulance Model 26, was mounted on a White chassis. This particular coach was sold to White & Son of Newcastle, Penn. It had an open driver's compartment and rather plain styling.

This open pallbearer's car was also built by G.A. Schnabel & Sons on their own assembled chassis. Notice the one man top and the radiator directly in front of the windshield similar to styling used on the prestige Renaults.

1914

Operated and built by the National Casket Company of Boston, this White hearse had extremely high body lines to accommodate several caskets at one time.

The Hornthal Company of New York City was a large livery that supplied both hearses and limousines to local funeral directors. Their main supplier of vehicles was the Peter Kief company which built this mammoth carved hearse on a White chassis. Note the area in which the carved draperies are contained and the shape of this enclosure. Once again all of the wood carving was done by hand, right down to the details on the tassel and the fringe.

White built this limousine style ambulance belonging to the Charles Stumpf Undertaking Co. of New York City. White was building a wide variety of their own coaches and these were quite popular within the trade.

Operated by the St. Laurence Hospital, this Riddle ambulance was mounted on a White chassis and was complete with hand held lamps mounted on the runningboard, red colored carriage lamps, and roll-down side screens for the driver's compartment.

The Birmingham, England, firm of John Marston Carriage Works built this little hearse on the chassis of a current model English-built Ford Model T. Note the long narrow rear side window, the crest in the side door and the open driver's cab.

Another White ambulance of yet another style. The rear window is of an unorthodox shape, with a valance above it, as are the driver's compartment side windows. The overall body style is quite pleasing and modern for 1914.

1914

The Peerless Motor Car Company of Cleveland built this interesting four-cylinder combination hearse, casket wagon as an integral part of their funeral service vehicle line.

One of the most interesting of the Peerless vehicles in this line was the Peerless hearse, featuring carved torches and wreaths on the rear body sides. This design was a vast departure from the traditional carved draperies and had a special dignity. As on other Peerless models, the carriage lamp is quite small. The coach itself is built on the Peerless four-cylinder passenger car chassis.

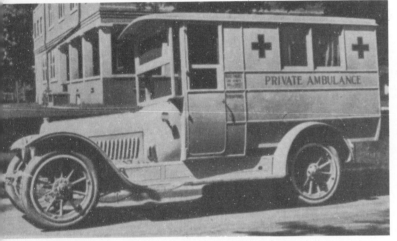

Mounted on a Hudson 6-54 chassis, this ambulance was especially built in a local shop for Theodore W. Abens of Aurora, Ill. It featured a fully enclosed body and a rather modern style that was copied from units mass produced by the big manufacturers.

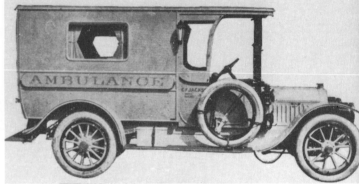

This Peerless ambulance was owned by C.F. Jackson and featured rather plain styling, an open driver's compartment, small modest carriage lamps and an interior fitted with hot and cold thermos bottles, a folding cot and attendant seats for two.

For most funeral directors the motor hearse was still a very expensive vehicle and some makers tried to announce lower priced models to get this business. This model made by the New York and Brooklyn Casket Company was called the Economy Hearse and was offered with a handsome yet strikingly dignified appearance. The body is quite plain with very little ornamentation other than four carved columns at the corners of the casket compartment.

In this era many funeral homes also owned and operated a local furniture store and had to operate a furniture delivery vehicle. Some firms recognized this fact and the additional fact that the funeral director needed an all-around utility vehicle to carry rough boxes, chairs, flowers, and other sundries to and from the funeral sight. Schelm built this combination funeral coach, furniture delivery car on a Ford Model T chassis that had been extended considerably. This utilitarian vehicle was to find many uses around the funeral home and many other companies were to adopt this idea and in the future offer similar models.

1915

The first Meteor funeral coach appeared on the scene in this year and was destined to become one of the most famous vehicles of this type ever constructed. The price quoted on the Meteor combination funeral coach/ambulance was F.O.B. Piqua, Ohio. The company had no dealers or road salesmen and, by eliminating these middlemen, Meteor was able to sell a "$2,800 car for $1,750." Most of the selling in these days was done through the mails by way of monthly literature and bargain sale notices. The advertising matter on this new Meteor car stressed the fact that the car was not priced low because of low quality, in fact the bodies were of high quality and compared favorably with more expensive units. As the bodies were built in high quanitities they were able to offer the low price and by building only one style, they were able to get better quality. At the same time, Meteor was offering its chassis to the public for $990. This was the same chassis that it was using on its own car and the same chassis that it had been selling to other funeral coach manufacturers for $988. The other manufacturers would then sell the car with their own body on it for $2,500 to $3,000, realizing a good profit. Meteor had previously been in the passenger car business and it seemed only normal to take this step as their chassis were becoming quite popular.

America's first 12 cylinder car was introduced by Packard in this year and was soon to make an impression on the funeral coach and ambulance field. Another first was declared in this year when the Miles Funeral Home in Toronto, Ontario, became the first owners of a commercially manufactured funeral coach in Canada.

The war was beginning to cause a shortage of materials in some quarters and some funeral coach and ambulance makers began to construct ambulances for shipment to the allied powers overseas.

Meteor entered the funeral coach and ambulance field with this coach and a new low price that did for this industry what the Model T did for the passenger car. Employing the Meteor six-cylinder chassis, this combination coach sold for $1,750 plus $35 for handling. The company did not have any road salesmen or any agents, and handled all sales directly through the mails, telegraph, or by phone. They also arranged liberal time payments to help all funeral directors to afford motorized vehicles. The Meteor Motor Company, destined to become one of the industry's big names, was located in Piqua, Ohio.

In a class by itself was this handsomely styled limousine, the S&S Model 1025 pallbearer's coach. It was rapidly convertible for use as an ambulance.

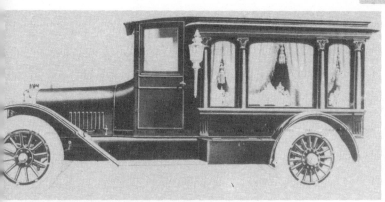

The Luverne Automobile Co. of Luverne, Minn., built this eight-column glass-sided funeral coach. Among its features were fringed draperies with tassels, a fully enclosed driver's compartment, and large ornate carriage lamps.

This ornate S&S hearse was delivered to W.C. Rigby of Medina, Penn. It was designated the Model 1525. The magnificently carved arches enclosed within their frames highly detailed carved wooden drapery. This coach featured the famous twelve column design originally introduced by S&S many years earlier on horse-drawn equipment. The car was powered by a Continental engine, with all of the body parts being made at the S&S plant in Cincinnati.

1915

The A.W. Miles Funeral Home of Toronto, was the first funeral firm in Canada to own and operate a completely manufactured funeral coach. This Cunningham was manufactured from components assembled from the largest suppliers in the U.S., such as Continental, Borg & Beck and Eaton. The Miles establishment is still located in Toronto and is one of the larger, more prominent funeral homes in that city.

This Cunningham ambulance, Style 992 was delivered to Garrett & Co. of Los Angeles. It featured the limousine styling that was becoming increasingly popular in this era. This car has a warning light mounted on the front windshield pillar on the right hand side.

This stylish pallbearer's coach was built for the William Tickner & Son funeral home of Baltimore by Cunningham and was a supreme example of this type of car.

"Built to maintain a reputation of 75 years," was the slogan used by Cunningham. The Cunningham Model R, Style 994 funeral coach was finished in silver gray. The car featured a fully enclosed body, now the dominant style, and an eight column carved casket compartment. All of the interior fittings were done in natural finish mahogany.

Another Cunningham Style 1017 combination casket car, hearse. This coach has very little carving and is finished in silver gray. Rosa & Hignell of Beacon, N.Y., took delivery of this coach.

William Sardo Undertaking of Washington, D.C., operated this elaborate Cunningham Style 981 funeral coach. The carvings are carried over to the front door tops. The double mosque deck type roof was continuing to be used on motorized coaches of the era.

1915

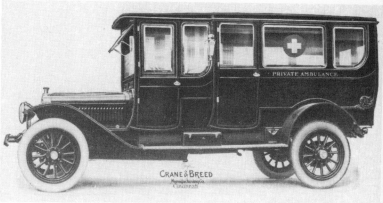

The Crane & Breed Model 1117¼ FW was offered in two versions, both on the Winton Six chassis. When equipped with the 33-horsepower motor, this coach sold for $3,900. With the 48-horsepower version it sold for $4,750. Although identical in exterior detail, these cars offered the funeral director a choice of prices for the same model. Once again the driver's compartment was open for the southern and middle tier states while the northern states were offered this model with a fully enclosed cab. The 1915 models featured a new shorter hood and dash cowl and different headlamps.

Designated the Crane & Breed 230-A closed-front limousine style ambulance, this model was of a standard design and considered to have been the ultimate in style. The mechanical specifications for this coach were the same as those for the hearse this year.

Mounted on a Velie chassis, this Rock Falls Model 88 funeral coach had an open driver's compartment and a highly carved draped casket compartment. Rock Falls would build a coach for a customer on either the Velie chassis or their own assembled one.

The Rock Falls Model 102 on a Velie chassis reflects the more modern limousine style with carvings limited to the belt line and leaded stained glass windows just below the roof line. Notice the full draperies and the tasteful way in which the carvings are handled.

This Rock Falls on a Velie chassis was designated a Model 99-B combination casket wagon, hearse. Once again the coach lamps are draped and the drapery carvings are found only above the belt line with the notable exception of the four corner pillars. This car was owned by Crawford S. Miller.

A service car, sometimes called a casket wagon, was primarily used as a utility vehicle around the funeral home. They were the workhorses that carted chairs, flowers, stands and caskets from the railroad station back to the funeral home. This is the Rock Falls Model 104 service auto mounted on a Rock Falls chassis. The styling of this car is purposely plain, as were most service cars, and features one window on each side with ray textured drapes.

1915

This funeral omnibus was offered by the J.C. Brill Co. renowned for its line of trolly cars for city use and for railroad equipment. Mounted on a gasoline powered chassis, this coach was painted a rich dark green overall, thus permitting its use for purposes other than funerals. Drapes were of black silk and tied back with black silk tassels at each corner.

An interesting hearse built on a Jeffery chassis. It shows very clearly the etched initials on the driver's compartment quarter window. The carriage lamps are small and the carving is found only on the four corner pillars. The large plate glass window features full drapery.

P.G. Katz & Son of Verona, Penn., bought this Schnable ambulance body and had it mounted on a used chassis. The body style is very different from anything seen before. The chassis looks like a 1910 Oldsmobile Limited.

The Owen Brothers of Lima, Ohio, built coaches from the funeral director's own designs, or the customer could choose from standard models. The company lengthened passenger car chassis especially for the bodies and specialized on Cadillac, Buick and Cole. This Model S-294 was mounted on a Cadillac chassis and features a small oval beveled plate glass window just behind the front doors. This car is a standard offering of the Owen line and is shown here finished in light grey.

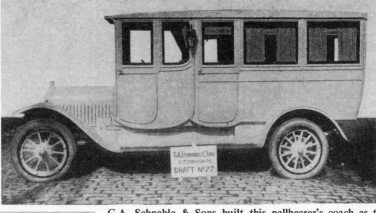

G.A. Schnable & Sons built this pallbearer's coach as the Model 27 and incorporated the latest limousine body style. The drapes are not the fancy tied back type, which hints that this car was convertible to an ambulance.

Another and quite unique offering from Owen was the Model C-82 on the eight-cylinder Cadillac chassis. In the center panel of the rear compartment, a window with a ray textured drape has been installed in place of carvings. That drape features a large cross that lends a whole new look.

This is an interesting view of a Seaman ambulance built on a Cadillac chassis for the Receiving Hospital of the city of Detroit. It clearly shows some of the interior appointments of this vehicle. The suspended cot and the plain interior of the car were typical of ambulances of this era. Another cot is shown ready to be loaded.

Seaman built the body on this Jonas chassis. The ambulance body was painted white overall and had gold lettering.

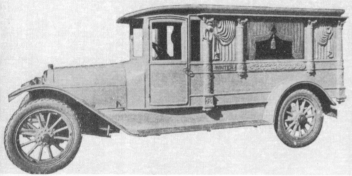

The Kunkel Model 170 combination casket car and hearse was built especially for the Winter Funeral Home. It featured a light gray finish and carved drapery panels between four full-length pillars and four shorter ones. The drapes in the large single windows on each side were the same type as seen on the Model 164 combination.

This massive carved vehicle is more unique than it looks. Built on a National chassis, the coach was designed to carry two caskets at one time. The overall size of the car is easily seen in relation to that of the gentleman standing beside it.

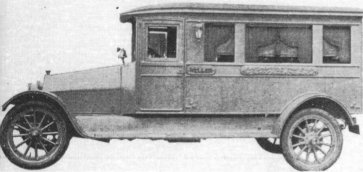

Kunkel built this Model 164 combination car on its own chassis for H.I. Weller of Cuyahoga Falls, Ohio. The car was rapidly convertible from an ambulance to a full hearse and featured fringed cut-away draperies.

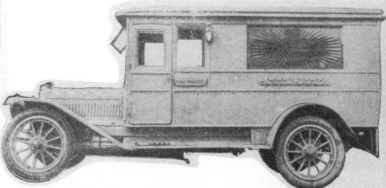

Kunkel utilized the Winton Six chassis for this casket wagon built for C.J. Ware Undertaking of Chillicothe, Ohio. The body was designated Model 165. The coach styling follows that of most casket cars of the era with a single window on each side of the casket compartment and ray textured drapes. The carved panel below the side window was used on the other Kunkel cars built in this year, and depicted strings of ivy.

The William Erby & Sons Co. of Chicago began the production of funeral coaches and ambulances this year. This hearse is one of the first vehicles to leave the plant. It features a small amount of draped carving on the rear body sides.

1915

An example of the progress of the motor hearse overseas is seen in this large Australian built oval windowed hearse. Style-wise the car reflects the styles prominent in North America with the addition of a large flower rack on the roof and a more highly developed casket table. This coach reportedly cost $5,700 new in Australia in this year.

This attractive combination casket delivery coach-hearse was built by the A. Geissel & Sons Co. of Philadelphia. It was mounted on the Buick C-4 chassis. Selling for $2,384, the coach featured an elaborately carved casket compartment, an enclosed driver's compartment, and beveled plate glass in all side windows.

Mounted on a Packard chassis, this Blue Ribbon glass-sided hearse left the casket almost completely open to view. Blue Ribbon Auto & Carriage Co. was located in Bridgeport, Conn., and built coaches in any design to order.

Columbia Body Co. of Detroit was most famous for little funeral service vehicles mounted on the Ford Model T chassis. This little casket wagon has a uniquely styled body with colonial type windows and a mildly redesigned front end.

Another Ford-based vehicle, this one was built by the William Pfeiffer Auto & Carriage Works of Omaha, a firm established in 1888. This hearse was sold to C.E. Johnston of Wanneta, Neb., and featured a light gray paint job and a rather high plain body.

The fact that the manufactured, assembled motor hearse was still a fairly expensive item that not all funeral directors could afford is best illustrated here with this Model T Ford that was used as a hearse by Ibbonbay of Cutbank, Montana. Equipped with a pickup bed, the casket was set across the back of the car and thus transported to the place of burial. Notice also how casket styles have changed since 1915.

Hornthal's hearse livery in New York City had for years used Peter Kief vehicles, but this year began the construction of their own coaches. This model, built on a Ford Model T chassis, was the standard casket delivery and service car.

1916

The twelve cylinder engine came to the funeral coach and ambulance industry in this year with the introduction of the Meteor "Twin Six" combination pallbearers' coach and ambulance. The twelve cylinder engine used by Meteor in this car produced 72 horsepower and was made specially for Meteor by Weidley.

"The twin six or 12 cylinder motor is not a fad," announced Meteor. "Such well known automobile manufacturers as Packard, National, Pathfinder, and H.A.L. Company have seen its worth and necessity in large cars," the advertising copy continued. Meanwhile, Meteor was selling its Combination for $1,750 at a fantastic rate. In the months of October, November, and December the company sold 200 cars alone. This would be a heart-warming event today for any coach maker. On January 1, 1916, its production was all sold out till March 10, as well as about half the production for April and May. This rush order surpassed even Meteor's fondest expectations, and the company offered to accept 30 orders for the month of March delivery, 50 cars for April and May, or about 80 cars up until June 1, 1916.

"These 80 cars will clear up the material we have contracted for and we will, no doubt, be forced to advance our price to cover the additional price of material. This is fair warning. Don't blame us if you have to pay a hundred dollars more for your Meteor after these 80 cars are sold." This was the type of flamboyant advertising that sold cars in this era. In the meantime, the six cylinder Meteor chassis had risen in price to $1,050 and was advertised complete with hood, fenders, and everything ready for mounting a body.

The fact that Meteor, a one-year-old funeral coach maker, could attain such an overnight success illustrates the fact that the motor hearse was here to stay and the demand was astronomical. At the same time, some of the smaller manufacturers began to copy the Meteor formula for success. J. Paul Bateman Co. of New Jersey began offering a body built on a Studebaker chassis for as little as $1,650. Although too small to assemble its own cars, Bateman used a wide variety of chassis this year, including such marques as Crow, Cadillac, Buick, Studebaker and Reo. There were many other firms scattered over North America engaged in the construction of these coaches and more entering the business annually. One of the most famous of the new arrivals for 1916 was the John W. Henney Co. of Freeport, Ill. Production of funeral coaches began in Freeport about mid-year with about 60 Henney coaches produced in 1916. Henney assembled his own chassis, and the bodies were wood frame covered with metal shaped by hammers in the hands of Finnish metal workers. At this point the paint was applied, a process that required three weeks with 21 coats of paint being applied. This was the process at Henney and at most all of the major manufacturers at the time.

Meteor announced the industry's first 12-cylinder funeral coach with this Model 85 combination pallbearer's – funeral coach. The car featured the limousine body style on a 148-inch wheelbase, 35x5 cord tires and Timken axles.

Seen here is a rear view of the Model 85 as a combination pallbearer's car – ambulance. This car, equipped with the Twin Six engine, sold for $2,150. It featured clean lines and two-tone paint jobs.

This Meteor Model 75 came with a six-cylinder engine as standard equipment but could be ordered with the Weidley 12-cylinder at extra cost. The coach sold for a base $1,750 in its six-cylinder form and $2,050 with the 12 cylinder. Note the limousine styling and the strange side window mouldings and the leaded glass windows above the large windows.

1916

When Meteor introduced its low-priced Model 75 last year sales zoomed. It taught the Meteor people that there was a real market for low priced, high quality coaches. They continued to sell the 75 for $1,750 but introduced a new Model 80 that sold for $1,700 at the beginning of the year, only to be lowered to $1,650 later in the year. When they lowered the price of this model, they gave the funeral director an added incentive to buy Meteor motorized equipment by offering to sell the first 100 cars ordered for only $1,575.

The instrument panel and control levers of the Meteor Model 75 are clearly seen in this photo. All controls were set far forward to enable the driver to get out of the right side of the car without interference. The entire front compartment was trimmed in the same material as the seats and doors. The steering column was bracketed solidly to the cowl with all control levers within easy reach of the driver.

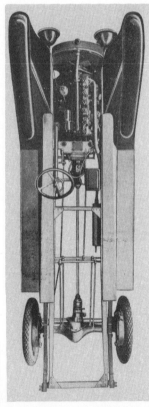

This is the engine called the Meteor Twin Six that was built for Meteor by Weidley. It produced 72 horsepower. The 12-cylinder engine was not a fad, as several passenger car manufacturers already offered this type of engine. Among these were Packard, National, Pathfinder and H.A.L. This engine featured a 3-inch bore and was ideally suited for funeral coach work.

The interior of the Meteor Model 75 had wood trim genuine solid mahogany and drapery made of soft silk po of a deep purple. The flat removable flower rack is clea seen just above the front seats. This car could also be orde as an ambulance or as a casket wagon. When ordered as ambulance, a set of plain silk curtains were furnished ar cot and two attendant seats were placed in the compartment.

Meanwhile, the Meteor Motor Company offered its chassis complete with hood, fenders, and everything ready for mounting of any desired body finished in lead coat for only $1,050. This was the same six-cylinder chassis Meteor used on the ambulances and hearses.

Another view of the Meteor combination coach that sold for a record low price of $1,750. This coach had a six-cylinder Continental engine that developed 45 horsepower, a 148-inch wheelbase, Timken axles, 35x4½ Goodyear tires, a Delco starter, Stromberg carburetor, and a tire pump. The car came complete with two sets of drapes, one plain for ambulance work, and the others (seen here) of high grade silk.

1916

43

Sayers & Scovill announced the new Model 1745 with elaborate carved draperies and the famous S&S twelve column design. The coach was finished in a light gray with highlights of dark gray around the carved sections. This coach also featured a nickel plated radiator with a moto-meter.

Shown here is the S&S Model 1025 limousine pallbearer's coach finished in black. This coach featured the same frontal treatment as seen on the Model 1745. All windows were of heavy beveled plate glass and there were roll-up blinds in addition to drapes.

From the rear the S&S 1745 was also beautiful. The carving was carried over to the swelled rear doors as was the two-tone grey finish. Note the folding step to assist in loading.

The Alliance Manufacturing Co. of Streator, Ill., entered the field this year, offering a line of coaches of high quality at a minimum price. All models in the Alliance line were standardized and built on their own chassis. Using the 40-horsepower Lycoming engine, this Alliance Standard carved panel hearse Model 179 sold for $1,795.

This was the Alliance standard limousine funeral coach also using the 40-horsepower six-cylinder engine but this model sold for $1,765. This difference in price is attributed to the fact that the limousine lacked the hand carved panels that adorned the Model 179 and that the limousine style was less expensive to construct.

Hornthals had begun to build its own coaches last year and in this year offered this stylish White hearse with rounded body lines. This company was also a leading hearse livery in New York City.

1916

All of the materials that went into the body of this Bateman eight-pillared carved hearse were claimed to have been the best. This coach is the J. Paul Bateman Model 36-S mounted on a Cole 8 chassis and sold complete for $2,480, or only $800 if the customer supplied the chassis.

The Bateman company received a letter from J.P. Rearick of Martinsburg, Penn., when he arrived home after taking delivery of this model 36-R casket wagon mounted on a Studebaker chassis. The letter stated that he was extremely pleased with the coach and would not hesitate to purchase another Bateman built coach. This vehicle cost the owner $1,650 as Bateman supplied both the chassis and the body.

This handsomely carved hearse body in white, ready for paint, was constructed of only the best materials and offered on any chassis sufficiently long enough to accept it. Marketed as the hearse body Number 0337 by the McCabe-Powers Carriage Company of St. Louis, this was one of a line of carved hearses, limousine hearses, combination, ambulance, and casket wagon bodies.

The J. Paul Bateman Company of Bridgeton, N.J., offered a full line of body styles on any chassis. The normal charge for constructing a body on a supplied chassis was $800 with a coach costing $2,400 or more if Bateman bought the chassis. Naturally, the buyer had to purchase the chassis. Shown here is the Bateman style 36-R with the body in the rough ready for paint. The chassis is a Hudson Big 6.

This Bateman funeral coach is seen mounted on a Crow chassis. It was designated a Model 36-R. The large plate glass windows of the Model 36-S were replaced with carved wooden ones on this model.

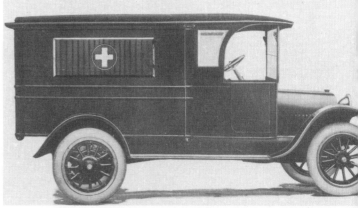

This McCabe-Powers combination ambulance and casket coach was marketed as a Model 380 and is here mounted on a Dodge chassis. The driver's compartment is open and beveled plate glass is used in the rear window.

This handsome car was placed in operation at the Harlem Branch of the New York & Brooklyn Casket Company, its maker, and was offered to the public as an Economy Casket Wagon.

1916

This attractive hearse was built on a Buick D-55 6-cylinder chassis by the William Pfeiffer Auto & Carriage Works of Omaha, for David. D. Reavis of Falls City, Neb. This firm concentrated on only a few styles and this enabled them to produce a product of very high quality at a suprisingly low price. This car was a style 31-R.

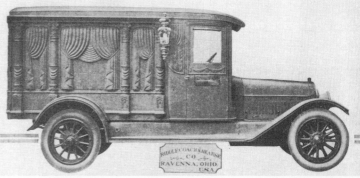

The Riddle Coach and Hearse Company of Ravenna, Ohio, offered this carved panel hearse on its own chassis. The Riddle Company offered a full line of coaches in all of the popular body styles.

Utilizing the Continental engine, Riddle also offered its chassis to the trade. A funeral director could purchase a chassis of his choice and take it to any body maker and have a body of his choice built on it.

Pfeiffer would mount a body on any chassis the customer preferred as seen on this casket wagon mounted on a Dodge chassis for Bernar & Peters of Sloan, Iowa. The Pfeiffer plan allowed the customer to buy a chassis from his local dealer, thereby assuring him of good service for his motor hearse.

The rebuilding of horse-drawn equipment on a motorized chassis was becoming quite popular and most firms would undertake this type of conversion. The Owen Brothers built this coach utilizing a horse-drawn body and a Buick chassis. This car was called a Model 308.

This picture illustrates the Owen Brothers Model 740 combination pallbearer's coach-ambulance on a Cadillac Eight chassis. This model featured six removable seats so that it could be rapidly converted to its ambulance duties.

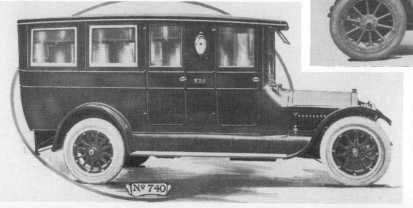

Designated the Owen Brothers Model 742 this funeral coach featured carved wooden panels with the center one being made available in glass, as a window, if desired. This car is mounted on a Cadillac chassis. Owen lengthened chassis and remodeled bodies on any new or used vehicle or supplied complete cars on both the Buick Six and the Cadillac Eight chassis.

1916

Setting the standards of the trade, Crane & Breed announced its Model 1152 for $3,785. Crane & Breed cars were advertised as the best coaches available for the price, with the Winton Six being a much more reliable power unit than any four on the market and offered twenty times longer a life span than any four-cylinder engine available.

Priced at a rather steep $4,300 this Crane & Breed Style 1153 utilized a Winton Six chassis with a 6-cylinder engine that developed exactly 33-75 horsepower S.A.E. rating. All prices quoted were F.O.B. Cincinnati and were for factory drive away.

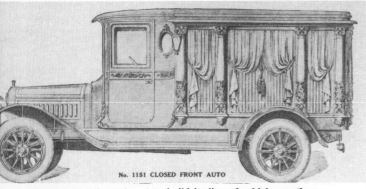

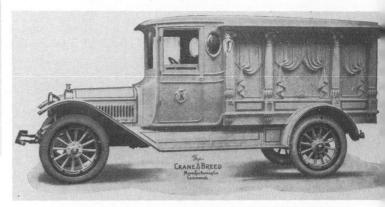

Crane & Breed continued to build its line of vehicles on the Winton Six chassis. This 33-horsepower version was the Model 1151 and was the first one delivered. It went to R.R. Bringhurst & Company of Philadelphia. The exact design designation of this car was called a closed front model.

This Crane & Breed Model 1150 sold for $3,825 and was delivered to Konatz Undertaking Co. of Ft. Scott, Kansas. The company quoted no asking prices, only the selling price. Most of the Crane & Breed coaches lasted for a full 10 years of service without any major repairs.

Calling themselves the "one price house" and "the house of quality," Crane & Breed offered a wider variety of models and ideas than any other firm. This Model 1155 ambulance was delivered to the Van Metre Undertaking Co. and was finished in white with gold trim. The extremely clean limousine lines are well fitted to this chassis and the car looks many years ahead of its time.

William Erby & Sons of Chicago, built this ambulance for the Iroquois Memorial Emergency Hospital. It was presented to the hospital by the Chicago millionaire phialnthropist T. Crane. Erby claimed that the fact that their cars were given preference spoke for itself of the quality of the bodies which were built with special care and attention to detail.

1916

Looking like a modern police paddy wagon, this Kissel Kar Worm Drive ambulance was the epitome of refinement in its day. Notice the lack of lights, the rear compartment vents, and the open driver's compartment, not to mention the stationary step at the rear doors.

The Kissel Motor Car Co. of Hartford, Wis., entered the funeral coach and ambulance market place in this year with the Kissel Kar Worm Drive vehicles. This Kissel Kar hearse reflects the styling prevalent in open cars in this year.

Shown here is the Michigan Model 1017 carved funeral coach. Michigan was one of the trade's larger makers, offering a line of vehicles built on their own assembled Michigan six-cylinder chassis.

Introducing a new Light Six line of funeral cars and ambulances, Michigan promised that these cars would be priced at under $2,000 and would be ready for inspection at the showroom in Grand Rapids no later than the 20th of April. The coach features the new clean limousine style that was becoming increasingly popular throughout the industry.

G.A. Schnabel & Sons built this rather plainly styled hearse on their own chassis. The driver's compartment is open and the exterior finish is painted black overall.

Another Schnabel, built on a Ford T chassis, is this little casket wagon. The styling of this car closely followed that of the other Schnabels.

1916

This Geissel funeral coach is on a Buick D-4 chassis. If the customer were to supply the chassis, the coach would cost $1,200, but its price was $2,395 if ordered direct from Geissel.

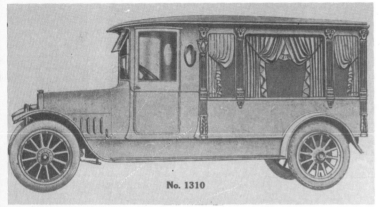

Offering coaches on chassis from Studebaker, Overland and Cadillac, the Champion Wagon Company of Owego, N.Y. offered this stylish Model 1310 carved panel funeral coach. The company took the chassis and lengthened them, then mounted its own custom designed bodies.

The John W. Henney Company of Freeport, Ill., was formed many years earlier and in this year announced its first motorized line of vehicles. This Henney carved panel funeral coach was mounted on an assembled Henney chassis and used the Continental engine.

Another maker to enter the funeral coach and ambulance field this year was the Houghton Motor Co. of Marion, Ohio. This attractive coach was the Model 400 and was offered on Houghton's own chassis.

Fred Groff of Lanchester, Penn., was the proud owner of this carved panel Cunningham Model V hearse. Cunningham retained the large ornate carriage lamps and the exceedingly ornate carvings on the casket compartment.

Offering both horse-drawn and motorized coaches, the Kunkel Carriage Works of Galion, Ohio, offered this complete auto funeral car Model 143½ for $2,475.

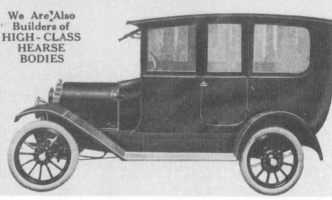

Mounted on an Overland chassis, this attractive Millspaugh & Irish six-way combination features center door styling. The casket rode beside the driver's seat and the car could be used as a hearse, casket delivery car, ambulance, or a pallbearer's car. This company also built high class hearse bodies.

1917

America entered World War I and the auto industry began to make even more material for the war. This was also true of the funeral coach and ambulance manufacturers who had been building small numbers of military ambulances prior to this year. Most of the military ambulances were built on a manufactured chassis from a production line passenger car or truck to make front line servicing of these vehicles easier. Cadillac, White, Ford and Buick were among those to see their chassis serving at the front equipped with ambulance bodies.

On the home front, the funeral car saw an increased price due to a war tax levied against all automobiles. Funeral directors in some localities continued to use the horse-drawn equipment, even though there were more modern types of funerary transportation. Styling of the cars of this era reflected a more modern approach to the carved side car while still maintaining the dignity of the older horse-drawn vehicles. The carved side style was maintained because the funeral directors wanted a coach that was distinctively a funeral vehicle and could not be mistaken for any other type of vehicle. But a new style loomed on the horizon in the form of the limousine coach — a hearse that reflected the style of the now common bearers' coach.

Meteor continued to offer the twelve cylinder models in an extended range that now included a carved side funeral coach. The enlarged model line-up featured models with prices both above and below the original $1,750. The Meteor Model 80 carved funeral coach used the Continental six cylinder engine that produced 45 horsepower. The coach sold for $1,700. The model 80A with the Weidley twelve cylinder powerplant sold for $2,000.

The A.J. Miller Co. of Bellefontaine, Ohio, began producing its own complete cars. Previous to this, it had built only funeral coach and ambulance bodies and mounted them on the popular Meteor chassis. The first Miller funeral cars had Continental Red Seal engines (like their Meteor counterparts), Borg & Beck clutches, and Eaton rear axles. Because these cars were run only 1000 to 2500 miles per year and at speeds of from 10 to 25 miles-per-hour, the new Miller assembled coaches, and all assembled cars for that matter, gave splendid service.

The A.J. Miller Company was founded by Amos Miller in 1853 in Bellefontaine, Ohio, to build buggies and carriages, but they did not venture into the hearse and ambulance field at this time. The company built its first hearses in 1870 and then only to special order. In fact, these vehicles did not become part of the firm's standard line for several years. It was in 1912 that A.J. Miller entered the auto hearse business, but then only on Meteor chassis.

This is the Meteor Model 85 or 85-B. The only difference between the two was that the Model 85 used the Weidley V-12 and the 85-B used the 45-horsepower Continental 7-N engine as did all the other six-cylinder Meteors. Shown above is the Model 85 combination ambulance-pallbearer's coach.

Continuing to offer the Weidley V-12 as an option, Meteor entered the year with a Model 80. When equipped with the 12-cylinder engine, this coach sold for $2,000, but in its standard form the car was equipped with a 45-horsepower six-cylinder made by Continental and sold for $1,700. Both cars rode on a wheelbase of 148 inches.

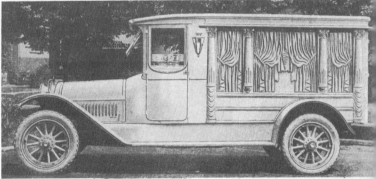

The Model 80 was also offered in a carved hearse style in addition to the limousine style shown here earlier. This car was also available as either a 45-horsepower six or a 72-horsepower twelve and sold for the same prices as the limousine style.

Meteor added two new models to its offerings this year with the Model 92 and 82. Both cars featured identical exterior appearance with some variation in their specifications. Both used six-cylinder 45-horsepower Continental engines and rode on wheelbases of 148 inches. The Model 92 used Columbia axles and sold for $1,650 while the Model 82 used Timken axles and was priced at $1,750. The tires for both cars measured 34x4½.

A.J. Miller of Bellefontaine, Ohio, began the production of its own funeral coaches and ambulances this year. Building its own chassis and using the Continental 52 horsepower six-cylinder engine, this Model 40 sold for $2,187. In the center panel, carvings depicted draperies while the two outside panels featured carved urns.

This Miller features more subtle styling and was designated the Model 42. Notice the small oval window behind the driver's area.

Shown here is the A.J. Miller Model 41 Colonial with swelled center panels that were removable to show a large plate glass window. It sold for $1,790. The chassis specifications for this model were the same as for the Model 40. Both cars used Goodyear balloon tires, size 34x4½.

Building ambulances for military service was not restricted to the manufacturers of civilian ambulances. The passenger car makers began to supply these vehicles to the military on their own chassis. Here is one such vehicle built by Hudson.

In some cases the A.J. Miller company mounted its bodies on other chassis as is seen here with a Miller Model 60 on an International chassis. This body style is unusual because of the large shield panel carved in the center of the carved drapery casket area. All Miller coaches featured small coach lamps and a small oval window just behind the driver's compartment.

The United States was now involved in World War I and the makers stepped up their production of ambulances for the military. Seen somewhere in France in October of this year is this American-built military ambulance in a scene that had become all too common during the war.

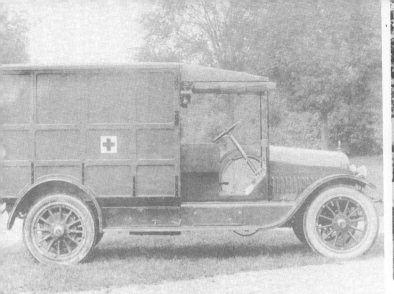

In contrast to the rather utilitarian appearance of the Hudson ambulance, this military ambulance was built by Cunningham and featured fully enclosed rear sides and a roof over the rear. Cunnungham built a large number of these vehicles and they were reported to have served notably.

Lewis E. Smith was the proud owner of this Cunningham Model 23-R and the coach was used in this firm's Middletown, N.Y., funeral home for quite a few years. This style features carved drapery panels in the four smaller areas beside the center window, with this main window featuring beveled plate glass and heavy draperies.

The U.S. Navy was a large customer for ambulances and in this year bought a few from Cunningham. The Navy set the specifications and the body style for the cars it ordered and they turned out looking like this one. With tires on the roof and two flag staffs for Red Cross pennants, this car is still quite attractive. The bell on the running board was produced by Van Duesen.

This Cunningham Model 10-A was built especially for A.J. Barron of Franklin, Penn., and featured the more common body styling with carved draperies on the rear body sides.

This beautiful Cunningham hearse is fitted with the latest body styling and is finished in white. Note the arched windows and the huge center window. Draperies in this car were of purple broadcloth or velvet and were hung in an entirely new fashion.

August Schubert Wagon Company of Oneida, N.Y., built this uniquely styled hearse on an unidentified chassis. This car was a converted horse-drawn vehicle and this is easily seen with a close examination of the top of the carved area. Note how this doesn't blend evenly with the roof line.

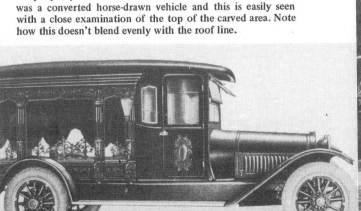

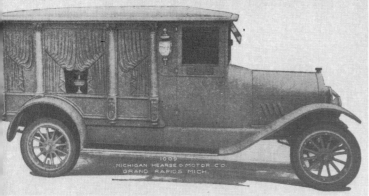

Mounted on their own chassis, the Michigan Model 1009 hearse featured styling similar to that shown on the Model 1010.

Michigan introduced a new design called the 1010 this year. Featuring a more modern approach to the carved panel hearse, this car used the Michigan chassis with a 45-horsepower Continental six-cylinder engine.

Michigan cautioned buyers to look beyond the price for a quality funeral coach. Offered was this Style 1016 on their own chassis with carvings limited to the drapery panels and the belt line.

Utilizing their own assembled chassis, Michigan offered this version of a limousine combination funeral coach to the trade. It was designated the Model 1032.

Houghton built cars in only one size and in two different styles with six different model details. Houghton built its own chassis as well as complete cars.

Houghton Motor Car Co. of Marion, Ohio, built this handsome little carved hearse on their own chassis with their own engine. They also marketed a line of combination cars, ambulances and casket wagons.

Houghton built this utility lightweight body on a Buick chassis. This high, narrow body was designated as a casket wagon and had one window on each side with ray textured draperies.

1917

Another ambulance offering from Rock Falls was the Model 45 mounted on their own chassis with large blinds for the rear windows.

Rock Falls also assembled their own chassis and offered this combination pallbearer's coach-ambulance in this year. The coach featured six removable seats and came complete with attendant seats and cots. The coach was styled like some of the more expensive passenger limousines of the era and presented a very dignified appearance.

Featuring bolder carvings and large carriage lamps, this Rock Falls hearse was considered the top of the line. It was mounted on the assembled Rock Falls chassis.

Rock Falls also offered some of the industry's most elaborately carved hearses as can be seen in this example, with moderately carved draperies in the casket area.

Announced as a new model, this Bateman Style 6017-A featured more carved surface with large carriage lamps and an oval window behind the driver's compartment. The car is shown here on a Cadillac V-8 chassis.

A customer in North Carolina took delivery of this silver gray hearse on a Studebaker chassis built by J. Paul Bateman of Bridgeton, N.J. Designated the Model 12016 this coach featured six large plate glass panels and a minimum amount of carving.

1917

The Utility Car Co. of Cincinnati, Ohio, built one coach that could be changed into four separate body styles for four different uses. Shown here is the car in its carved hearse version. The carved panels were removable.

With the smaller columns placed back on the car, it made it the first marketed flower car. Note that the draped panels are still demounted.

The Utility Car Co. ambulance. When converted to its ambulance duties, the car featured the draped panels and the short columns demounted. A blind was equipped on this car that could be pulled down to read "Private Ambulance," thus converting the car to its ambulance form.

This picture shows the same car in its casket car form. The plain panel with the wreath carving is mounted in the center in front of the window and all of the carved columns are removed. Note that this car in all of its various versions maintained a dignity and style not found on most cars of the era. With the flower car version, Utility was to market the industry's first car designed to haul flowers and be designated as a flower car. All of these cars feature beveled glass in the driver's compartment doors, and a unique versatility not found in any line of coaches or one coach since.

The Sayers & Scovill Model 1025 was built especially large to accommodate six pallbearer's or to meet its duties as an ambulance. Sayers & Scovill marketed more models of a wider variety of styles than any other maker in the field and kept their cars in line with all of the progress made in the passenger car industry.

This S&S Style 1695 featured beveled glass in the doors and restrained carvings on the rear body sides. The drapery in the center rear window was of hand carved wood.

1917

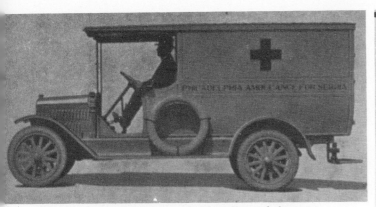

The city of Philidelphia sent this Kissel Kar ambulance to Serbia. The styling of this vehicle is very similar to that seen on earlier Kissel Kar ambulances and was powered by a Kissel engine and drive train.

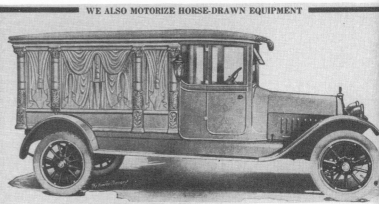

Velie built a special duty chassis just for such vehicles as this Eureka hearse. Eureka called this the Model 20 and the overall style of the car was somewhat lower than competing coaches. Eureka was located in Sterling, Ill.

Kunkel offered this carved drapery funeral coach as the Model 50. The car features elaborate draped carvings on the rear body sides and eight carved pillars.

William A. Carroll Co. of New Beford, Mass., succeeded the Peace Co. and offered this hearse with beveled plate glass and urn carvings on a Buick D-6-55 chassis.

The economy funeral coach and ambulance was becoming quite popular and there were quite a few built by several little makers. This casket wagon was built by H.H. Babcock of Watertown, N.Y., and was mounted on the economical Ford T chassis. The car weighed 1,900 pounds complete and rode on an extended wheelbase of 130 inches with a body length of 96 inches from the back of the driver's seat. This car sold for $699. The company also built hearses and ambulances. They claimed that this car was suited to country as well as city uses.

The A.J. Volk Funeral Home of Hoboken, N.J., owned this Geissel built combination casket wagon-ambulance. Many ambulances of the era were called invalid cars. This car has both drapes and roll-up blinds.

1918

The rapid increase in the use of the automobile and the general manufacturing "boom" of the war years was felt by the funeral coach and ambulance makers, and the market expanded to new heights. Because of the war, accessory and automobile manufacturers offered devices for prolonging the life of cars. Carburetors designed to use low grade fuel and even kerosene were introduced to help conserve fuel. The fully enclosed funeral coach and ambulance was here to stay and practically every coach made was of this style.

Another new field opening was in the area of coach rebuilding. The old horse-drawn coaches were getting harder to sell, with everyone switching to the newer motorized vehicles, and some smaller makers began to specialize in rebuilding the bodies of these former horse vehicles and mounting them on a motorized chassis. Some makers would do the job for as little as $800 if the customer supplied the chassis. Other makers, having made their success with coaches, began to construct passenger cars especially for the funeral profession. These cars were built in the form of limousines or large sedans and were used primarily by liveries or large funeral homes.

Patriotism in the United States was at an all-time high and this Cunningham hearse reflects that feeling with the American flag serving as drapes for the casket compartment. This coach was owned and operated by J.A. Still of Bradford, Penn. It was a Cunningham Style 41-A.

Some models in the Cunningham line now offered a Continental eight-cylinder engine and this Model 1204 was among them. Notice how the carved draperies are restrained to the window area above the beltline.

The American Red Cross operated a fleet of these dependable Model T ambulances at the front during World War One. Notice that the body is constructed of boards with a canvas roof.

With World War I raging in Europe, the American fighting-man that died on the front was to have his body returned to the U.S.A. for burial. In this military funeral in New York City, an armored motor car is used as a hearse to bear the body of a fallen soldier.

On the home front this was the type of ambulance popular with the trade. This particular car with an unidentified make of body was mounted on a Buick chassis and was operated by the Buick Motor Company in Flint, Michigan. Most all professional cars of the era, be they hearses or ambulances, utilized the double rear door style seen here.

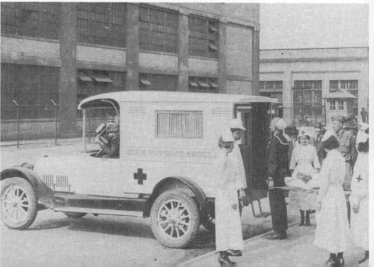

1918

Sayers & Scovill continued to offer one of the industry's largest selections of carved funeral cars. The arched areas on this S&S Model 1745 are filled with draped carvings and the car is of the famous S&S twelve-column design.

Another S&S with the twelve-column design was the Model 245 finished in white. The carvings on this model were more ornate than on some of the other S&S carved offerings. All S&S units were completely assembled vehicles built as professional cars from the ground up.

For the man who demanded a car of more utility, S&S offered this Style 1025 combination pallbearers' coach - ambulance with attractive, practical limousine styling.

This sedan styled combination ambulance and casket wagon was built on a Packard chassis for August Eickelberg of New York City by A. Geissel & Sons in this year. The car had blanked out rear quarters and plain black blinds on the rear door windows.

Philip Herwig of Baltimore was the owner of this Giessel glass-sided hearse with massive, heavy draperies. This car retained the large ornate carriage lamps.

This combination hearse-casket wagon was built by A. Geissel & Sons of Philadelphia for P.A. Murphy of Pittsburg. It featured both carved panels and a ray textured drape behind the casket compartment's center window.

1918

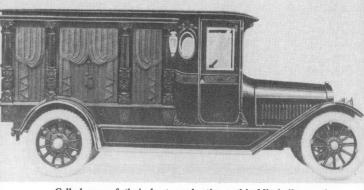

Called one of their best productions, this Mitchell was the Model 375. It was built by the Mitchell Hearse Co. of Ingersoll, Ontario. The carved panel hearse features elaborate carved drapery panels and driver's cab trim with small oval windows behind the driver's seat. Heavy beveled glass is used in the side windows.

Rock Falls built this casket auto, hearse combination, with full draperies and a minute amount of carving on the belt line. Built in quasi-limousine styling, this car had a section of leaded glass windows above the regular large windows, which gives the car a higher look.

Another manufacturer to market undertaker's limousines in this year was Rock Falls, with this seven-passenger car. It utilized the Rock Falls' chassis and all driving components.

The A.J. Miller Company marketed this stylish Model 40 again this year and used the same 45 horse Continental engine found on similar Meteor models. Once again the carving and styling of this coach were unique.

This beautifully restored 1918 Rock Falls carved panel furneral car is the pride and joy of Paul Latham of Oakland, Cal. The car is in completely original condition and wears a coat of white paint that was identical to the original.

Another firm that built a special chassis for funeral coach and ambulance work was Velie. This chassis utilized the Velie six-cylinder engine and was strengthened to accommodate the heavier bodies that would be mounted on them.

1918

Meteor discontinued offering the V-12 engine in favor of Continental engines. This Model 95 carved funeral coach used a 45 horsepower Continental six cylinder. It rode on a wheelbase of 148 inches. The car featured ornate carvings and a swelled rear door treatment, and sold for $2,250.

This Meteor Model 58 sedan ambulance seated ten persons in its sedan or pallbearers' form, or the seats could be removed to use the car as an ambulance. It also had an invisible type rear door at the back to further enhance its sedan styling. Utilizing the 45 horsepower Continental Six, this model sold for $2,050.

This Hoover ambulance was mounted on a Cadillac chassis and has an open driver's cab and only two small windows in the rear compartment. Obviously this was a genuine emergency vehicle and not a show car.

Meteor continued to lead the industry in the field of low priced, high volume funeral service vehicles. In May of this year the employees of the Meteor Plant in Piqua, Ohio, participated in a Liberty Bond parade, and this coach was one of the entries. This white finished Model 47 carried a sign on the roof displaying sentiments of the campaign while another sign inside the coach depicted the Kaiser.

Mounted on a Reo 3½ ton truck chassis, this utilitarian looking Hoover ambulance has an open driver's compartment, a large ambulance sign over the windshield and two small windows in the rear compartment. The inset picture illustrates the fully equipped patient compartment with a capacity to handle two injured or sick people and have two attendants ride along. The electric lights on the roof and the medicine cabinet on the side wall were standard.

This Ford T chassied Hoover service car has sunburst or rayed carvings where one would ordinarily expect to see a window, and elaborately carved wreath and column patterns on the rear body sides. Note the interesting curvature of the cab and the corner windows.

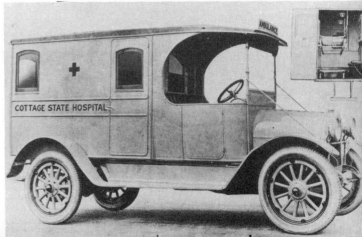

1918

Mounted on a large Pierce truck chassis, this funeral omnibus is an example of the type of vehicle that was tried in New York City. These monsters accommodated the casket, the pallbearers, and a full entourage of mourners. This type of vehicle was not in any great demand, even though a company called the Fifth Avenue Coach Co. was formed to build them. They were found to be too cumbersome and impractical.

H.H. Babcock constructed hearse and casket wagon bodies on both extended Ford and Dodge chassis. Seen here is an example of their combination casket wagon-hearse body mounted on an extended Dodge chassis. The oddly shaped windows of the driver's compartment contained beveled glass.

This Babcock Dodge reflects the company's train of style thought on casket wagons — very plain styles without any frills or ornate carvings. This car was also mounted on an extended Dodge chassis and had beveled glass in the side windows. These windows had stylish arched tops.

Another coach on the extended Dodge chassis is shown here with this hearse built by H.H. Babcock of Watertown, N.Y. It was designated as a Model 145 and had small carved panels set into the area beside the main center windows that featured beveled plate glass without drapes. The small oval window just behind the driver's area appears on this coach, too.

The body on this carved draped panel hearse, used this year by the Albemy J. Janisse Funeral Home of Windsor, Ontario, is by an unknown builder. Note exterior mounted crucifixes in the center panels of this eight-column car, and liberal use of wreath ornamentation.

E. M. Miller, sometimes called Miller Quincy, of Quincy, Ill., was also building bodies on extended chassis of new cars or used and extended chassis. They had a full line of differently styled bodies of which this is the Style 1830. It is mounted on a Dodge chassis.

1918

Advertising that superior performance was built into all Kissel hearses and ambulances, the company offered this Kissel Kar hearse. This body style could accurately be described as a limousine with the extra long windows in the rear compartment. Heavy beveled glass was used in these windows.

Keystone Vehicle Co. of Columbus, Ohio, built this funeral car Style 309, which sported unusual window treatment in the casket compartment. This car is shown mounted on a Dodge truck chassis.

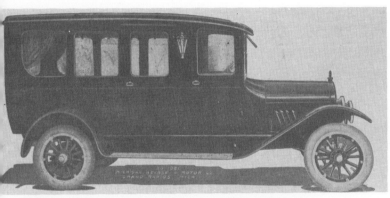

Michigan's latest offering was this combination pallbearer's coach, mourner's car, child's hearse and ambulance designated as the Model 1081 and called a "real undertaker's coach." This car was built on Michigan's own chassis with their own engine.

Mounted on a Dodge chassis, Keystone offered this carved panel hearse. Keystone was a master of the extended chassis car and mounted bodies on both Ford and Dodge.

Alliance offered this large body on this Ford TT commercial chassis. The car reflects the more modern limousine styling and the draperies are ornate, ruffled silk, blind type fixtures. Note the minute carving below the window line.

This coach features a four-column horse-drawn body that was cleverly converted to an eight-column motorized hearse by the August Schubert Wagon Co. of Oneida, N.Y., for F.C. Kipp Jr. & Son in Ossining, N.Y. Schubert's main business was this type of conversion job that transformed horse-drawn vehicles into modern auto hearses.

1919

The Michigan Hearse Co. introduced a new model this year. Called the Model 1012, it was of the twelve-column style. The angled louvres give the car something of a sporting look.

Crane and Breed, one of the real premium hearse builders, continued to produce its beautifully carved hearses on the Winton Special Six chassis.

The funeral coach and ambulance market had expanded to include literally hundreds of manufacturers. The carved style car that had been prominent among hearses since the beginning of the motor age for these professional vehicles was starting to decline in its popularity. Every year the number of manufacturers involved in the construction of these vehicles had grown and new ideas had been slow to develop and take hold. The limousine style was slowly making headway within the industry and each year there were a few more of this style appearing from different makers.

Another new model was announced in this year by the James Cunningham, Sons and Company of Rochester, N.Y. This new car was of carved style and called the model 43A. It could be ordered with either glass or wood panels between the carved columns that decorated the rear sides. Selling for $4,500, plus 5% war tax, the car was considered one of the finest carved funeral coaches available. Cunningham, one of the pioneers in the auto hearse and ambulance field, was established in 1838 to build high quality carriages, buggies and, of course, hearses and ambulances. In 1909, when Crane & Breed announced the first commercially built auto hearse, Cunningham was close behind them with one of their own. In fact, Cunningham offered a motorized ambulance at the same time that Crane & Breed introduced the first commercially-built auto hearse.

The Meteor model 80 carved funeral car sold for $2,200 in this year and the Meteor model 95 Deluxe carved funeral car was offered complete for $2,475. These cars were powered by the Continental Red Seal 9-N 50-horsepower engine equipped with a Stromberg carburetor and a Delco lighting, starting and ignition system. Clutches were Borg & Beck and the transmission was a Covert three-speed with reverse and Hyatt bearings.

Shown here is a Cunningham carved-drape panel hearse, eight-column style. Note the ornate carriage lamps, owner's initial escutcheon on the door, and the rearview mirror.

James Cunningham Son & Co. Style 43-A hearse offered either glass or wood panels. The price of this unit was $4,500, plus war tax. Of the classic eight-column style, this was a high-quality model in every respect.

1919

One of Meteor's finest products was the Model 95 deluxe funeral car. A massive, carved-drape hearse, the Model 95 listed at $2,475 and was powered by a Continental Red Seal 50-horsepower motor.

Meteor's Model 80 funeral car carried a $2,200 price tag. This one, finished in light gray, has an eight-column body ornamented with carved drapery panels.

Meteor Motor Car Co. of Piqua, Ohio, built this Model 85-B combination ambulance and pallbearers' coach. Price was $2,300 plus 5% war tax.

Sayers & Scovill Model 195 carved-panel hearse was offered in standard black finish or the light gray which was now becoming increasingly popular with the trade.

Sayers and Scovill "Clifton" eight-passenger sedan-limousine was designated Style No. 50. Many funeral directors frequently rented cars like this to competitors.

Sayers and Scovill built this impressive eight-columned motor hearse. It was advertised as the "Puritan," Style 145.

Keystone Vehicle Co. of Columbus, Ohio, built this Model 1100 limousine on a Dodge chassis.

1919

No, this wasn't Lorraine's basic model — but it shows the sturdiness of the chassis, billed by Lorraine as "our latest type."

This is a Lorraine carved-panel hearse of what was known as the 12-column style. The columns at the front and rear of the body are doubled up. The car is finished in light gray.

Lorraine described this 12-column carved hearse as being finished in "Three Tone Black" — obviously light, dark and medium black! You figure it out.

Owen Brothers, of Lima, Ohio, did many horse-drawn hearse conversions. This is a typical transformation, mounted on the rugged Reo 143-inch wheelbase chassis. It is a standard eight-column, carved-drape style coach.

Rock Falls Manufacturing Co. pioneered a bold art form for funeral car advertising, utilizing pastel shades against elegant backgrounds. This carved-panel Rock Falls motor Hearse is depicted with wire wheels.

"Not Expensive — But Mighty Good." That was Owen Bros. claim for its big Model 8-765 mounted on a Reo Six chassis 143-inch wheelbase. The manufacturer claimed this car was "built very low, thus placing the center of gravity for utmost stability on the road." Carvings were of solid wood.

The Albemy Janisse Funeral Home of Windsor, Ontario, was the proud owner of this two-tone, carved-drape funeral car. Chassis and body builder are unknown. It was photographed in Windsor Grove Cemetery.

1919

Light, harmonious lines characterized this carved-panel hearse built by the Streator Hearse & Body Co. of Streater, Ill. The chassis appears to be a Velie.

"Tell Us Your Hearse Troubles," Williams Wagon Works of Macon, Ga., urged its customers in this ad for its open-front No. 310 hearse body on a Ford chassis. Like many other manufacturers, Williams would mount its bodies on any chassis specified by the customer.

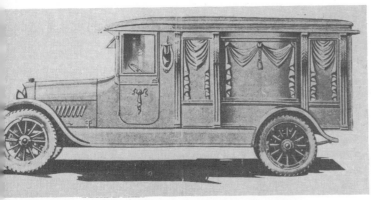

Reo Atlanta Co., of Atlanta, Ga., marketed this modestly carved funeral car on the Reo Speedwagon motor chassis. Note simple columns and dignified oval window behind the front door.

Another newcomer to the business this year was the Cole-Stratton Co. of New York City. This is an artist's rendering of an eight-column, glass-side hearse body mounted on a Dodge chassis. Note the uniform arch shape of side windows and doors.

This photo shows a 1919 Hoover carved-hearse body mounted on a 1913 Cadillac chassis. Carriage lamps were quite small on some models. This medium-priced hearse was finished in light gray.

This is a Babcock Model 148 funeral body, glass-side style, mounted on a 1919 Dodge chassis. Styling is of the familiar eight-column type.

The Rhode Island Vehicle Company, of Providence built this handsome eight-column glass-side hearse for a C. M. Page. Note the heavy, tasselled draperies, and filigree-type urn ornaments in each of its six side windows.

1919

Geissel and Sons built this elegant, raised-center, glass-side hearse for William F. Harding of New Brunswick, N.J. Note the heavy, formal drapes, and light colored paint job.

This is a 1919 Geissel casket wagon built on a White chassis for New York City funeral director Frank E. Campbell. An eight-column body style, it featured rayed "sunburst" panels between the columns.

Shown here is a Eureka carved-drape hearse, eight-column style, set against a rather artistic background of classic statuary.

Schelm built this combination funeral car-casket wagon to order for H. M. Kilpatrick, of Elmwood, Ill., on a Model T Ford chassis. A real dual-purpose unit, when it wasn't hauling caskets, it carried and delivered furniture for its owner.

Auto Top and Body Works of San Francisco, offered this closed-curtain funeral limousine on a Pierce-Arrow chassis. This was a high, massive car but it sported graceful, pleasing lines.

Fitz Gibbon & Crisp, of Trenton, N.J., built this huge funeral onmibus on a Pierce-Arrow truck chassis. Designed to eliminate the funeral procession from urban traffic, it could carry a driver, casket and 19 mourners and pallbearers.

1920

The golden age of the funeral coach and ambulance had begun. Gone were the evolutionary days when the motor hearse and ambulance had to prove itself. The automobile and its respective cousins, the coaches, had proven themselves in both peace and war service and were now considered the normal mode of transportation.

Styles were beginning to become more liberal, with more innovative differences between the various makers throughout the country. The largest manufacturers in the industry were Henney, Sayers & Scovill, Rock Falls, Meteor, Riddle, A.J. Miller and Cunningham. These were the volume makers and they set the dominant styles for the rest of the industry to follow.

The limousine type body style was becoming somewhat more common and there were even some quite interesting variations on this theme. Some tried to combine the older carved side style with the full greenhouse type limousine style by placing carved wooden columns between the full plate glass windows. More glass area was favored in this era and every maker seemed to have his own way of approaching the situation. Carved draperies were becoming minimal and less ornate on some models while other makers went to real draperies of heavy velvet behind the new large windows. The combination car was becoming a better buy for some of the smaller funeral directors. These cars could be rapidly changed from a funeral coach to an ambulance or a service car by simply changing the interior fittings and nameplates. Bearers' coaches were losing their market as more and more individuals bought their own vehicles and could drive to the funeral and the cemetery in their own cars.

Meteor's Model 85-B combination ambulance and pallbearers' coach was a dual purpose unit that sold for $2,600. Note the dainty coach lamps and pull-down shades.

Among Meteor's premium models was the massive Model 97 funeral car, which used plate glass side windows in place of the carved drapery panels of the popular Model 95. Nameplates were mounted on the upper cab door side glass.

The Meteor Model 95 Funeral Car Deluxe had richly carved drapery panels and eight columns. Note the tastefully complimentary ornamentation on lower body panels and doors.

The Meteor Model 62 carried a hefty $3,000 price tag and boasted a three-piece windshield and "oversize" tires. This was a heavily ornamental casket car of the classic eight-column design.

William A. Carroll of New Bedford, Mass., designed and built this eight-column, glass-side funeral coach mounted on a Velie chassis for Joseph P. Marfing, of Brooklyn, N.Y.

1920

The Sayers & Scovill "Puritan" carved drapery hearse was among the more popular funeral car models of 1920. Styling was heavy but pleasingly proportioned. Columns were handsomely ornamented and framed rich, matte-finished carved drapes.

The Sayers & Scovill Style 1025 combination accompanying text states . . . "the beautiful interior of this limousine and ambulance combination shows the artistic possibilities of a dual-purpose vehicle."

This handsome motor ambulance was built by Cunningham for J. T. Hinton and Sons of Memphis. The ambulance's owners had no illusions about the wisdom of their purchase. They called it "The Finest In The World."

This elaborately-carved hearse is a Cunningham finished in a light gray. Note the center arches and richly carved drapery panels. Side lamps matched the drapery panels in elegance. This was one of the finest funeral cars of its type ever built.

The Lorraine Hearse was built by a firm known as the Motor Hearse Corp. of America, in Richmond, Ind. This Lorraine carved car sported a three-piece windshield, twelve-column body and finely detailed drapery panels.

This is a rear view of the Lorraine hearse. Note how carvings were extended to the double rear doors.

1920

This Rock Falls hearse utilized an interesting combination of carved drapery panels and bevelled, etched glass. The designer appeared to be seeking the best of two styling worlds — the traditional carved hearse and the up-and-coming limousine.

Rock Falls built this good-looking ambulance, which had such up-to-date styling features as a two-tone paint job and leaded window panes. Note the individual rear compartment door.

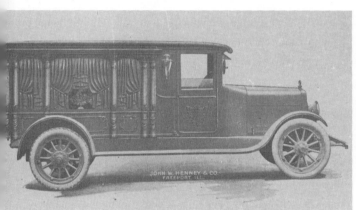

A relative newcomer to the funeral car and ambulance field was the John W. Henney Company of Freeport, Ill. Shown here is a Henney twelve-column carved panel hearse.

This is a Rock Falls pallbearers' coach. Note the similarity to the limousine style funeral coach which was gaining in popularity with funeral directors who were beginning to tire of the tall, ornate carved cars of the day.

This Hoover carved hearse, built on a Cadillac touring car chassis, used carved drape panels in the outboard sections and plate glass center windows behind which was a rayed panel.

Another Hoover body, mounted on a lengthened Cadillac touring car chassis, is of the formal carved style with eight-columns. Carved drape panels on this model were symmetrically proportioned.

1920

The Keystone Standard "Satisfaction Plus" combination body is mounted on a lengthened Dodge Brothers commercial chassis. It is seen in its ambulance trim.

The Keystone Body Co. called this medium priced hearse "The Undertaker's Ideal Motor Car." The modestly carved body is mounted on a Dodge chassis, one of the more popular funeral car chassis of the day.

This is the 1920 Keystone combination hearse and ambulance, showing the demountable equipment which could be quickly put in place to prepare the car for either one of its double lives. Shown standing next to the car is the flower rack with the casket table lying in front of it and two ambulance attendant seats on the ground.

This carved panel hearse was built by the E. M. Miller Co. of Quincy, Ill., on a Dodge Brothers' chassis. Cars built by E. M. Miller were often called Miller-Quincy units to differentiate them from the products of the A. J. Miller Co. of Bellefontaine, Ohio.

Riddle built this handsome eight-column carved hearse, which had a two-tone paint scheme. Note similarity to Sayers & Scovill models of the same era. Ornate side lamps were nearly four feet high.

This is Columbia's No. 12-A ambulance, built on a Model T Ford chassis. Columns were used to ornament the body corners.

1920

William Schelm, of East Peoria, Ill., was the builder of this very professional looking combination casket and furniture wagon, constructed on the rugged Model T Ford chassis. Note the full cab and column replicas on the vehicle's body.

William Merz of Philadelphia operated this 30-passenger funeral bus built on a 3½-ton Mack Bulldog truck chassis. For some reason, funeral omnibusses never won any measure of public acceptance despite their favorable effect on urban traffic patterns.

Called the Wolverine Sedan, this combination pallbearers' coach and ambulance had a gracefully styled body with unusually thin pillars, and a rakishly angled windshield.

A. J. Diefenderfer of New York City, was the builder of the limousine-inspired hearse which carried the unusual model name of "In Favour." The body was basically of the four-column style with half-columns separating the rear side windows.

This is an eight-columned carved car built by Owen Brothers of Lima, Ohio. Carved center panels were available in place of the glass center panels shown on this particular model.

The Michigan Hearse and Motor Co. of Grand Rapids offered the "Michigan Six" which had, among its features, worm drive. Note the finely-detailed carvings and trim.

1920

William Erby and Sons Co. was the builder of this combination car on a Cadillac chassis. Note the graceful, sweeping body lines and generous glass area.

Serious attempts at styling were evident in the lines of this ambulance built by William Erby and Sons Co. of Chicago. The oval-shaped side window is unique.

The Mitchell Hearse Company, of Ingersoll, Ontario, built this combination casket wagon and motor hearse on a Reo chassis. An optional extra was a carved-panel side which could be quickly bolted into place to convert the car from a plain casket car to a formal carved hearse.

Simplicity was the keynote of the design of this funeral car built by the August Schubert Wagon Co. of Oneida, N.Y. Notice the unique window treatment. This would have to be categorized as a four-column model, despite the half-columns framing the side windows.

Undertaker George Andrews poses proudly with his new auto hearse. This was the latest in auto funeral equipment in use in Australia at the time.

English hearse builder Woodall-Nicholson of Halifax, England, constructed this motor hearse on a 30-horsepower Armstrong-Siddely chassis. Flowers were carried on the roof of this car.

1921

With the popularity of the sedan and the flexibility of the combination coach, Meteor announced a new 8-passenger combination sedan-ambulance. The car was side-loading and from the outside could not be distinguished from any normal sedan. Equipped with the Continental Red Seal engine and the, by now famous, Meteor name and chassis, the car sold for $2,650 plus 5% war tax.

In its death throes, the ornately carved funeral coach hit its peak with the new 1921 models. The epitome in ornate carvings and ornamentation came with a car built by A. Geissel & Sons of Philadelphia, for Charles J. Cristinzio, also of Philadelphia. Mounted on a Reo chassis the car was, from the driver's seat back, one huge mass of carved wood and plate glass. Featuring eight columns carved in the form of praying angels and five more large angels on the roof, the car was the most elaborate built since the days when horse-drawn vehicles were in their heyday.

Michigan Hearse and Motor Company of Grand Rapids, announced the "latest achievement in funeral car design" in this year also. Called the "Gothic," this car was mounted on the Michigan Worm Drive Six chassis which had proven itself quite satisfactorily over the last seven years. The name given to this car was very well chosen, as the car resembled a gothic cathedral in many respects. The body design was pure gothic, right down to the highly detailed carvings and the leaded stained glass windows that featured, "rich autumnal tints, soft and glowing, imposing yet subdued, in complete harmony with the light-tone grey finish of the body." The interior was genuine mahogany and the appointments of their own exclusive design. Michigan claimed that the draped hearse would give way to the march of progress and that this coach was the wave of the future.

One of the most expensive cars in the S&S line was this Model 245 "Masterpiece" of the twelve-column body style. The coach was mounted on the S&S chassis that was especially engineered for funeral work and was built complete in the S&S factory in Cincinnati.

Meanwhile, the sedan had become the most popular style of passenger automobile among those able to afford a closed car. There was a demand for a passenger car designed especially for the funeral director, a car that could be used for everyday chores about town as well as on the job. This car would have to reflect dignity and versatility and would have to be dependable.

Sayers & Scovill realizing this demand introduced a new "professional car for the moritcian" in two basic forms. The Sayers Six Roadster was a sharp looking little convertible with wire wheels and a two tone body. The roadster filled every requirement for funeral work. Ample space was provided in two compartments for all equipment needed including a cooling board. The roadster was powerful, speedy, comfortable, and had dignity that befitted the profession. The Sayers Six five-passenger sedan or limousine was another of the passenger cars designed by S & S with the funeral director in mind.

In this year S&S offered a complete line of sedan type professional cars. This Model 55 Clifton seven-passenger funeral limousine was quite popular with both liveries and funeral directors alike.

Sayers & Scovill presented their 1921 line as one of the most complete ever offered, with many models and body styles. With arch-like coves containing draped carvings, this two-tone gray S&S Westminster was a new model for the year. All of the intricate carving work on this car was done by hand while the coach itself was mounted on an assembled S&S chassis.

1921

Looking for all the world like a normal passenger car, this S&S sedan was designated the Samaritan and for some very good reasons. This car is a combination sedan-ambulance and was rapidly converted from a seven-passenger sedan to its ambulance duties in only five minutes. The only exterior give away to its dual purpose is the one door handle on each side.

This photo shows the S&S Samaritan in its converted ambulance state. Note how the doors open to allow easy loading and how the cot rides beside the driver's seat. In this form the little jump seats come in handy for attendant seating.

The Sayers Six roadster by S&S filled every requirement for funeral work. Ample space was provided in two compartments for all the equipment needed, including a cooling board. All three large doors were fitted with Yale locks and provided easy access to the compartments. This roadster was powerful, speedy, and had dignity befitting the profession.

The other passenger car in the Sayers Six line was this five-passenger model that was transformed in two minutes from a limousine to a normal five-passenger sedan. These cars utilized the S&S assembled chassis with a Continental six-cylinder engine.

The limousine style was becoming the prominent type, and this S&S Kensington ambulance adds a new styling touch to even this new body type. With a small rear compartment window and leaded glass, this car features a two-tone paint finish, pull down shades, and a small carriage lamp. Note the sturdy looking Westinghouse shock absorbers on the front and the placement of the "parking lamps" on top of the headlights.

The August Schubert Wagon Co. of Oneida, N.Y., built this carved drapery hearse for Amedee Archambault & Sons of Lowell, Mass., on a Packard chassis. The styling of this car was rather archaic when compared with some of the coaches being turned out by the larger makers.

This Meteor Model 96 featured a central glass window flanked by carved panels and eight columns.

Designated the Model 86, this was a Meteor ten-passenger, side-loading sedan ambulance with a concealed rear door. This car could be used as either a pallbearer's coach or an ambulance. Note the blinds. This model was equipped with Houdaille shock absorbers and 35x5 cord tires and sold for $2,850. All Meteor funeral coaches and combinations shown here were powered by a Continental Red Seal 50 horsepower six-cylinder engine, had Columbia axles, Warner transmission and clutch, Delco lighting and starting systems, and Stromberg carburetors. The cars rode on a wheelbase of 144 inches and had heavy pressed steel one-piece fenders and crank window lifts.

Meteor continued to offer the trade a wide range of medium and low priced vehicles mounted on their own chassis, using a Continental engine. Through this and the following series of pictures, many of the models offered by Meteor are shown and the model differences are obvious. This Model 97 funeral coach featured the traditional eight-column design coupled with a massive expanse of glass. This was called a glass-sided funeral coach.

Like S&S, Meteor also offered a sedan ambulance with this Model 65. This was a side loading car and rode on a 127-inch wheelbase. Other specifications were the same as for the rest of the Meteor line. This car sold for $2,850 plus 5% war tax.

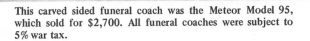

This carved sided funeral coach was the Meteor Model 95, which sold for $2,700. All funeral coaches were subject to 5% war tax.

This coach represented one of Meteor's combination units. It was the Model 98. By removing the drapes and pulling down the shade, this car was quickly converted to an ambulance. The car was the same size as the Model 95, with the only difference between the two being that seen in the pictures. For its ambulance duties, the car was fitted with a Bomgardner cot and an attendant's seat. Priced at $2,600, this car was available in black, dove gray, or a new color called silver molton which was strongly recommended.

1921

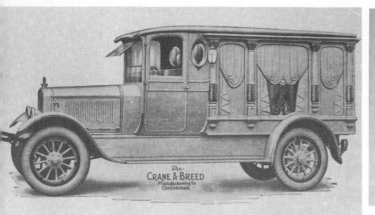

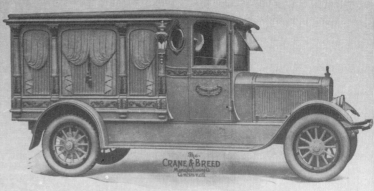

This Crane & Breed Winton shows some model differences from the others. This car has a glass window in the casket compartment and less ornate carvings. This is one of the Crane & Breed models that reflects a clean approach to the carved car style.

Crane & Breed continued to offer a line of high quality funeral coaches mounted on the Winton Six chassis. This model features a fully carved casket compartment and oval windows behind the driver's seat. The cars were available in gray, black, or dark green showing that there was a trend developing away from strictly black funeral vehicles.

Crane & Breed built this combination coach to appeal to the lower price bracket. Note that with the exception of the pillars on the side of the car and the drapery carvings in the windows, this car is quite plain. The carved panels were removable to quickly convert the car to an ambulance, or a hearse. It is seen here in its service car form. The name plaque signifies that this car was owned by the Townes Funeral Home.

Cynthiana Carriage Co. was established in 1876 and built bodies on several makes, including Ford and Pierce Arrow. This Cynthiana-Ford features the latest carved coach styling and modern lines. Notice how the size of the carriage lamps has been generally decreasing.

Featuring etched windows just below the roof line, this stylish Rock Falls carved drapery hearse was the epitome of style through these years. It was mounted on Rock Falls' own assembled chassis.

A small coach company in Kentucky built a line of bodies that they would mount on any extended chassis. Cynthiana Carriage Co. of Cynthiana, Ky., built this Model 31 on an extended Dodge chassis. Notice the carved drapery panels and the blanked off window areas in the casket compartment. They offered complete cars on the Dodge and Dixie chassis and bodies for both Pierce Arrow and Ford.

1921

Michigan claimed that the draped hearse would give way to their latest achievement in funeral car design. With a patented design called the Gothic, Michigan believed that they had introduced the funeral coach of the future. The Gothic body design was pure Gothic even to the carved, leaded stained glass windows colored in rich autumnal tints. Interior fittings were done in genuine mahogany with the appointments of Michigan's own design. The chassis was the same worm drive six-cylinder that Michigan assembled for the rest of the line, and had used with satisfaction for the past seven years.

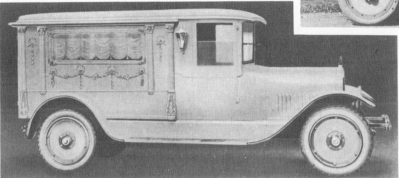

Michigan also offered this subtly carved hearse as the Model 1075-B at $2,550, in the same choice of colors.

This Michigan Model 1070-B was a service car, a flower car, or an emergency hearse. It sold for $2,500. All Michigan coaches used an assembled chassis with worm drive and a six-cylinder engine.

To achieve quantity production of the new Gothic, Michigan offered the other models in the line at lower prices. The standard carved panel, Model 1078-B, funeral coach shown here sold for $2,600 and was available in three-tone gray, two-tone black, or silver bronze.

In this, the "golden age" of the funeral coach, the carved car reached its peak of ornamentation and ostentation. This car was the most elaborate carved funeral coach ever seen in the motor era. It was built by Geissel for Charles J. Cristinzio of Philadelphia. Built on a Reo chassis, this car was covered with carvings of angels and cherubs and was topped off by a large carving of Gabriel on a pedestal. All of the carving was done by hand. Never would there be another model built by any firm as lavish and ornate as this.

A. Geissel & Sons of Philadelphia built this stylish carved panel hearse. Note the carved drapery panels beside the center window and the unique drapery within this main window.

Hoover of York, Penn., built a wide variety of coach styles and types on any chassis the customer would select. This Model SC-511 carved panel hearse was mounted on a Packard chassis and featured arched carving areas.

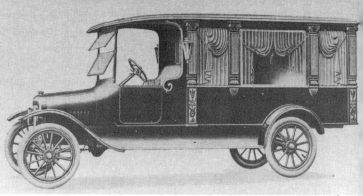

Hoover's open front offering on the Ford chassis was this Model D-15661 with the carvings limited to the area above the belt line. Note the slightly modified radiator shell on this model.

Hoover also offered ambulances like this open front Model SC-483 on a Dodge chassis. Styling is quite subtle and the carved work is restrained to the area around the driver's seat.

Hoover offered a complete range of models on the Model T chassis, with this carved drapery panel hearse, Model SC-501, as the fully enclosed hearse in the line.

This stylish little car was designated a Hoover Model SC-580 combination casket wagon and furniture car. The low deck and open driver's compartment are integral parts of this type of styling. The rear deck could accommodate the largest rough box, casket, or a full complement of chairs or flower stands and baskets. These vehicles were also used as flower cars to deliver flowers.

Two fully enclosed versions of service cars or casket wagons were offered by Hoover on the Ford T chassis. This was the more ornate of the two and was designated the Model D-10348. It had carved wreaths and a carved ray or sunburst area in the center on the body sides. The other service car, called a Style D-10248, was without any decoration behind the driver's compartment.

This carved Hoover is shown mounted on a Dodge commercial chassis. Called a Model SC-458, this eight-column carved drapery car has beveled glass in both the driver's windows and the oval window behind the driver's compartment.

1921

This Riddle carved funeral coach was built on their own chassis and features ornate filigree type carvings with columns and a hand carved drape around the rear window.

This Riddle all glass funeral coach is shown in front of the tomb of President William McKinley in Canton, Ohio. It was purchased by Gordon & Hollinger of Massillion, Ohio. Riddle was located in Ravenna, Ohio.

Another Riddle, this limousine style sedan ambulance loaded from the side. The hinges for the extra-wide rear door can be seen at the rear of the small third window. Notice the pull-down shades.

The Tiamons-Von Hook Funeral Home of Chico, Cal., owned this coach that was built especially for them by Riddle. The coach features a body from a horse-drawn vehicle mounted on a Ford Model T chassis.

Riddle also built special limousines for the funeral profession. This Riddle Model 1074 limousine was designed for either public or private service, with a roomy body that seated eight persons comfortably. The car had a glass partition that could be raised or lowered to transform the car from a limousine to a regular sedan. Built without the partition and with a double side door, the car could also be used as an ambulance.

Designated the Model 14-A, this Babcock hearse was mounted on a Dodge commercial or truck chassis. This coach featured one window with ray textured draperies and four mildly carved columns on each side.

1921

Mounted on a revamped Reo Speed Wagon chassis, this E.M. Miller style 848 combination coach and ambulance also was a dual purpose vehicle. A full-width rear seat indicates that the car could also be used as a passenger sedan. This coach was also available on E.M. Miller's own Model K six-cylinder chassis.

E.M. Miller of Quincy, Ill., offered this Model 855 built on their own six-cylinder chassis. Note that the carved drapery casket compartment is finished in two-tone gray.

Another E.M. Miller creation was this seven-passenger limousine for the funeral director, built on their own special six-cylinder chassis. This type of car was at its height of popularity in this era and was widely used by the profession.

Another E.M. Miller creation was this combination seven-passenger limousine and side door loading ambulance mounted on a Dodge Brothers' business chassis and designated as Style 2175. This body was available on any chassis and could be used as a dual purpose car, passenger sedan, or ambulance.

Seen here is the same Mitchell model with the carved panel in place. This panel combined the traditional drapery carvings with that of a dignified shield and wreath. The columns were carved into the panel. Mitchell coaches were built in Canada but were available in the U.S.

O.J. Mitchell of Ingersoll, Ontario built this combination casket wagon, hearse and ambulance on a Reo chassis. The car came complete with removable side panels that changed the casket compartment's exterior appearance from this plain casket wagon to a carved drapery hearse.

1921

In business since 1900, the J.M. Karwisch Wagon Works of Atlanta, built hearse and ambulance bodies on extended chassis. This carved panel hearse is shown mounted on an extended Ford T. chassis. The carved drapery panels were executed with great skill and all were hand carved. Karwisch called this a high quality funeral car without the high price.

This interesting sea plane was the first flying hearse. It carried six passengers, the casket, pilot, and the co-pilot. Built especially as a hearse by Aeromarine Plane & Motor Co. for the Fifth Avenue Memorial of New York City, this craft used two 400 horsepower Liberty engines and had a cruising speed of 100 miles-per-hour.

This uniquely different hearse was offered by Stratton-Bliss Co. of New York City for $3,250. The arched window tops are carried over to the driver's compartment door windows of this Dodge chassied coach. The glass in these windows was called French plate and was a heavy beveled type.

This British Ford TT chassis was fitted with a convertible body. The car could be converted from a hearse as shown, to either a limousine or a delivery truck. This type of car was placed on the market this year in London.

This is the type of motor hearse that had become common in some parts of England within the last few years. Notice the low deck that was used for carrying flowers and the ornately etched glass side window.

Biehl's Wagon and Auto Body Works of Reading, Penn., built hearse bodies for over fifty years. This is an example of their product line for 1921.

1922

Both Meteor and A.J. Miller began producing the new limousine style funeral coach — a style that was slowly gaining favor with more makers taking it up every year. Variations on this style had become common with the ambulance in previous years. Some of these had leaded stained glass in their side windows. The ambulances of the day carried such items as chair-cots, baby baskets, thermos bottles for both hot and cold water, individual attendant seats, and came equipped with drapes for the large side windows. These cars were designed with lightness, speed, flexibility and dependability, to rush the sick and injured to the hospital. With the new sedan ambulance the funeral director had an extremely versatile seven or eight passenger limousine or sedan that was quickly convertible into a practical and useful ambulance. Several more makers added these vehicles to their regular lines and among them was Sayers & Scovill (S & S). With only one door handle visible on either side, the S & S Samaritain sedan and invalid car (ambulance) was an extremely attractive vehicle.

Another of the unique cars of this era was the low deck casket and first call wagon. Resembling the modern pick-up truck these cars were used to transport the body of the deceased from the place of death back to the funeral home for preparation and then back to the home for the funeral. They doubled as the workhorses of the funeral profession and were sometimes called service cars. They had the capacity to carry several caskets or rough boxes and were also handy for transporting chairs, pedestals, baskets and flowers. This type of vehicle had been in use by the profession in one form or another since the days of the horse-drawn vehicle.

The carved cars continued in their popularity although a new type of coach style was on the way. This view of the S&S Corinthian shows the car in a 3/4 front view that enhances both the body and the frontal ensemble. Once again S&S offered one of the most complete lines of carved funeral coaches available.

Another offering in the vast S&S line of carved funeral cars was this Model 575 Westminster shown finished in gray. This car retains the twelve-column style for which S&S was famous.

This magnificent lineup of Cunningham funeral limousines was owned by the Walsh Funeral Car Company of New York City, a funeral limousine and hearse rental service. Cunningham, in addition to building high quality hearses and ambulances, built a whole line of passenger cars that were highly regarded in their day.

Mounted on a Pierce-Arrow chassis, this stylish coach was built by the house of William Erby of Chicago. It had intricate drapery carvings on the rear body sides. Notice the horn mounted on the lower cowl.

Called the "Heart of America" this Dodge chassied hearse was built by the Holcker Manufacturing Co. of Kansas City and was designated as a Model 55. The use of two-tone paint set this car apart from some of the other coaches built on the Dodge chassis, and lent this car a special styling distinction.

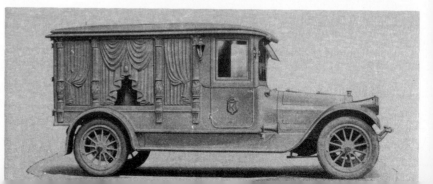

1922

This seven-passenger sedan was instantly convertible to a practical and desirable sedan ambulance. Built by A.J. Miller this car was mounted on a Studebaker six chassis and sold for $2,250. A.J. Miller was located in Bellefontaine, Ohio.

A.J. Miller mounted their line of vehicles on Studebaker chassis and offered this carved style hearse as the Model 62 which sold for $2,300.

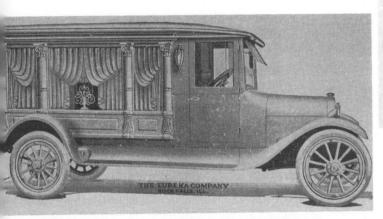

Eureka of Sterling, Ill. were builders of fine undertakers' vehicles, and this carved panel Dodge hearse is a fine example of their work.

This twelve-column carved hearse was built by the A.J. Miller Company and was mounted on a Dodge commercial chassis. This chassis was lengthened to 138-inch wheelbase and this lengthening was approved by the Dodge Company and was fully guaranteed by them. Selling for $1,815, this Model L Dodge funeral car had carvings and columns made of solid popular with interior appointments of genuine mahogany. The driver's seat was upholstered in leather. The space behind the driver's seat was 7 feet 10 inches long, 46 inches wide, and 56 inches high above the removable casket table.

This carved-side hearse was the Meteor Model 100 when equipped with the 50 horsepower Continental engine and Model 200 when it used the more powerful 70 horsepower version. These cars sold for $2350 as a Model 100 and $2,700 as a Model 200.

This model, called a Style 205 side loading sedan ambulance, was marketed by Meteor. It had a concealed rear door. In this form the car cost $2,850, but was also available as a 10-passenger sedan ambulance without the leaded glass windows in the rear compartment. In this form the car was designated as a Model 204 and sold for $2,850 plus 5% war tax. These cars were mounted on Meteor's own Series 200 chassis and were powered by a Continental Red Seal six-cylinder engine.

1922

Keystone built this limousine type ambulance as their Model 801-B on a Cadillac chassis.

Called a Style 54, this Keystone casket wagon had a low rear deck with a removable top. Mounted on a Dodge chassis that was lengthened, this car could carry one to four rough boxes and was also handy for carrying chairs, pedestals, flowers and baskets. Keystone offered a complete line of high grade motor funeral equipment.

This ambulance was mounted on the latest Dodge Brothers business car chassis and the body was built by E.M. Miller of Quincy, Ill. as their Model 1823. This body was offered on any chassis and Miller also built their own six-cylinder chassis. This car came complete with a nickel-plated Bombgardner Chair-Cot, thermos bottles, and two attendant seats.

Smith & Smith of Newark, N.J. owned this uniquely styled Giessel funeral coach on a Cadillac chassis. The extreme rear quarters of this car are blanked off in service car fashion while the rear door windows contain frosted glass that is set into leaded frames.

This Kunkel Model 307 carved funeral coach is seen here mounted on a Cadillac chassis and the accompanying copy read, "A thing of beauty is a joy forever."

Kunkel built this ambulance with a limousine body style and leaded glass in the rear windows. Notice the bell on the cowl.

1922

Complete with a cord operated bell and leaded glass rear compartment windows, this McCabe-Powers ambulance is striking in appearance with its two-tone paint finish. The glass in the rear compartment windows is frosted to lend privacy to the patients.

McCabe-Powers of St. Louis built this carved panel hearse on an unknown make of chassis. Although a small firm, McCabe-Powers built a full line of vehicles for funeral use.

The Pilot Motor Car Co. marketed a line of Pilot funeral cars called Lorraine. Shown here is the Lorraine hearse finished in silver with dark gray highlights and a carved casket compartment.

Shown in the "front yard" of the W.H. Combs Co., in the Land of Flowers of Miami, these two Lorraine funeral cars were of two models. The one on the extreme left was a standard carved hearse while the one on the right was a combination hearse, sedan, ambulance. Both cars were finished in the special silver and gray paint job Lorraine was famous for.

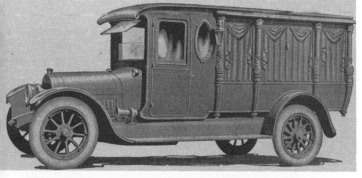

This combination body, mounted on a Cadillac chassis, was constructed by Hoover and was a Model 55. This body had a massive appearance with swelled sides, hand-carved pillars and panels, and was painted gray overall. This car was on display at the York, Penn., plant.

Hoover built this carved funeral coach on an unidentified chassis. The draped carvings and four columns per side were carved from ash. Hoover said that their coaches invited comparison.

1922

The Owen Brothers of Lima, Ohio, introduced this new Model S-294 carved drapery funeral coach this year. This was strictly a funeral coach of distinctive design with heavy draped carvings on the rear body sides, an oval window and large carriage lamps. The chassis of this car is unidentified.

Called the Style 910, this hearse was built by the Williams Carriage, Hearse and Auto Co. It featured an arched center carved drapery section that gave this particular car a special dignity. The wreath on the door was a traditional symbol of memorialization and funerals.

Mounted on an older model White chassis is this unusual Japanese hearse used in Yokohama this year. Notice the Shinto-type architecture of the casket compartment and the traditional garb of the two inscrutable Japanese funeral directors.

In Canada, Dominion Manufacturers of Toronto was building a line of high grade coaches on extended chassis. This combination pallbearers' coach and ambulance reflects the more modern limousine type styling. It was designated a Model 114.

Loaded with a casket, this hearse was of the type that was in daily use on the European continent this year. With a coach-type casket compartment without any windows, but with intricate gilded carvings, this coach is mounted on an Austro-Fiat chassis and was built in Austria. Notice the high casket lids styling and the overall shape of this receptacle.

This Dominion carved funeral coach was built on a Canadian Dodge Brothers' chassis. The carved drapery panels are all of solid wood. The car is of eight-column design.

1923

The golden age of the funeral coach and ambulance continued with several more new makes entering the fold. Meteor Motor Car Co. of Piqua, Ohio, introduced a new four model range as a sister line to the Meteor. The idea behind this model line-up was to enable the smaller funeral directors and hospitals which couldn't afford the price of the large Meteor to obtain Meteor quality at a lower price. Called the Mort, the new cars sold for approximately $1,000 less than the larger Meteor line, while having all of the quality, workmanship and reliability of the Meteor Company behind it. The name Mort meant "little" with reference to the ambulances; but for the hearse, the name, which is French for dead, had something of a bizarre quality about. Advertisements carried the logo, Mort Motor Co., Piqua, Ohio. These cars used the same Meteor chassis and the same Continental engines but were marketed separately as a make to themselves.

Meanwhile, in neighboring Bellefontaine, Ohio, the A.J. Miller Company had begun to build coaches on the Studebaker chassis instead of assembling their own. Competitive advertising was saying things like this: "Funeral car chassis must be assembled by the funeral car builder especially for funeral coach and ambulance work. Standard manufactured chassis built by pleasure car makers are wholly unsuited for hearse and ambulance service." This launched a competitive war between Meteor and Miller that was to last over three decades. Miller's advertising was saying things like this in retaliation: "The assembled funeral car is suffering the same fate as the Dodo bird. The Dodo passed out because it couldn't keep pace with progress. Neither does the assembled funeral car chassis. It doesn't meet the needs of the modern mortician. So it is destined for oblivion. One of these days we are going to find that it is as dead as the Dodo!" Not being able to comment on Miller's knowledge of bird life, it is, however, interesting to note that within the next few years most of the makers would switch to the manufactured chassis.

The Superior Motor Coach Body Company was organized in Lima, Ohio, by a group of mechanics, coach craftsmen, and businessmen to build wooden passenger bodies on the chassis of the Lima-built Garford truck. These deluxe passenger or bus bodies were the beginning of the largest funeral car and ambulance maker in the U.S.

In Lima, Ohio, a new name appeared that within the next two years would make an imprint on the history of the funeral coach and ambulance. Superior Motor Coach Body Company was organized to manufacture deluxe wooden passenger bodies for the Lima-built Garford truck chassis.

Two new sedan ambulances were introduced to the trade in this year. One emanated from the house of S & S in Cincinnati and was called the Brighton, while the other was introduced with the Mort. The Mort Model D six-passenger side-loading sedan ambulance was said to be the latest and best adaption of the sedan type car to the various uses of the funeral director.

The limousine coach style was gaining more followers in this year and there was an increasingly insistent demand for this style. It was called the "car of the day" as it could not be confused with other new designs of the era. This one body style breathed an air of refinement into the field and offered the funeral director a new and more modern style from which to choose if he no longer desired a carved car.

This striking limousine style hearse was built by Cunningham and was designated as a Model 102-A. This particular vehicle was operated by the Cunningham & O'Connor Funeral Home and was painted a two-tone gray with purple velvet draperies.

Crane & Breed continued to market a line of high quality coaches mounted on the Winton Six chassis. This carved model was a standard Crane & Breed offering and this picture was used for several years in succession in the company's advertising. Crane & Breed was beginning to de-emphasize coach production and concentrate on a strong line of high grade caskets.

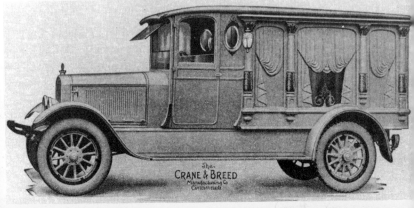

The S&S Arlington hearse was a new addition to the S&S range this year. It featured the long narrow rear quarter window that was popular on the Kensington ambulance. This window had French plate beveled glass. The car was of the increasingly popular limousine style.

Built for the personal or business uses of the funeral director or the livery, this S&S Brighton seven-passenger limousine was a very attractive car. This particular model was not convertible into an ambulance as was the S&S Samaritan. The car was powered with a Continental Red Seal six-cylinder engine.

Powered by a Continental Red Seal six-cylinder engine, this was the new S&S Olympian carved side funeral coach. This was a deluxe model with nickel plated radiator, bumpers, headlamp casings, and wheels.

Equipped with large Westinghouse shock absorbers, this S&S Kensington ambulance was one of the best vehicles of its type on the market. In limousine style with a modified rear quarter window scheme, this car had leaded glass in the rear quarter windows. Notice the chrome plated disc-type wheels.

Mounted on E.M. Miller's own medium weight, powerful chassis equipped with a Continental Red Seal six-cylinder engine, this coach was designated a Model 1957. It was also called a Model Q hearse. It was equipped with American Bosch ignition, five cord tires, and disc wheels if preferred. Notice the two-tone highlights on the carved sections of the casket compartment.

Woodall-Nicholson of Halifax, England, built this coach on the chassis of a 1923 Austin. The body was transferred to this chassis from an older horse-drawn hearse and reflects the general English hearse styling of the era. The arched roof had carved frames around it and flowers were carried on it. Notice the massive expanses of glass and the lack of any drapery or other ornamentation.

This E.M. Miller ambulance, shown on a Cadillac chassis, was a Model 891. This car has the leaded glass in the rear window with the addition of a stained glass cross.

1923

Meteor added two new model series to their extensive range in this year. Called the Models 200 and 300, the cars were identical except for the power train. The selection of engines governed the price and the model designation. The series 200 was powered by a 70 horsepower Continental Red Seal six-cylinder engine and sold for $2,800, while the Model 300 was powered by a 50 horse Continental engine and was priced at $2,450.

The limousine styling that was gaining popularity on all coaches was carried through on this Mort Model V funeral coach. Priced at $1,950 this car was also powered by the 40 horsepower Continental engine.

The Meteor ambulance line in limousine styling was also listed under two model designations — the 204 and the 205. Both were limousine style sedan ambulances that were side loading and had concealed rear doors to further enhance their sedan styling. The 205 shown here had stained leaded glass windows in the rear compartment, was powered by a 70 horsepower Continental Red Seal engine, and sold for $2,850. The model 204 was equipped with a 50 horsepower Continental engine and went for $2,500. Notice the bell mounted in the center of the radiator.

Meteor introduced a new line of cars that were lower in price and somewhat smaller in overall size this year. Called Mort, these cars were built along side the regular Meteor line at the Piqua plant. This Mort Model X hearse features classic twelve-column styling and carved draperies. This car closely resembled the regular Meteor line. This particular model was powered by a 40 horsepower Continental Red Seal engine and sold for $1,750.

Meteor offered this limousine styled hearse this year for the first time. It too was available in two model power choices at two prices. The standard Model 206 limousine hearse was powered by the 50 horsepower Continental engine and was priced at $2,350. The Model 207 was identical in appearance to the 206 but was powered by the 70 horsepower Continental unit and sold for $2,700.

This model was unique to the Mort line and was designated as a Model D side loading sedan and ambulance. The car in its sedan configuration was a full six-passenger car with a full leather interior and was suited to duties as the funeral director's private car or for use as a pallbearers' car. It was converted in a matter of minutes to ambulance duties. The body was framed in staunch ash and sheathed in steel. This car sold for $1,500 and was powered by the same 40 horsepower engine used in other Mort units.

Studebaker chassis served as the bases for many coaches built by A.J. Miller this year. Shown here is the A.J. Miller Model 75 sedan ambulance on the Big Six chassis. The photo shows the car looking like any other passenger car of the era but the inset picture shows the interior converted for ambulance work. The car sold for $2,250. This body could also be mounted on other chassis.

A.J. Miller also offered a choice of powerplants on coaches built on their own assembled chassis. This Model 65 was mounted on Miller's own chassis and was available equipped with a 55 horsepower six-cylinder Continental Red Seal engine as standard at $2,200. On the same chassis, at extra cost, a 70 horsepower Continental Red Seal six-cylinder was available. This version cost $2,400.

This A.J. Miller carved funeral coach is shown mounted on their own chassis. It was a Model 67. Miller also used the Continental 70 horse engine. This car sold for $2,400.

The A.J. Miller limousine hearse Style 95 was also available on different chassis, with this one being on their own. This car, with strange draperies, sold for $2,750.

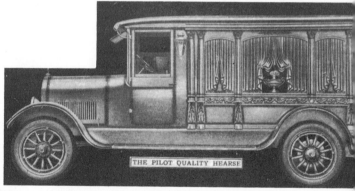

Marketed as the Lorraine, these Pilot funeral cars were in a wide variety of styles and body choices. Pilot Motor Car Co. was located in Richmond, Ind. Notice that this car is finished in the silver finish similar to that on the Banks coach.

With the new year and a new style of coach, the limousine, A.J. Miller announced this uniquely styled landau limousine coach. This was the first limousine introduced by Miller and was designated a Model 90. The car was equipped with a 70 horsepower Continental six-cylinder engine and sold for $2,500. Limousine styled coaches began to be offered by most every manufacturer large or small. This limousine hearse, mounted on a Cadillac chassis, was built by the Owen Brothers of Lima, Ohio. This car featured a 36-inch rear door opening, silk roll-up curtains or draperies and was offered as a straight hearse or as a combination.

1923

Using a Ford engine and transmission, these Banks carved hearses looked very similar to those produced by Lorraine. This example is finished in a color called "molten old silver" with darker shading. This color was said to have been very durable and easy to keep clean. Lighted shading and other color choices where available. Banks Motor Corp. was located at Louisville, Ky.

Built by the Davis Funeral Car Company of Raleigh, N.C., this Davis Dodge carved hearse looks surprisingly like a Meteor. Davis, like so many other makers, was a small and rather localized firm that never really made it on a national scale but had some success in their particular area of manufacture.

This Reo wears a body built by Mitchell of Ingersoll, Ontario. The carved panels were removable to reveal a plain paneled casket wagon type coach. This was the Mitchell & Company's Style #3.

McCabe-Powers built this ambulance, which features an arch-type rear compartment window with all rear windows being fitted with leaded glass. Notice the shaping of the roof and the rear end.

Offered on their own light six chassis, Henney marketed this carved hearse without any glass area in the rear compartment for $2,160.

Built on an older Packard chassis and designated Style SC-861, this Hoover carved funeral coach has one very large window in the casket compartment on each side, and shows that the traditional type of funeral coach was slow to give way to the all new limousine style.

Endicott-Johnson Shoe Company operated these two industrial paneled ambulances built by August Schubert Wagon Company of Oneida, N.Y. on Larrabee "Speed Six" truck chassis. The size of the coaches is easily seen when compared to the men standing next to these cars.

1924

This attractive and light colored service car was offered as part of the Cunningham line this year. Notice the plain, straight-through styling and the absence of ornamentation or carving. Cunningham built their own chassis and bodies, assembling their cars complete.

The limousine of full greenhouse style was the up and coming trend for these professional cars and every maker rushed to introduce new limousine models. This limousine hearse was a new model in the S&S line and was designated the Columbus. Once again, all S&S cars used a Continental engine and a completely assembled chassis.

S&S continued to market their stylish Samaritan sedan ambulance through the year. These sedan ambulances were very versatile vehicles and enjoyed a considerable amount of popularity with the trade.

The limousine style funeral coach finally captured th majority of the coach market and every manufacture offered some variation on the limousine theme. With thi new style the designers were faced with a problem of how to make their firm's coach appear different from thos offered by the competition. This was undertaken by th stylists with great skill in many cases. Some firm retained a small amount of carving and placed this above below and around the body side windows. The Hoove Body Company of York, Pa., was particularly adept a this in their model 1033 hearse body that appeared ir this year.

Other ways of making the new limousine appear to be more formal came with the leather back style. As the roofs of the cars were made of wooden slats covered with a leather or other like material, some companies placed padding under this surface covering and covered the back portion of the roof with this material. Landau bows were many times placed on the extreme rear portion of the roof sides to lend the coach a more formal effect Sometimes a small oval window was placed here instead of the landau irons. These windows were sometime etched with crosses or other figures of memorialization Still other companies used the landau bow and the ova windows together to form a totally new effect.

Another variation on the limousine theme was intro duced by Sayers & Scovill this year. The Arlingtor funeral coach was an outstanding development in the limousine style. These coaches featured a long, narrow rear side window with beveled glass. The S & S Kensington ambulance had this same type of window but with leaded glass in this area, while the window itself wa somewhat smaller.

McCabe-Powers Auto Body Co. of St. Louis offered this Model 536 combination pallbearers' coach and ambulance this year on any suitable chassis. It is shown here on an earlier Dodge chassis. The doors are open, giving a good view of the car in its ambulance form.

1924

This stylish limousine hearse was a Meteor Style 216, which sold for $3,150. All sales were direct through the factory and prices were F.O.B. Piqua. All makers advertised prices in this manner and delivery of the vehicles was made at the plant.

One of the larger Meteor carved models was this Model 210 that carried a price of $3,050. This car was equipped with four-wheel hydraulic brakes and a Continental Red Seal six-cylinder engine that was rated at 70 horsepower.

Unlike the Meteor line, Mort coaches were offered on chassis other than the assembled Meteor-Mort standard offering. This carved Mort Model A is shown here mounted on the Dodge business car chassis. With the Dodge chassis the Mort model A was referred to as a Mort-Dodge and sold for $1,750.

Looking very much like its sister coach, the Meteor, this Mort Model S was a less expensive version of the Meteor line. Utilizing the same chassis and running gear as the Meteor, this car sold for $2,150. It was powered by a 50 horsepower Continental Red Seal engine, had Lockheed hydraulic four-wheel brakes, and a Borg & Beck transmission.

A new model in the Mort line was this combination car designated as a Mort Model T. Selling for $2,150 this car featured a novel combination body with a small leaded glass window on each of the rear compartment sides. The framing around this window was done in frosted glass. Specifications for this car were the same as for the Mort Model S seen earlier in this chapter.

Redesigned and offered on their own chassis, the Mort sedan ambulance re-appeared this year. This stylish sedan was referred to as the Mort model 327 side loading sedan ambulance and was offered on either a 50 horsepower chassis utilizing the Continental engine for $2,450 or with the more powerful 70 horse version of the Continental. In the latter form, the car carried a price of $2,700. These cars were equipped with Firestone balloon cord tires and an overall style that made the car hard to recognize from any normal sedan.

1924

With their own scientifically constructed chassis with a spring suspension that assures the greatest possible riding comfort, Henney offered a line of coaches with the bodies being built by Meritas. These bodies reduced the interior vibrations and kept drumming to a minimum. This Henney Model 1162 limousine hearse boasted a light weight of only 3700 pounds and a low maintenance cost.

With a Meritas made body, a 70 horsepower Continental Red Seal six-cylinder engine, and unique styling, Henney offered this new limousine-landau model as the Style 162. The graceful curves of the roof line further enhanced the cars modern appearance. The addition of a fabric covered top with landau bows on the rear quarters offered yet another style to the still new limousine type coach. Close examination of this picture reveals the flower tray (filled with floral tributes) that rode above the casket high enough to easily display the flowers through the windows. All coaches came complete with a flower tray.

Featuring sedan styling and surprising versatility, this Henney Model 1110 combination car also had a Meritas made body. This car combined the utility of the sedan ambulance with sedan landau styling. The small picture shows this car in form for its ambulance duties. Without a doubt, this Henney sedan ambulance was one of the most attractive vehicles of its kind available.

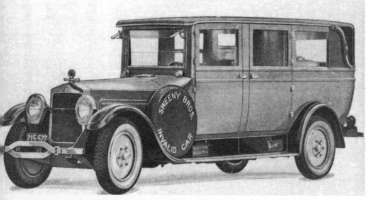

Introduced this year as a 1925 model, this is the A.J. Miller Model 95 mounted on the Studebaker Big Six chassis. This car features limousine styling with a fabric covered top and landau bows at the rear quarters.

This carved coach was built by the A.J. Miller Company and was their Model 65. It is shown mounted on their own chassis that utilized the 70 horse Continental six-cylinder engine and sold for $2,400. A.J. Miller was one of Meteor's closest competitors.

The A.J. Miller Model 95 is seen here as it was mounted on a Studebaker chassis. Notice the limousine style and the strange draperies. Small inset pictures show inside view and a good view of the back of the coach.

1924

Called the Model 200 special limousine burial coach, this car is shown on a Dodge chassis and was built by Eureka. The body design was patented and featured a wide landau bow area and a limousine style.

Shown here is the standard Eureka limousine funeral coach with a smaller landau area and mounted on a Cadillac chassis. Notice the pull-down shades that Eureka used.

This hearse, finished in Pilgrim gray, was owned by Clarence R. Huff of Huntington, N.Y., and was driven for over 7,300 miles without any mechanical problems. The body was built by Eureka and the chassis was that of a Buick.

With the limousine style being relatively new and with variations on this theme being offered by nearly every maker, this Hoover Type 1033 hearse is unique in its own way. The narrow rear compartment windows are augmented with beveled glass and heavy draperies. The belt line of the rear compartment is highlighted with carvings while the front doors carry carvings of the ancient Egyptian winged disc.

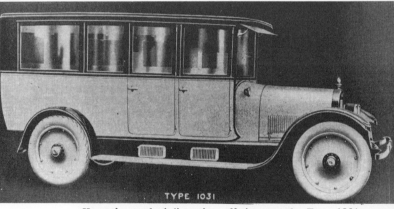

Hoover's standard limousine offering was the Type 1031 hearse without any beveled glass or carvings. The car follows a clean line with a fresh approach to the limousine style.

Despite the attractive limousine offerings from the firm, Hoover continued to produce the carved drapery style of coach. This Model 3C-1285 hearse features a twelve-column design and a crucifix in addition to the carved draperies and carvings on the front doors. Notice that the carved coaches retained the carriage lamps while most of the new limousine style coaches dropped this feature.

Seen here on a Dodge chassis is a hearse built by the Williams Carriage & Auto Company of St. Louis. This model was designated Style 912. It featured sunburst or ray type draperies in all of the rear compartment windows, minimal carving work, and eight mildly carved columns.

1924

Referred to as a sedan type of coach, this Kunkel Model 405 ambulance had limousine styling with all frosted leaded glass windows. The car was a side loader, and the extra width for loading was obtained by making the area behind the rear side doors open with the doors.

Another variation on the limousine theme was this unusual hearse from Kunkel. Called a Model 630, this car featured a small landau bow in the rear quarters and a small vertical window with leaded glass. This coach retains the coach lamps.

This E.M. Miller limousine style hearse is on the Cadillac chassis, and was designated a Model 743-A. Notice the front window treatment and the flower tray that is prominent in the rear compartment.

Mounted on the Cadillac V-63 chassis, this attractive E.M. Miller, Miller-Quincy, limousine hearse was one of the firm's finest offerings. The body could be mounted on any dependable chassis including E.M. Miller's own assembled chassis. This hearse is a rear loading vehicle only. Notice the absence of rear side doors.

This E.M. Miller Model Q, Style 1827 ambulance is shown on the assembled Miller-Quincy chassis. This chassis utilized the Continental engine and hydraulic four wheel brakes. Interesting features of this coach are found in its leaded glass windows, single drape in the door, and carved columns at the body corners.

Designed for Harry T. Pyle of Brooklyn, this Geissel ambulance was equipped with five electric dome lights, running water and a basin, electric fan, a swinging cot with air matresses, cabinets for linen and medicine bottles, two thermos bottles arranged on the rear compartment side, a heater on the floor of the rear compartment and two attendant's seats. Riding on a Stearns-Knight chassis, this attractive body boasted all leaded glass windows in the rear compartment, a two-tone paint finish and small carriage lamps.

This extremely high looking Pierce-Arrow wears a body made by Geissel. This limousine body accentuates the lines of the Pierce-Arrow chassis and comes off as a very attractive unit.

1924

An Owen Brothers' limousine type hearse, called the Model 782, is shown here mounted on the 50 horsepower Studebaker Big Six chassis. The car itself features limousine styling with a blanked off area behind the driver's seat, pull-down shades and fender mounts.

Owen offered this sedan ambulance on the Nash chassis. The car was again, a side loading style with only one door handle showing on the side. Both side doors opened to facilitate loading.

Owen offered this pleasing Model 745 straight ambulance in this year. The car came complete with a Bomgardner cot, heater, medicine cabinets, compartments for two thermos bottles, and an electric fan. Shown here on a Cadillac chassis, this car also had leaded glass in the rear compartment windows.

Riddle built this carved hearse this year and finished the car in white. Notice the chrome wheels and the wide white wall tires on this vehicle. Carvings were still done entirely by hand, of high grade hardwood.

The Bridgeton Hearse and Ambulance Co. of Bridgeton, N.J., offered a complete line of vehicles again this year, with this Dodge chassied car being their combination model.

Cynthiana Carriage Co. moved to Covington, Ky., in this year and built this carved hearse on a Dodge chassis as their Model 31 funeral car. The small oval window and the restrained use of carving work highlighted this car's style.

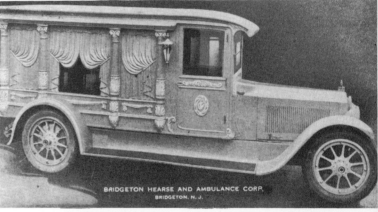

Another Bridgeton, this time a carved hearse with all the trimmings, featured eight columns, massive carved draperies and detailed carving work on the doors.

Called the Model 61 combination coach, this Cynthiana product reveals much more restrained use of carvings and a small window on each side of the rear compartment. Even though rather plainly styled, the car is still attractive. Once again, the chassis is Dodge.

1925

The depression and the keen competition had taken its toll among the funeral coach and ambulance manufacturers in business through the previous years. With the considerable shrinkage in the field, it was unusual to see two new companies joining the ranks in this year. Two years earlier the Superior Motor Coach Body Company had been established in Lima, Ohio, to build bus bodies for the locally made Garford truck. In 1925, the name of the firm was changed to Superior Body Company and the production facilities added to embark upon the funeral coach and ambulance field, building the new bodies and mounting them on a manufactured chassis, as was the fashion now. The first coaches were mounted on Cadillac chassis. The new Superior funeral coaches and ambulances were stylish indeed. Funeral cars were of the limousine leather back style, while ambulances were all of the straight limousine style.

Another firm to enter the funeral coach and ambulance field this year was The Flxible Company of Loudonville, Ohio. Flxible had been established in 1912 in nearby Mansfield, Ohio, by Hugo H. Young to build sidecars for motorcycles. Established as the Flxible Sidecar Company, they built sidecars for motorcycles in a flourishing little business until 1921, when Henry Ford reduced the price of the Model T below that of a motorcycle with a sidecar. The popularity of the motorcycle then began to fade, and as it became increasingly hard to compete with the automobile, the Flxible Company decided to enter a new field of manufacturing. They began to build busses and sold their first vehicle to E.L. Harter who operated a bus service between Ashland and Mount Vernon, Ohio. The following year, it was decided that the highly-developed coach-craft skills found in the new Loudonville plant could be put to use in yet another transportation field. Thus, in 1925, Flxible entered the funeral coach and ambulance arena. The first coaches were built on Chevrolet truck chassis and were of the limousine style.

Mounted on a Chevrolet truck chassis, this little hearse was the first such vehicle to come out of the Loudonville, Ohio, professional car facilities of the Flxible Co. seen through the rear windows are the flower tray and the shades that were of the pull down variety. Flxible built about 16 of these coaches this year.

While most of the manufacturers switched to the manufactured chassis, some held firm to the assembled car. Among them were Cunningham, Sayers & Scovill, Henney, Riddle, and Meteor.

The Sayers & Scovill Company of Cincinnati built a new style of limousine coach to meet the new demands that emanated from funeral directors for simplicity and dignity in the funeral procession. The limousine style was now well entrenched and the newer landau type of limousine coach was also called the leather back car. The roofs of these cars were covered in fabric and the rear quarters featured a landau bow. This S&S landau type car was called the Norwood and had a small oval window in the rear quarter panel in addition to the landau bows.

Continued without any major change was the S&S Samaritan combination eight-passenger sedan and ambulance. This type of car continued to enjoy a considerable amount of popularity.

This style of coach was a uniquely S&S type of car. A variation of the limousine style, this S&S Washington featured a long narrow window in the rear compartment that was filled with heavy beveled plate glass. A small flower vase was mounted on the partition wall just inside the rear compartment.

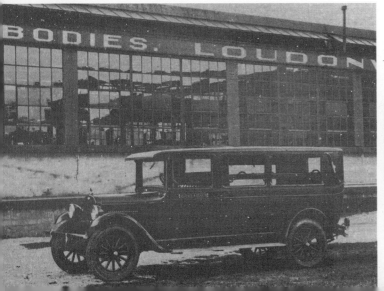

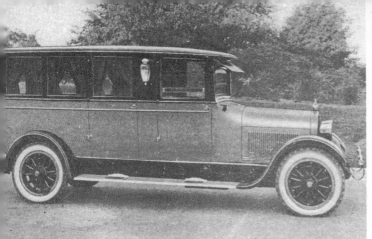

This stylish two-tone limousine hearse was built as a Model 743 by the Owen Brothers of Lima, Ohio. Notice the white wall tires and the bumper. This car retained the small carriage lamps and had velour draperies of deep purple in the rear windows.

Called the Eureka Style 194 Deluxe, this car is mounted on a Rock Falls chassis. The model was available in two versions. The standard 194 did not have the stylish oval windows in the rear quarter that are clearly seen here on the deluxe model. Also available on the deluxe model were the etched windows in the rear compartment, in the form of stars. Other etchings on the windows in this car consist of an ornamental border around the stars. Note the crucifix in the small oval windows.

Eureka introduced this attractive town car style funeral car in this year on a Lincoln chassis. The car is seen here with a new device that Eureka was developing, called the three-way casket table. Note the wire wheels and the landau back styling.

New to the scene this year was a line of professional vehicles from Studebaker. They had bodies built by the Superior Body Co. of Lima, Ohio, and were sold through Studebaker's own, already established dealer network. This car was mounted on the Big Six chassis with a wheelbase of 158 inches and four wheel hydraulic brakes. The engine produced 75 horsepower and the complete unit sold for $3,550.

The Superior Body Company of Lima, Ohio, organized in 1923 to build bus bodies for the Garford truck, this year expanded plant facilities to include the building of a line of hearses and ambulances. This first Superior funeral coach was mounted on a Cadillac chassis and was of the landau leather back style. The flower tray can be seen through the rear windows as can the sliding glass partition of this car. The front seats were appointed in leather. The disc wheels were standard on this car as was the side mount.

Along with entering a new car in the funeral car field, Superior offered this ambulance. The car was finished in white, and featured a spotlight mounted on the windshield pillar, frosted glass in the rear quarter windows, and a sidemount. This car, like the funeral coach built by Superior this year, was mounted on a Cadillac chassis.

1925

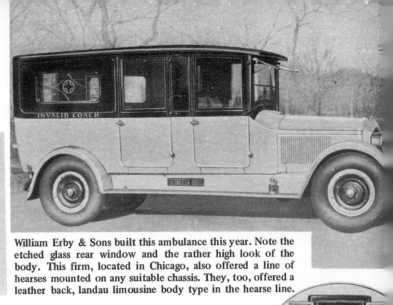

Earlier in this book were some hearses built on the Ford model T chassis by Columbia of Detroit. This year they offered the trade this small car mounted on the Ford chassis. Called a Funeral Director's Special First Call Car, the vehicle was equipped with all the room needed to carry all of the essentials of the trade. It must be remembered that these were the days of holding the funeral in a private home, usually that of the deceased, and the funeral director had to carry all of his tools out to the house to do his work. Columbia offered this special bodied car complete for just $525.

William Erby & Sons built this ambulance this year. Note the etched glass rear window and the rather high look of the body. This firm, located in Chicago, also offered a line of hearses mounted on any suitable chassis. They, too, offered a leather back, landau limousine body type in the hearse line.

A.J. Miller began to build coaches on chassis other than Studebaker in this year and offered this Model 96 limousine funeral coach on a Nash Advanced Six chassis. Styling was quite plain with little to distinguish these vehicles.

A beautiful leather back landau limousine hearse was this model offered by the James Cunningham Co. Featuring both the landau bow and a small oval window combined with a two-tone paint job and white wall tires, this car was one of the most attractive offered this year.

Cunningham also built some of the finest ambulances offered. This model was supplied to Dickinson-Streeter Co. of Springfield, Mass. It featured a siren on the running boards, leaded glass in the rear windows and a very small carriage lamp.

Not only did Cunningham offer some of the industry's most attractive coaches, they also built some rather nice looking passenger cars. This Cunningham Style 133½-A 8-passenger limousine features some of the same styling seen on the hearse. Notice the landau or leather back roof with landau irons and small oval windows.

A fleet of Cunningham hearses like this one was delivered to Hornthal & Co. of New York City. This firm was a large livery for hearses and limousines and did a brisk trade with the city's smaller funeral homes. Notice that this coach retains the carved glass side style of earlier cars.

1925

Meteor offered this Model 226 leather back hearse for $3,150. It was powered by a 70 horsepower Continental Red Seal engine. The large Meteor cars were powered by larger six-cylinder plants than those offered on the sister Mort.

Mort continued in production. This Mort Model 325 limousine ambulance was powered by a 50 horsepower Continental engine and sold for $2,450. The car featured a combination of both frosted and stained glass in the leaded rear windows.

This Mort limousine style funeral coach, designated a Model 326, had four wheel hydraulic brakes and a 50 horsepower Continental engine. It sold for $2,450.

This car was the Meteor Model 226-A limousine funeral car. It was powered by the same engine used on all other Meteor models and used the same chassis. The difference between this car and the standard 226 are easily seen in body styling. This car features no landau bows or leather back styling. The coach sold for $3,150, same as the landau version. This car is seen with Gordon tire covers on the side mounts. Gordon of Columbus, Ohio, was a prime supplier of spare tire covers with special inscriptions.

Henney continued to construct their own chassis and offered this DeLuxe Six ambulance this year. It featured leaded glass in the rear compartment windows and a landau bow in the rear quarters.

This stylish Henney coach was named the "Chicago Special" for some unknown reason. It is of limousine leather back styling with curvaceous landau irons and uniquely etched windows for the rear compartment, the lower half of which was comprised of stained glass sections. Note the Gordon tire cover on the side mount.

1926

A new and distinct triumph of the coach-builders art was realized in this year when Eureka Manufacturing Co. of Rock Falls, Ill., announced the side loading funeral coach. Previously, all funeral cars were loaded through the rear doors. But this year an application for a patent was applied for and granted to Wilber S. Myers, who had joined Eureka in 1921 as a body builder, for his table design on the Eureka side loading hearse. Eureka was founded in 1871 and first incorporated in 1887. The first products manufactured by Eureka were school desks, teachers' desks, chairs, and portable blackboards. Some time later, the firm began building carts for the government to be used as rural mail delivery wagons; and by 1900, the Eureka Company was listed as one of the largest carriage builders in the United States. Shortly thereafter, the company took up building hearses and ambulances and in 1910 began to build motorized funeral cars and ambulances mounted on passenger car chassis.

The side loading hearse was to become yet another "wave of the future" that was to last to this very day. In this year the Studebaker Corp. decided to enter the funeral coach and ambulance field, and the newly established Superior Body Company of Lima, Ohio, gained the contract to construct the bodies for these cars that were to be marketed strictly under the Studebaker name. With this decision, Studebaker withdrew its chassis from the coachmakers; and the A.J. Miller Company, that had been using this chassis exclusively, now turned to Nash and Packard chassis for its hearse and ambulance work. Miller's new streamlined limousine body was designed especially for the new chassis they were now using. Known as the models 101 and 102, the A.J. Miller Nashes and Packards became quite popular and were widely distributed.

While Studebaker entered the business in this year, Cadillac also made an effort to design their own coaches. These new coaches used the Cadillac 90-degree eight-cylinder engine and sat on a 150-inch wheelbase.

Instant approval was accorded Henney's new low priced "Light Six" funeral coach introduced in this year. Meanwhile, Henney produced a line of large coaches that wore a body made by Meritas Body Company and finished with Duco paint. Their model number 210 sedan ambulance was a very attractive landau back car.

The E.M. Miller Company had been established in Quincy, Ill., in 1856 to build high quality carriages, and thereafter entered the hearse and ambulance field. When switching over to the motorized chassis, E.M. Miller built their coaches on existing manufactured chassis although they did assemble a few of their own chassis through the years. The products of the E.M. Miller Co. were called Miller-Quincy and were said to have been "designed and built along conservative lines, free from fads and freakish features that soon made vehicles look out-of-date." This was an implied reference to the carved coaches that were still being built in limited numbers by some manufacturers, while Miller-Quincy had ceased to build this style within the last couple of years. This year saw the last of the Riddle hearses and ambulances that were built in Ravenna, Ohio. In the decline of the number of funeral coach and ambulance makers, Rock Falls had also gone out of business a year earlier.

A car worthy of note from this year was the special S & S Golden Anniversary Car. Celebrating their golden anniversary in 1926, Sayers and Scovill announced this deluxe version of the popular Washington casket coach. All of the exterior mountings on this "golden car" were finished in gold. The car's distribution was strictly limited to one such model in any city, and it carried a price tag of $10,000 while being available to special order only. S & S also introduced a smaller version of the popular Washington in this year, called the Junior Washington. This car was to appeal to the funeral director that longed for an S & S Washington but could not meet the price that the full-sized coach carried.

Built by the Knightstown Body Company of Knightstown, Ind., this stylish leather back landau funeral coach had very large oval windows in the rear quarters and smaller landau irons that wrapped around the oddly shaped windows. Knightstown built bodies to fit any chassis with the capacity to carry them. This car is mounted on a Packard chassis.

Built by the Holcker Manufacturing Co. of Kansas City, this Buick chassis coach featured the leather back landau style with small oval windows in the rear quarters as well as a large landau bow.

1926

Albert Alt of Grand Rapids, Mich., was the receiver of this handsome Eureka leather back, landau hearse this year. Crucifixes are mounted in the rear most windows. This car was mounted on a Chandler chassis.

The triumph of the coach builder's art was realized and a totally new era in funeral cars was ushered in with this model from Eureka. This was the first side-loading funeral coach. Eureka designed and patented the table that allowed a casket to be loaded from either side or from the rear as had been the custom. This revolutionary car is shown here as it appeared in the announcement advertisements, mounted on a Cadillac chassis. The rear side doors opened from the same pillar thus allowing room for a casket table to swing out.

This Eureka Model 194 limousine funeral coach is shown mounted on a Lincoln chassis. Eureka was one of the major makers supplying professional bodies for the Lincoln chassis. The smart looking coach has no landau irons or carvings, but was finished in two-tone paint. This car is not a side loader.

Designed and built by the Kelley Manufacturing Co. of Bucyrus, Ohio, this special casket and vault delivery body is shown mounted on a Ford TT chassis. The body was all steel and it was said to have been a mighty convenient rig for delivering caskets to the residence, rough boxes or vaults to the cemetery, or for everyday hauling around the funeral home. It had a special compartment at the top of the rear body to carry paraphernalia.

Shown mounted on a Cunningham chassis, this Eureka DeLuxe side-loading hearse has a landau area and a leather back roof style. This car is also equipped with the small oval windows in the rear quarters and pull down shades. The fact that this car is on a Cunningham chassis is not unique, because the makers that assembled their own chassis sold these products alone or with bodies mounted.

Eureka continued to mount bodies on any suitable chassis with this straight ambulance body being on a Stearns-Knight chassis. This particular car was the deluxe ambulance and featured leaded glass in the rear windows with a stained glass cross in the centers. This car went to Wymer Funeral Temple in Cedar Rapids, Iowa.

1926

With S&S celebrating their golden anniversary this year, it seemed fitting to introduce a special model. They did this with this car called the "Golden Anniverary Golden Car," which was first exhibited at the National Funeral Directors Convention where it was the center of attention. Basically a DeLuxe Washington hearse, this car was trimmed in outstanding materials that made it really outstanding. All exterior mountings were of gold with thousands of dollars worth of this material going into the construction of the car. The interior mountings were all of gold and old ivory, with the woodwork being of East Indian rosewood inlaid with white mahogany. Sayers & Scovill was established in 1876.

Geissel offered a line of high grade coaches mounted on any suitable chassis and this limousine hearse was but one of their body styles. Mounted on a Cadillac chassis, the car has clean limousine styling with a small coach lamp on the B-pillar and silk draperies in the rear compartment windows.

This S&S 8-passenger limousine for funerals and livery work was new for 1926 and was called the S&S Newport. This particular car was owned by O.J. Peterson of Chicago and featured leather back landau styling with a small oval window in the rear quarters. This was not a combination car, but a straight 8-passenger sedan that could be used by the funeral director as personal transportation, during funerals for transporting the family, or pallbearers.

The S&S Norwood was re-introduced in a refined and more sophisticated version. This S&S DeLuxe Norwood looks like it had an extremely high waist line. The car has a leather back landau roof treatment with oval windows in the rear quarters of the roof.

Specializing in the reconversion of older motorized coaches to a more modern style car, the L & M Manufacturing Co. offered this body style seen here on a Nash Advanced Six chassis. This stylish limousine body was designated a Model 76 by L & M and they also offered several other limousine type body models. The leather back is prominent on this car. L & M was located in Dayton, Ohio.

This stylish L & M Model 77 combination pallbearer's car and ambulance is shown mounted on a Cadillac chassis. L & M converted old carved type cars to the modern limousine style and used only the chassis of the carved car. Some of their conversions were carried out on cars as old as fifteen or twenty years. This Dayton, Ohio, firm used dried hardwood for the body framework and all of the joints were mortised and tenoned. Interior woodwork was solid walnut and chase mohair or leather was used for trimming throughout.

Cadillac's marketing department never missed a potential area for sales and they offered this custom built leather back suburban as a combination livery car and a pallbearers' coach. This car was owned by C. Kampp & Sons of Chicago. This was only one model of a whole range of new models aimed at the funeral profession.

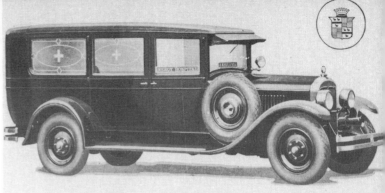

Cadillac also offered a line of hearses and ambulances on a wheelbase of 150 inches. The bodies were most likely built by Detrich of Superior to Cadillac's specifications and designs. Cadillac claimed to have designed all of the interior appointments, including the casket rollers and hardware. Equipment included on these cars consisted of custom headlights, spring covers, moto-meter, Lorraine spotlight, two fresh air heaters, 33x6.75 tires, the 90 degree eight-cylinder engine, windshield wiper, and a choice of either disc or spoke type wheels. Notice the frosted glass windows in the rear part of this ambulance.

Built primarily for the funeral and livery trade, this Case Model Y 7-passenger limousine was offered to the trade in this year by the Case Motor Co. of Racine, Wis.

This fleet of Cunningham limousine style hearses was delivered to Zabriskie & Scott of Paterson, N.J. Cunningham made many such mass deliveries of coaches, and made no bones about advertising that fact with pictures such as this.

Patrick Boland of Chicago took delivery of this eight-column glass sided Cunningham funeral coach in July. The large ornate carriage lamps and the carved body, attractive as they are, seemed somewhat antiquated beside the firm's more modern looking limousine offerings.

Hoover's Type 862 casket body is shown here mounted on a Dodge truck chassis. The body was devoid of any ornamentation other than the four carved columns at the body corners and the carriage lamps. Notice the colonial style side windows.

Designated a Model SC-1399 limousine hearse, this Hoover featured the most up-to-date limousine styling with simplicity of construction and durability that made these cars last for many years. This special hearse body was designed to express the highest degree of refinement and dignity required by the funeral director and was available on any suitable chassis.

This Meteor Model 236 offered a Meteor chassis that was powered by a 70 Horsepower Continental engine with a three joint propeller shaft, Timken axles, Lockheed 4-wheel hydraulic brakes, Firestone balloon cord tires, Watson stabilators all around, as well as improved front and rear fenders and bumpers. This model sold for $3,150 while the Model 236-A, without the landau, leather back styling, sold for the same price. There was no difference in specifications. The casket compartment in both machines was done in Velmo mohair and Wilton carpeting with a removable arched flower rack riding above the casket.

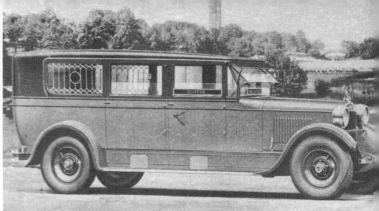

This ambulance style was also offered in both Meteor and Mort versions with both being identical with the exception of the engines. The ambulance in the Meteor line was referred to as a Model 235-A with the Mort version being designated as the Model 335-A. The prices for these cars were the same as those for the limousine hearse shown earlier in this chapter. Chassis details for both cars with the exception of engines are as follows: Timken axles, Continental engines, Lockheed four-wheel hydraulic brakes, five Firestone balloon cord tires, four Watson stabilators, front and rear bumpers, and an interior upholstered in genuine leather that was hand buffed. Ambulance equipment included a cot, folding attendant's seats, fan, and Kysor heater. The car was designed to load from either the rear or the side. They were built as shown or with the landau leather back styling.

Available with either a six or an eight cylinder engine, this Kissel hearse and ambulance was quite a stylish car in its day. These coaches were custom built to the same high standards that Kissel set for their passenger car line, and were built at the Kissel Motor Car Co. plant in Hartford, Wis. Notice the cycle-type fenders and the landau, leather back styling.

Available in either the Meteor version or in a Mort car version, this limousine style coach was a standard offering from the Meteor-Mort Company of Piqua, Ohio. In the Meteor version, this car was designated as a Model 236-A and sold for $3,150 while the Mort was referred to as a Model 336-A and was $2,650. The Mort version was powered by a 50 horsepower Continental Red Seal six-cylinder engine while the Meteor cars were equipped with a six-cylinder engine made by Continental that produced 70 horsepower.

Called the Model 666 "Ideal Limousine" funeral coach, this Kunkel hearse had very straight forward styling, no frills or ornamentation. It could be loaded from either the rear or from either side and was available on any suitable chassis.

"The country has gone Nash," was the headline of one A.J. Miller advertisement this year. This was one of the Miller limousine models mounted on a Nash chassis. Miller was one of the major manufacturers to make the switch from the assembled chassis to that of a manufactured vehicle this year.

The older carved coaches had now begun to take a back seat to the newer and more modern looking limousine coach, but this, by no means, meant that this type of car had become extinct. This ornately carved hearse was offered by August Schubert Wagon Co. of Oneida, N.Y. It featured intricate carvings of cherubs in the panels along side the center plate glass window in the casket compartment.

1926

107

Auburn aimed a model at the funeral profession with this 7-passenger sedan-limousine. Priced at $2,495, this was designated as the Auburn Model 8-88 and featured all of the stylish lines of the other Auburn models in a vehicle especially designed for the funeral director.

This combination ambulance, pallbearers' car was also marketed by Studebaker. The car is fitted for its ambulance duties. As an ambulance there was ample room for the driver, the attendant and the cot carrying the patient. This combination pallbearers' car and ambulance was powered by the Big Six Studebaker engine and was available through the Studebaker organization of 3,000 dealers.

Studebaker, now manufacturing their own complete funeral cars, offered this leather back, landau hearse in this year. Studebaker was the only large manufacturer to offer funeral cars, ambulances, and service cars through their nation-wide dealer organization. Superior Body Co. of Lima, Ohio, built the bodies for all of the Studebaker vehicles for professional use. These cars offered three major advantages over some of the assembled coaches available. These were, quality with a Superior built body, service through Studebaker's dealer outlets, and price. This coach sold for $3,550 F.O.B. Southbend.

The Studebaker combination pallbearers' car and ambulance is shown here as a pallbearers' car. In this form the car could accommodate seven passengers. The passenger's seat in the front was removable when the car was converted to an ambulance. The car had an automatic spark control, and the lights were controlled by a steering wheel switch. The upholstery was of genuine mohair. This car, complete with Bomgardner cot, sold for $2,875 and could be changed into an ambulance in less than two minutes.

The motorcycle had hit its popularity peak around this time and there were several of these vehicles pressed into service as ambulances. Their superiorities could be easily seen — speed, versatility, economy of operation, as well as their drawbacks — lack of full time attendant care, lack of proper space for medical supplies, and lack of comfort for the patient, just to name a few. This Harley Davidson motorcycle ambulance was operated by the Wilhelm Funeral Home of St. Petersburg, Fla. this year.

This Miller-Quincy, denoted a Model 85 limousine funeral coach, is shown mounted on a Studebaker Standard Six chassis. The Model 85 was the company's newest and lowest priced coach and was of limousine leather back, landau styling.

This E.M. Miller, Miller-Quincy, leather back landau funeral coach features the style of the popular limousine coach with the oval rear quarter windows. E.M. Miller, commonly called Miller-Quincy, was founded in Quincy, Ill., in 1856, and was one of the makers of fine professional car bodies. They mounted bodies on any suitable chassis and also assembled their own chassis. This Model 74 is shown mounted on a Cadillac chassis.

1926

This new low-priced Henney was called the Henney Light Six, and was accorded the instant approval of the funeral directors. The car was Henney's entry into the lower priced segment of the field and it retained the style used by all Henney products.

Henney offered this combination sedan-ambulance this year as the Model 210. This car, mounted on Henney's own assembled chassis, had a Duco finished all-metal body or one with a Meritas-made body.

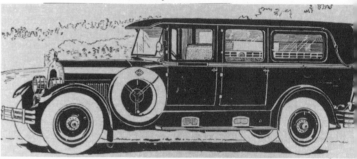

Called the Henney Model 162-A limousine hearse, this car was also available with a choice of types of body makes, Duco finished all-metal or Meritas-made, and featured the landau, leather back styling so popular in this year. The removable flower tray is easily visible through the rear windows.

This Henney Model 160-A limousine hearse was also mounted on an assembled Henney chassis and was available with either an all-metal body or a Meritas-made one. Note the sunburst etching on the rear quarter windows and the screened lower parts of these rear windows.

Mounted on a Dodge truck chassis, this rather boxy looking Cynthiana Model 81 had the popular limousine styling with pull-down shades in the rear compartment windows and a two-tone paint finish. Cynthiana Carriage Co. was now located in Covington, Ky.

In service in Port-of-Spain, Trinidad, this hearse had a low deck that accommodated flowers on the roof and had eight carved columns and large glass windows on the rear compartment sides. The chassis appears to be that of an early Cadillac.

One of the first motorized hearses to be used in Vienna, Austria, was this open-sided coach, used for the first time in January of this year. Note the high top casket and the over head opening rear door.

1927

Eureka's introduction of the side-loading or three-way funeral coach took the industry by storm. This innovation was shortly in fantastic demand and Eureka saw the profit in it. They began constructing these new tables for their competition as well as for their own use. No longer was it necessary for the funeral director to be limited to a rear loading coach. He could now purchase a three-way car and be able to load from either side as well as from the rear. Eureka was to be the supplier of these tables to the competition for the next ten years and this was to open a new chapter in the story of the funeral coach. As more makers contracted with Eureka for the new table, the customers began to be offered a larger selection of coach types. There were the straight-end loaders, three-way cars and, for a time, straight side-loading coaches.

"For the funeral director who knows the frequent inconvenience of rear loading," Henney introduced its NU-3-Way coach to the trade using the patented Eureka table. The three-way car was to acquire more converts with each succeeding year.

With the advent of the limousine body type, coachbuilders began their evolution away from the all-wood body construction. At first frames were made of seasoned hard woods — oak, maple, ash — and outer panels were of sheet steel. Except for some "hammered" rear quarter panels, the steel was flat or "wrapped," with the result that the first limousine hearses and ambulances were square and box-like in appearance. In this year new metal working machinery was acquired by A.J. Miller. The first hydraulic presses used in hearse and ambulance construction were converted "tire presses." With the use of pressed steel panels, hearse and ambulance designs that could not be attained with earlier types of metal panels became possible. Miller's hydraulic presses gave the Miller designers a head start over the rest of the industry in the area of "streamlining."

Another new style that began to appear in this year was the town car funeral coach and ambulance. A style borrowed from premium models of passenger cars of the era, this type of coach added a new dimension to the limousine style.

Offering their own combination pallbearer's-ambulance coach, Studebaker built this stylish sedan. Utilizing all Studebaker chassis components, this car was instantly convertible to ambulance duties. The car sold for $2,875 and was available through any Studebaker agency.

Another of the Studebaker offerings was this DeLuxe Funeral Car that was priced at $3,550 F.O.B. South Bend. The body was built by Superior of Lima, Ohio, and was a very cleanly styled limousine type. The chassis was that of a specially engineered Studebaker and service was available through any Studebaker dealership. This car rode on a 158-inch wheelbase.

With both side doors open and the center door posts removed from their special clamps, the Studebaker combination pallbearers'-ambulance provided ample side loading space. The car is seen here in its ambulance form complete with cot and patient.

This Studebaker offering rode on a wheelbase of 146 inches and was called the Arlington. Priced at only $2,986, the Arlington was destined to become one of the most popular cars in the Studebaker range of professional cars. Other models in the Studebaker line include a DeLuxe Ambulance priced at $3,550 and a Bellevue Ambulance at $2,985. The Bellevue used the same chassis as the Arlington series funeral coaches.

1927

A.J. Miller built this DeLuxe funeral coach on the Packard chassis. Note the handsome contour of the lines and the balanced appearance of the complete ensemble. The interior was complete with an oval flower tray and hardwood highlights. The price of the complete unit was $3,350 on the Packard six-cylinder chassis and $4,400 on the Packard eight chassis with the same body.

This interesting shot shows the interior of an A.J. Miller ambulance of this era. The car was completely appointed with running water and a sink, medicine cabinets, linen storage cabinets, and a heater. Folding attendant seats line the right side of the compartment while the cot would be placed on the left hand side and held by the catches seen here. Note too, the folding cot on the roof above the right side windows.

A.J. Miller offered this rather plain looking service car on the Nash chassis this year. The car carries absolutely no decoration or ornamentation whatsoever and is painted black overall. The car has disc wheels and a small window in the rear compartment. This was very much a utility vehicle.

A.J. Miller also offered their bodies on Nash chassis, with the one shown here being on the 1927 Nash Advanced Six. The body was rather gracefully styled with the window ways being of stamped steel without joints or seams. The car as seen here sold for $2,500. When this body was mounted on the Nash Special Six chassis it sold for $2,250.

Meteor continued to offer one of the largest lines of assembled cars available. They made significant reductions in price for their cars in this year and offered six models. The Mort line had been dropped and now Meteor cars were available with Continental engines in six-cylinder form that delivered 50 and 70 horsepower and a Continental eight-cylinder plant that produced 72 horsepower. Shown here is the Meteor Model 846 landau back funeral car with the eight-cylinder engine. The price of this car was reduced a full $600 to $2,550. The same body style was offered in the other horsepower variations.

Built on the same chassis, this Meteor ambulance was designated a model 246-A metal back ambulance and was powered by a 70 horsepower Continental Red Seal six-cylinder engine. This engine used a Stromberg carburetor, a Delco electrical system and a Willard battery. In this form the car sold for $2,350.

1927

Henney introduced their NU-3-WAY coach in this year, incorporating the Eureka patented side-loading table and building it under license. The three-way coach was rapidly becoming one of the more popular types with almost every maker offering a variation on this theme. This Henney coach features the landau, leather back styling and a large oval flower tray over the casket in the rear compartment. Henney built cars on their own assembled chassis.

From the front it is easy to see that these Henney coaches were completely assembled. The radiator shell and the fenders were all made by suppliers, to Henney specifications, and both the bumpers and the radiator shell bear the Henney emblem. Note the moto-meter and the spotlight.

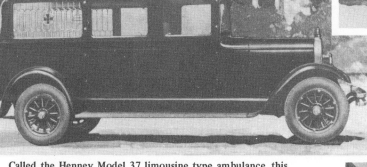

Henney coachwork on this Model 370 DeLuxe landau hearse was unsurpassed for beauty of design and appointment. This car was designed to meet the newer demands of the funeral director and was a rear loading coach only.

Called the Henney Model 37 limousine type ambulance, this model was also mounted on the Light Six chassis and met the demand for a small, light-weight ambulance in a lower price range. The leaded glass in the rear compartment windows was complemented with a red stained glass cross in the center while the body had the leather back, landau styling.

Henney continued to market their successful Light Six models through this year. This Light Six Model 27 limousine type hearse incorporated all of the standard Henney features into a smaller size car at a lower price.

Looking more like a bus than a funeral coach, this hearse was built by August Schubert Wagon Company of Oneida, N.Y. With a rather long body and straight limousine styling without any drapes, the car does look somewhat like a bus. The little bulge on the roof is the rear compartment vent. Chassis appears to be that of a Hudson.

The larger Henney DeLuxe models were also available in ambulance form. They followed the same styling line, with leaded and stained glass in the rear most rear compartment windows and the landau, leather back type roof styling.

1927

This Miller-Quincy limousine hearse was mounted on a Hudson chassis and was designated a Model 75. The plain styling with the pull down shades give this coach the appearance of being a combination ambulance-hearse.

Called a Model 85 limousine funeral coach, this Miller-Quincy was mounted on a Chrysler 90 chassis. Note that this car has a large landau area and rounded rear quarter windows. The removable flower tray is easily seen through the rear windows.

Miller-Quincy (E.M. Miller) built this stylish landau, leather back limousine coach called a Model 74 on a Cadillac chassis. Note the oval portholes on the rear quarters and the curved landau irons.

The Miller-Quincy Model 1823 ambulance, mounted here on a Dodge chassis, enabled the operator to serve the sick and injured in a manner that was remembered by grateful families. Notice the carved column styling and the carriage lamps. This company also offered a carved drapery hearse designated the Model 1957 that was just as ornate as some of the earlier carved coaches. But the carved car in these years did not enjoy the popularity it once had.

Riding on the Studebaker Big Six chassis, the Holcker straight limousine funeral coach bears styling that was quite plain and utilitarian. With four windows on each side and heavy velour draperies this car was the lowest priced offering from Holcker.

Building funeral vehicles since 1888, Holcker Manufacturing Co. of Kansas City, Mo., now offered a complete line of automotive professional cars mounted on any suitable chassis. Holcker styling was in line with the other makers and was clean and crisp. This limousine coach rides on a Hudson chassis and features side mounts and a leather back, landau style.

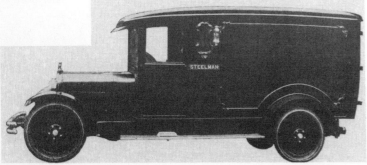

This large, plain service car was built by Giessel and mounted on a Dodge chassis. Note the elaborate window behind the driver's compartment and the carved outline on the rear body sides.

1927

S&S built this Kensington ambulance to the most exacting standards and with a style that was similar to the ever popular Washington. The rear quarter window area contains leaded glass and a cross insigna. This cross insigna is carried over to the windshield where it is repeated.

The Sayers & Scovill Washington funeral coach continued to enjoy popularity through this year with its unique styling features like the slim beveled plate glass rear quarter window and high waist line.

Many makers marketed their chassis alone, without bodies. This is the rugged S&S chassis. This chassis was especially designed for the work of a funeral car or an ambulance and was made to endure the stresses that these special vehicles have to take. The engine itself had many special features such as a heavy counterbalanced crankshaft, larger cylinder heads, and reinforced cylinder head gaskets. A heavy duty oiling system, valves of special heat-resisting alloy steel were additional features of the Continental engine used by S&S.

The graceful lines that characterized the S&S Washington are readily seen in this view of the rear quarter. Note the Gordon tire cover, the double opening rear doors and the rear quarter window.

This stylish little coach was built by the Pfeiffer Top & Body Co. of Omaha and was mounted on a Chevrolet chassis. The limousine, landau styling of this car adds class to the rather mundane chassis.

Building coaches on any desired chassis, the Knightstown Body Co. of Knightstown, Ind., offered this model as the Style #17 combination funeral coach and ambulance. With a gracefully curved rear quarter window and long curved landau bows, this car is shown mounted on a Willys-Knight chassis.

Another offering from Knightstown was this side loading hearse mounted on a Packard chassis. This coach was a side loading car only and there was no rear door. The spare tire was mounted in this place in sedan fashion

1927

Mounted on the Buick Master Six chassis, the Eureka Model 194 side loading burial coach reflects styling that was very popular in this era. The leather back roof effect is highly visible, and the landau irons complement the overall effect. Pull down shades were standard.

Another Eureka style was this coach built on the Buick Master Six chassis. Notice the S&S-type elongated rear quarter window with the beveled plate glass. All Eureka coach bodies were hand built, like most of the makers, with metal body pieces formed over a wooden body frame.

The first armoured ambulance ever put into private use was built by Cunningham this year for J.T. Hinton & Sons of Memphis and mounted on a Packard chassis. The entire body, fenders, front end, etc., were Cunningham and the car used boiler plate sides, bullet-proof glass, featured hot and cold running water, and forced air ventilation with hot water heating. The car cost $13,000 and was only the first of several Cunningham built ambulances of this type to be ordered by this firm. It is not really clear why they required an armoured ambulance, but it is obvious that Cunningham was willing to build it for them at a considerable cost.

Flexible of Loudonville, Ohio, continued to construct professional car bodies and mounted these on Buick chassis. Built on either the Buick Series 50 or 47 chassis, these cars were hand crafted limousine type coaches in either ambulance or funeral car style. In this year Flxible mounted 30 bodies on the Series 50 and 10 on the Series 47 chassis.

Like S&S, Cunningham continued to assemble cars on their own chassis and the name Cunningham was regarded as one of the finest professional vehicles on the road. This high, stylish Model 156A limousine hearse reflects the up-to-date styling Cuningham was then known for. Note the windshield styling and the two-tone paint finish.

Flxible built this Buick chassied ambulance in the finest limousine style for the North Wildwood Fire Department. The company built 30 of these cars mounted on the Buick Series 50 chassis and an additional 10 on the Buick 47 chassis, making a total of 40 cars for the year.

1927

The Kissel combination hearse-ambulance used innovative styling. The front fenders and runningboards, like the entire frontal ensemble of the car, were borrowed from the passenger versions of the Kissel. These cars were extremely low. This model has the leather back landau styling on a limousine body.

Looking like any other ordinary Kissel sedan, this combination pallbearers' coach, ambulance was built by Kissel. It featured side doors that opened from the center, thereby allowing maximum loading space. The three small pictures show the car in its various phases.

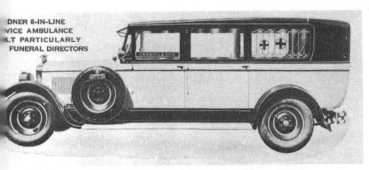

The St. Louis Coffin Company and the Gardner Motor Co. of St. Louis put their heads together and worked out this attractive Gardner ambulance and a complete line of other professional cars. This ambulance, like the other Gardner cars, was powered by the Gardner eight-cylinder engine and the chassis was designed especially for funeral or ambulance work. Note the leaded glass and cross in the rear compartment quarter window.

Claimed to be the only American built funeral cars with all the major components completely cushioned in rubber, the Kissel professional cars were derived from their line of passenger cars. Kissel offered a complete line of ambulances, funeral coaches, and combinations

The Montpelier Manufacturing Co. of Montpelier, Ohio, built a magnificent limousine body for this Pierce Arrow in this year and sold the complete coach for $1,995. It was reported to have been called the "most beautiful car at the show" with reference to the National Funeral Directors Convention display in this year.

The Juckem Company of Minneapolis built some extremely high quality coaches of which this limousine hearse on a Buick Master Six chassis was one. The car with rather straight, squared lines was priced at $2,250 and featured smooth mohair upholstery and Duco two-tone finish in Buckingham grey.

1927

The Owen Brothers of Lima, Ohio, built this handsome two-tone funeral coach, Model 782 on a Dodge chassis. The car features limousine styling with a dignified carriage lamp mounted on the extra wide "B" pillars.

Superior Body of Lima, Ohio, built a few cars with chassis other than Studebaker this year. This stylish Cadillac funeral car is an excellent example of the quality and style that Superior was to become famous for. Note the Cadillac crest on the Gordon tire cover, the disc wheels and the flower tray in the rear compartment.

The Pierce Brothers Mortuary of Los Angeles designed this unique town car hearse and had it built for them by Copple Auto Works of Los Angeles on a Pierce Arrow chassis. The car had a landau, leather back roof style, and large side doors. This car was equipped with a side-loading table.

Operated by Andrew Morales of San Antonio, this Harley Davidson emergency ambulance was unique to say the least. There were a few of these motorcycle ambulances around but by and large they were quite scarce and only in use in the warmer climates.

Built by either Dietrich or Superior, these Cadillac Custom Imperial Limousine Funeral coaches and ambulances were advertised by Cadillac exclusively and there was no mention of a body maker. The cars had the most advanced limousine styling and Cadillac claimed to have designed all of the interior appointments and hardware. In any event, these cars remain to be called simply Cadillac Custom Imperial coaches.

This stylish Lincoln wears a body custom built by Brownell & Burt of Taunton, Mass. The chassis had a wheelbase of 150 inches and the body is of straight limousine style with a leather back.

1928

Meteor was beginning to market funeral coaches and ambulances that were equipped with a Continental Red Seal 80-horsepower eight-cylinder in-line engine. The new eight-cylinder Meteor coaches were said to have had power to carry the car with a full load over any highway grade. Sayers & Scovill was, at this time, using the 75 horsepower Continental eight-cylinder engine in some models, and claimed to be the only maker using this motor at the time.

Three years earlier, S & S had introduced the now famous Washington Six, called the "masterpiece of casket coaches." They entered this new limousine style car into a highly competitive field, but were quite confident that funeral directors would welcome this new arrival from S & S. They were quite right in this assumption. The car was one of the largest selling models they had ever built.

Several more town car style coaches were introduced to the field in this year with Eureka leading the way with a 3-way town car model mounted on a Lincoln chassis. They built several more versions of this model on various chassis both as funeral coaches and as ambulances. In this year the Kissel Motor Car Co. of Hartford, Wis., which had been actively building coaches for several years, signed a contract with the National Casket Company of Boston, to build these cars under the name National Kissel. This was a marketing strike that was to meet with considerable success for both National and Kissel. The new National Kissel line was presented with a full range of ambulances, hearses and service cars. The cars were, like all Kissel cars, quite long and very low. They retained the famous bicycle type fenders that the Kissel passenger car had always had and added a new limousine body. The finished product was a very lithe funeral coach or ambulance that claimed a ground clearance of only 28 inches. This was supposed to be a smaller funeral coach, ambulance and service car line designed to meet the demand of funeral directors for a high grade coach at a moderate price.

Pierce Arrow made their model 36 available to the funeral profession complete with a set of folding jump seats and a glass partition. Many funeral directors looked to Pierce Arrow for their passenger vehicles and some even requested Pierce Arrow chassis for their funeral coaches. Most firms would mount a body on the Pierce chassis, producing extremely attractive machines.

Seen here is a row of A.J. Miller Nash funeral coaches ready for delivery. The first car is an ambulance while the following two are funeral coaches. Note the Model T in the background.

A.J. Miller was concentrating on a strong line of coaches on the manufactured chassis from Nash and Packard. This stylish Miller-Nash was designated a Model N-103 and was of straight limousine styling with a metal back. The flower tray is easily visible through the rear side windows. This car was finished in dark blue with cream pin striping on the body sides and wheels.

Called the Miller DeLuxe Model 105 Packard funeral coach, this attractive vehicle was finished overall in sand with black roof and side striping and a light cream pin striping.

1928

"Friendly local service at 3,000 points throughout the United States" was one of the main sales points of the Superior bodied Studebaker coaches. This Studebaker Arlington funeral coach with body by Superior of Lima, Ohio, was priced at $2,385 and could be serviced at any local Studebaker or Erskine dealer as the sign indicates. This was taunted as a major superiority over assembled cars as local service was readily available for cars with a manufactured chassis.

Brownell & Burt of Taunton, Mass., built this strikingly high ambulance on a Lincoln chassis for the City of Salem Police Department. The wheelbase of this Lincoln chassis was 150 inches and Lincoln supplied 24 such bare special chassis to body makers.

The larger Superior bodied Studebaker DeLuxe series, of which this example is the funeral coach, was mounted on a longer chassis and sold for $2,985. "Powered for the back road" was the slogan that bannered across the top of Studebaker advertisements for the DeLuxe series. These cars were powered by the Studebaker Big Six engine.

The Bellevue ambulance, built with a Superior body and a Studebaker 146-inch wheelbase chassis, was said to have been safe, fast, dependable and priced right at $2,560. The Bellevue rode on the same wheelbase as the Arlington funeral coach and was one of the line's most popular models. Note the ornately leaded rear quarter window glass and the leather back styling.

All DeLuxe models rode on a 158-inch wheelbase and this DeLuxe ambulance with a Superior built body was an aristocrat among ambulances and the top offering from Studebaker in the ambulance line. The white lacquer paint finish was optional at extra cost while lacquer was a standard finish. The white color was also not guaranteed to stand the test of time. This car sold for $3,185.

Utilizing the Arlington-Bellevue wheelbase of 146 inches, this Studebaker service car with body by Superior was the last word in a service car and was priced at only $2,285.

1928

When S&S introduced the Washington three years earlier, they knew that they were moving into a highly competitive field. Now in 1928, the S&S Washington was one of the trade's best selling coaches and gave a high value for the money. This car was built for a stately appearance and a long life and was completely assembled on S&S's own assembled chassis, powered by the Continental Red Seal engine.

Shown here is the S&S body frame as it appeared while setting in the alignment jig. The assembly was made of only the best hardwood, and was made additionally sturdy through the use of metal joints. This body is for a Washington casket coach.

This is the Continental Red Seal six-cylinder engine, used only by S&S this year. This motor produced 75 horsepower and was an S&S exclusive.

Robert Thompson of Los Angeles built several funeral coaches this year, of which this example was mounted on a Cadillac chassis. Thompson coaches were finished in DuPont Duco and were mounted on any suitable chassis.

Cadillac announced a new ambulance in their Imperial professional car line. This stylish coach had a two-tone paint finish with leaded glass in the rear quarter windows. Styling of this limousine was of the metal back variety.

The rear view of the Meteor Model 858-A service car shows the graceful curves that dominated this style and the ridge of the leather back top. This car continues to use double rear doors and was the first to become a standardized model offered to the trade. Note the leaded glass in the rear quarter windows.

Meteor also continued to assemble its own chassis for their cars. This Meteor Model 858-A service car was powered by the 80 horsepower Continental Red Seal straight eight engine and came complete with a patented extention sill roller in the rear door to facilitate loading. This car sold for $2,185 F.O.B. Piqua, Ohio.

1928

This Flxible ambulance was built on the Buick Series 50 chassis and in this year the firm built a total of 124 coaches on three chassis series — 50, 47 and 25. During the year they built only two cars on the series 25 chassis. Note the stylish stained and leaded glass in the rear quarter windows and the side mounts on this car.

Designated as a Buick Model 47, this Flxible limousine style hearse featured a leather back style and this model had optional fender mounts. In this year Flxible built 45 Model 47's and 77 Model 50's, all on the Buick chassis. Flxible was located in Loudonville, Ohio.

Cunningham continued to assemble their own complete cars, of which this limousine coach is a prime example. This car was delivered to J.W. McCormick of Columbia, S.C. It is one of Cunningham's special ambulances. Of interest are the side ventilators and the leaded glass rear quarter windows, two-tone paint finish, and small carriage lamps on the "B" pillars.

This fleet of Cunningham ambulances, hearses and service cars was operated by McGibbon & Currey of Liberty, N.Y. The ages of these cars span almost a full decade, with the oldest being the service car on the extreme right and the latest model being the limousine hearse on the right. The coach on the extreme right is a Cunningham 7-passenger pallbearers' limousine.

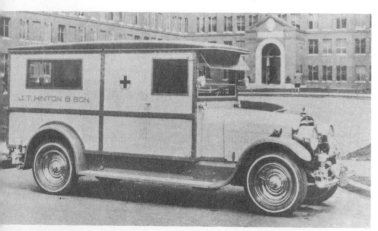

Designed in collaboration with Captain Frayser Hinton and built especially for J.T. Hinton & Son of Memphis, this is another armoured ambulance built by Cunningham. This car was built entirely on Cunningham running gear. Note the interesting Westinghouse shock absorbers and the thin white wall tires. The side plating was of thick boiler plate steel and the glass was bullet-proof.

The Montpelier Company of Montpelier, Ohio, continued to manufacture bodies of high quality in very low volume. This example was mounted on a Cadillac Model 61 chassis and is a very nice-looking vehicle. Montpelier, like so many of the smaller companies, was finding it increasingly difficult to compete with the industry giants and harder to make it in the sales race.

1928

This Lagerquist Model 141 is shown mounted on a Buick Master Six chassis and was delivered to the Jones Funeral Home of Boone, Iowa. Note the rounded rear roof lines and the rather high waist line of this car.

The Lagerquist Auto Company of Des Moines built this Model 141 funeral coach body and mounted it on a Lincoln chassis. Lagerquist was another of the smaller makers that did well in their own locality but could not survive nationally.

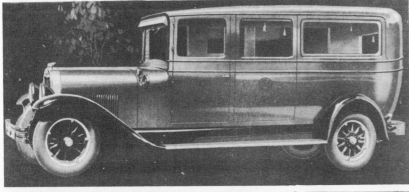

Henney ambulances could be depended upon, and there is no better proof of this than this car which was supplied by Henney to the Federal Government. The U.S. Veterans' Bureau officials demanded a showdown among ambulances from the leading makers, and as a result of this test conducted by the U.S. Bureau of Standards, a contract was awarded to Henney for twenty-three ambulances that were used by the U.S. Veterans' Bureau hospitals throughout the country. This car was one of these ambulances, and was used by U.S.V.D. Hospital Number 95 in Northampton, Mass.

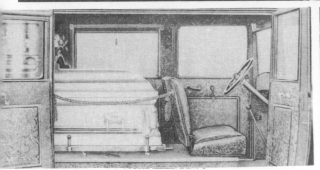

The interior of the Henney NU-3-WAY service funeral coach featured a roomy and comfortable driver's compartment. Note the easy access to the casket compartment. Both side doors swing wide open so that the casket and table can swing out within easy reach. Also of interest are the luxurious interior appointments and the oval flower tray that can be seen to the left of the interior.

Designated as the Henney Model 630, this stylish limousine funeral coach sold for $2,235 F.O.B. Freeport, Ill. It was built on a chassis totally assembled by Henney.

"Leadership" was the headline of a 1928 Henney advertisement for the Nu-3-WAY funeral coach and the ad was also done in pastel colors for added impact. Note how the casket extends out on the casket table in place for side service. Henney was using the Eureka table under an agreement whereby Eureka built the tables and Henney bought them. Most of the side servicing cars of this era used this table.

Henney completely assembled their own cars from the ground up. They used Continental engines and were offered in a choice of horsepowers of from 50 to 70 all in six-cylinder forms. This coach is a straight limousine hearse with a metal back style and pull-down shades.

1928

Combs Funeral Home of Miami, Fl., operated this Harley Davidson ambulance this year. It is the last seen and reported upon in the trade journals.

Public approval of the limousine body style made the type all the more popular. This coach built by the Knightstown Body Company of Knightstown, Ind., was mounted on a Packard chassis and has a two-tone paint finish and a fabric covered roof. Disc wheels were in the height of popularity. It is interesting to notice the bar type bumpers on this car.

Seen here is an attractive little Knightstown service car with a landau bow on the rear quarters, small windows on the rear body sides, and a two-tone paint finish. Note the step in the rear to facilitate rear entrance.

Knightstown also offered a side loading coach but did not use the Eureka table. The small circles on the rear compartment floor of this car are casket rollers to accomplish the movement of the casket through either side door or through the rear. Note the driver's compartment in this view.

Knightstown, Ind., supported two independent coach makers in these years. This car, built by Silver Knightstown, is shown mounted on a Buick chassis and is of limousine styling. Silver Knightstown claimed that their side loading funeral coaches had no tables to pull out or jam and that there was nothing to rattle.

When the Knightstown coach was used as a side loading vehicle, the partition folded forward as did either front seat to allow access to the casket. This was a much simpler system than that employed by Eureka and Henney. Some other makers were using similar methods to make side loading coaches while others used variations of this type of system.

This stunning Silver Knightstown combination funeral coach-ambulance was mounted on a Lincoln chassis and is a side loading car only. This is evident by the rear mounted spare. Silver Knightstown coaches seemed quite long in comparison to those made by some of the other makers on the same chassis.

1928

The Eureka side loading or "Side-Way" funeral coach was the trend setter in these years. Eureka of Rock Falls, Ill., was the first maker to introduce the side loading table and held the exclusive patent on this table. They supplied it to many other competitors for about 10 years. This Eureka coach is seen mounted on a Buick chassis.

Along with being an innovator, Eureka was a styling leader in this field. They built this attractive Packard town car style ambulance for Spencer-Sturla of Memphis. The car features such things as a fully padded fabric covered roof with landau bows and a large siren on the front.

Eureka built up a strong line of bodies all with their revolutionary side loading feature and mounted these on any suitable chassis. This Eureka side loading funeral coach is seen here on a Reo Flying Cloud chassis that was extended and strengthened by Eureka before the body was set. The extension was guaranteed by both Eureka and usually by the company supplying the chassis if it was new. Some of these Eureka side loading bodies were mounted on used chassis.

In February of this year Eureka announced a new concept in their revolutionary side loading funeral coach. The original Eureka side loaders worked three ways — through either side or through the back door — but this new model line had no back door and a spare was mounted in its place. This side loading Eureka is mounted on a Packard chassis and has small landau bows and a leather back roof style. It looks quite a bit like a large sedan of the era.

The Owen Brothers of Lima, Ohio, offered this model designated as a Model 946 funeral coach mounted on a Cadillac chassis. This model was completed as a regular rear loader but was also available as a side loader with either a regular floor with rollers or a swinging casket table. This car was also offered as a combination ambulance-funeral coach.

Owen called this the most beautiful funeral coach in America. Designated the Imperial, this limousine funeral car was available as either a rear or a side loader and was called a Model 945. Shown here on a Cadillac chassis, this new model was finished in Duco and featured intricate carvings around the top edges of the side windows, which were gracefully curved, and a small carriage lamp on the "B" pillar.

Miller-Quincy (E.M. Miller of Quincy, Ill.) said in this year that you could take any good chassis, new or used, and they would mount a new limousine body on it for a modest charge. This Miller-Quincy Model 85 landau, leather back limousine was mounted on a Velie hearse chassis and was quite an attractive vehicle. Velie built a special chassis for funeral coach and ambulance work as did several other makers of passenger cars.

1928

Kissel, who had been building funeral coaches, ambulances and service cars for several years, in this year joined up with the National Casket Company of Boston and marketed these cars as National-Kissel professional coaches. The entire line consisted of two hearses, one ambulance and a service car. This model was the new smaller Kissel funeral coach that was built to meet the demands of funeral directors for high grade cars at a moderate price.

The largest car in the National-Kissel line-up was this combination ambulance-funeral coach. "Bodies of artistic beauty" was the way in which they described this stylish hearse with rounded lines, bicycle fenders, and a long narrow piece of heavy beveled plate glass in the rear quarter windows.

The St. Louis Coffin Co. continued to offer a line of Gardner professional vehicles. This model was equipped with a Gardner straight eight 70 horsepower engine and was called a Model 77. A straight limousine coach this car sold for $2,495. Gardner supplied the chassis and the St. Louis Coffin Company built the bodies.

Another National-Kissel was this attractive service car with styling similar to the large National-Kissel funeral coaches and ambulances. National-Kissel vehicles were available through any Kissel agent or through the National Casket Co. of Boston or any of their traveling sales people.

Cynthiana called this car a Model 85 funeral coach, and this particular car is shown mounted on a Dodge Brothers commercial chassis. The company would also build a car on a chassis with a 60 horsepower Continental Red Seal engine, a special Studebaker chassis or the body could be furnished alone and mounted on the buyer's own chassis, new or used. This car features landau leather back styling with a small oval window in the landau area, and intricately etched rear compartment windows.

Wearing coachwork by W.H.F. Blume of St. Louis, with the body being built by St. Louis Coffin Company, this Gardner straight eight deluxe service car was very stylish indeed. Note the Gordon tire cover and the small carriage lamp on the rear body side. Service cars, sometimes called casket wagons or first call cars, were used by the funeral director as an all-round work horse to carry flowers, chairs, baskets and often to make the initial first call at the residence of the deceased.

1929

Sayers & Scovill rocked the funeral coach and ambulance field and the trade in general this year with the introduction of a new and exclusive model. Called the Signed Sculpture hearse, this new S & S town car hearse featured a large bronze panel on the rear sides of the car that depicted the "Angel of Memory." This coach, with new dignity and symbolism, put solemn meaning into the hearse that lead the procession. S & S had asked a world renowned master of sculpture to create a most sympathetic and understanding interpretation of death for them and to capture it in sculpture. The sculptor was Clement Barnhorn and the cast bronze panels that decorated the sides of this new coach bore his signature. Thus the name — Signed Sculpture. The production of this coach was made purposely limited and was offered only in selected cities and to only the recognized leaders in the profession. The sculptured design was itself copyrighted and patented and the coach that bore this enormous piece of artwork sold for $8,500, F.O.B. Cincinnati.

Several makers added 3-way or side-servicing cars to their ranges and among them was Silver-Knightstown and A.J. Miller. Henney scored another first in this year with the introduction of an airplane hearse. The idea was not new, but Henney saw aircraft as the wave of the future. Unfortunately for them, the craft was not too successful.

Called the "Signed Sculpture," this unique S&S funeral coach sparked a return to the carved type funeral car. This coach had a large cast bronze panel on the rear compartment sides that enclosed a carving in relief of the "Angel of Memory." Totally an assembled car, like all S&S coaches, this car was powered by a Continental Red Seal engine and was of the dignified town car style that had become increasingly popular in this field.

In this view of the Signed Sculpture funeral coach the front windows have been rolled up and the sculpture stands out somewhat better. This car was built in very limited numbers and there were two very good reasons for this — the price was $8,500 and there was a depression of some magnitude in this year.

This close-up of the panel on the Signed Sculpture funeral coach shows the intricate details that were part of this panel. The angel symbolizes the angel of memory and the scarves on her wings (at the top of the figure) signified that the activity of life had ceased and the wings were stilled. In the angel's hands are two very symbolic items. The right hand holds a cycus leaf, which has been a symbol of memorialization since the days of the Egyptians where it was found on the walls of Egyptian tombs as decoration, and was found again later in the same usage in the Etruscan era. Cycus leaves are still in use today in formal memorial wreaths. The other hand holds a wreath of memory.

The magnificent "Signed Sculpture" funeral coach got its name from the fact that Clement Barnhorn, a famous sculptor, designed and sculpted the bronze panel on this car. His signature was reproduced on each of these panels and thus the name. Here Mr. Barnhorn is at work in his studio at the Cincinnati Museum of Art, working on the sculpture.

1929

Producing one of the most complete lines of professional vehicles offered, S&S also built this clean looking Evanston ambulance. This was the most luxurious ambulance in the S&S line and was built on the same wheelbase and on the same chassis as the Signed Sculpture hearse. The Evanston, like the Fairview, shared the styling of the Washington with the long narrow rear quarter window.

This is an example of S&S coachwork in the ambulance field in this year. Designated the S&S Fairview this car had all of the unique styling features of the Washington seen earlier in this chapter with the addition of an exceptionally complete emergency equipment in the rear compartment. Notice that the long slim rear window has leaded glass placed in it instead of the heavy beveled plate glass.

The S&S DeLuxe Washington had an out-and-out quality look that made this coach one of the most popular vehicles in the field. All the power a person could ask for was supplied in this car by a Continental Red Seal eight-cylinder engine that produced 85 horsepower. The engine was fitted with a dual carburetor and intake manifold. The S&S chassis was fitted with Lovejoy shock absorbers, oversized axles, and four wheel mechanical brakes, all as standard equipment. The body had long beveled side glasses, a massive countoured radiator, airplane type fenders and new styled head lamps.

Uniquely styled, this Cunningham combination service and flower car was an extremely attractive vehicle. Cunningham assembled their own chassis, utilizing the Continental Red Seal engine. Note the low sculptured rear deck, disc wheels, carriage lamps, and gracefully curved front fenders.

This is the drawing room type interior of the S&S Washington funeral coach. All interiors were finished to customer order. Notice the spacious walnut medallion panels and the large walnut door belt panels. Rich damask drapes are fitted to all of the rear compartment windows and the upholstery was luxurious to say the least.

A.J. Miller designed and built this pleasingly styled body and offered it on either a Nash chassis, as seen here, or on a Packard. The graceful curve and highlights of this body quickly made Miller cars one of the most popular names in the medium price class.

Mounted on a Chevrolet chassis, this Lagerquist funeral coach was built by the Lagerquist Auto Co. of Des Moines. It was called one of the greatest values ever built. Note the graceful curves of the body and the arch type flower tray which can be seen clearly through the rear side windows.

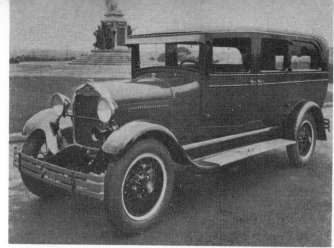

Another Lagerquist funeral coach is this unique little burial coach mounted on the Ford AA truck chassis. Lagerquist, a small mid-western concern, specialized in the extension of low or popular priced chassis for funeral coach or ambulance use.

ral Car
Model 866-A

Wearing the newer, more modern styling the 800 Series was the best looking line Meteor had ever built. Seen here is the funeral coach Model 866-A with a metal back. It was also offered with the leather back landau styling. Powered by an eight-cylinder Continental 80 horsepower engine with a Stromberg carburetor this car was very similar to the other Meteor line with the older type styling. These cars had rubber flow stabilators by Watson and four wheel Lockheed brakes. It sold for $2,250.

With a straight eight Continental Red Seal engine that developed 80 horsepower, this Meteor funeral coach retained the older styling while some of the other models in the line were restyled. This car could be had as either a landau, leather back, or as a metal back limousine. The vehicle was equipped with four-wheel Lockheed brakes and Watson stabilators and sold for $2,250.

lan-Ambulance
Model 867

One of the most attractive offerings in the Meteor 800 Series was this luxurious 867 sedan-ambulance. In its sedan form, it could carry nine passengers, but could be transformed into an ambulance within five minutes. Trimmed in mohair and fitted with a cot, heater, fan and driving light, this was an extremely attractive car. It, too, was powered by the 80 horsepower Continental engine with a Stromberg carburetor and sold for $2,250.

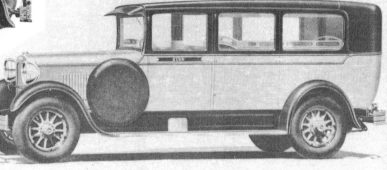

At the less expensive end of the scale, Meteor offered a new series called the 666, which was to represent the extreme limit of value. Utilizing a refined Continental six cylinder engine that produced 60 horsepower, the Model 666 funeral car had a deep narrow radiator, full crown fenders and many other newly designed body features. This car sold for only $1,885, F.O.B. Piqua.

rvice Car
Model 868

Another offering in the Meteor 800 Series was this Model 868 service car built on the standard eight-cylinder chassis with a Stromberg Duplex carburetor. The car was equipped with an 80 horse Continental engine, new style Watson stabilators, extra tire and wheel, and a full set of bumpers. It was fitted with a removable flower tray and rack and extension rollers, and was priced at $2,250. Note the leaded rear window.

1929

Built by Eureka for Balbirnie Funeral Home of Muskegon, Mich., this attractive Lincoln town car funeral coach features a fully padded top with large landau area and bows. This car was also used as an ambulance and the siren mounted on the front bumper gives this fact away.

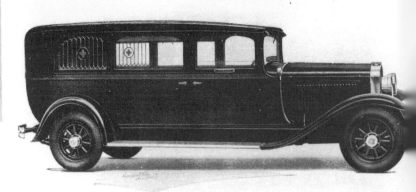

Eureka Model #29 was this body seen mounted on a Dodge chassis. This ambulance has rounded side window tops and leaded glass in the rear quarter windows. The Eureka body was quite well styled and at this time Eureka was reveling in the glory of having been the first to offer a unique 3-way coach.

Mounted on the Buick chassis, this hearse body was called the Eureka 3-way loading Burial Coach Supreme. After three years of use in all parts of the country the three-way table had proven itself durable and quite useful, and a desirable piece of equipment.

Mounted on the Windsor White Prince 8 funeral car chassis, this Eureka side loading funeral coach features yet another unique Eureka patented device. The Eureka patented casket clamp instantly anchored any casket firmly on the table and held it rigid with no loss of time by putting in bier pins.

Superior of Lima, Ohio, continued to build a strong line of bodies for Studebaker and offered this Arlington funeral coach for $2,375. This car was powered by a smooth, silent Studebaker 75 horsepower six-cylinder engine. In ambulance form this car sold for $2,560 and was called the Bellevue. Both funeral coach and ambulance rode on a 146-inch wheelbase.

One of the premium offerings in the Superior Studebaker line this year was the deluxe funeral coach. Selling for $2,985 this coach rode on a 158-inch wheelbase and was powered by the Studebaker Big Six engine. It had extra long springs and full balloon tires with powerful four wheel brakes. This car was also offered in ambulance form for $3,185.

A very early snowmobile. Built from a Model A Ford, this snowmobile hearse was used by M.J. Managan of Dunsieth, N.D. The unique body and running gear of this converted Ford was built by Gleason Garage of Bottineau, N.D. This has to be one of the most unusual vehicles ever built for use in this field, but one can see where it could come in handy in the area in which it was employed.

Building coaches on the Buick chassis only, the Flxible Company of Loudonville, Ohio, offered their cars on two Buick chassis series in this year. They built 91 cars on the Buick Series 50 chassis and another 173 on the Series 47 chassis. This Flxible ambulance on the Series 47 chassis was built for the DuPont Company for use in their Willmington, Del., plant. This car featured regular glass panes in the rear quarter windows and pull-down shades in all of the rear compartment windows.

Flxible also built a few ambulances for the American Legion with this unit going to the local in Flemington, N.J. The car was finished in two-tone gray and had ornate leaded glass rear quarter windows. Note the siren on the front bumper.

Henney continued to market their NU-3-Way funeral coach on a newly redesigned body that was still being mounted on their own assembled chassis. With a rather high waist line and a very long overall appearance, the Henney was considered one of the finest funeral cars in the country and Henney was the largest funeral coach and ambulance maker in the U.S.

HENNEY *Again is* FIRST

A new venture for Henney was this interesting airplane ambulance, offered in this year. With a 1,000 mile cruising range and a 100 miles-per hour cruising speed, this craft was built by Great Lakes Aircraft Corp., an associate company of Henney, on a Martin 74 airframe. The plane carried a medical crew of two plus the pilot, co-pilot and a stretcher patient. It was equipped with hot and cold running water and a complete array of medical supplies. Although Henney claimed to have been "first again" with a new and unusual offering in this field, the venture was not successful.

This Henney body frame would hold over 1500 pounds of weight directly over the wide door opening without sagging. This was aptly demonstrated when Henney placed eight men on this frame and the results were perfect. Heavy wrought-iron bracing was placed within the frame to reinforce an already rigid design. The arrow in this picture points out the bracing. This was the body frame of a Henney 3-way coach.

Henney contended that the truly well conducted funeral should have an atmosphere of reverence and a feeling of respect and that this was not possible to portray with an antiquated end loading hearse. Here we see the Henney Nu-3-Way funeral coach with the casket on the table swung out for easy removal. Notice the extremely wide opening rear side door.

This Chevrolet was the first child's hearse to be used in the South and has the typical white paint finish. The car was a rear loading vehicle. It is interesting to notice the special rear window treatment on this vehicle.

Operated by the East Undertaking Company of Texarkana, Texas, this stylishly elegant Lincoln ambulance was built by the Silver Knightstown Company of Knightstown, Ind. It was designated a Model 68 DeLuxe ambulance. Note the unusual leaded, stained glass work in the rear quarter windows and the dual spotlights on the windshield pillars.

Silver Knightstown also built this nice-looking ambulance on a Packard chassis. Once again the unusual rear quarter window treatment appears along with the spotlights and a center mounted red warning light and a siren on the bumper. This model was called a Style 88. It wears chrome plated disc wheels.

Silver Knightstown's own side loading funeral coach was protected by U.S. Patent 1680811 which covered the self-guiding of the casket into the side loading coach. This example of the Silver Knightstown three-way coach is shown mounted on a Buick chassis.

Operated by the Johns Undertaking Company of Birmingham, Ala., this Miller Quincy ambulance-hearse combination coach is mounted on a Lincoln chassis. Note the unique windshield treatment and the landau bow and small oval window in the rear quarter. This was the Style 79 combination coach, and features a cross decal in the rear quarter window and a siren on the front bumper with a megaphone-type gong along side the siren.

Another Miller Quincy funeral coach. This was the Model 77 funeral car with a limousine body, leather back styling with landau bows, and small oval windows in the rear quarters. Miller Quincy was a trade name for the E.M. Miller Co. of Quincy, Ill. It claimed "modern bodies by an old time coach builder." This example of their coach work is mounted on a Hudson chassis.

Mounted on a Velie special hearse chassis with a 60 horsepower-six cylinder engine, this Miller Quincy (E.M. Miller) Style 96 body features a straight limousine style with a metal back. Miller Quincy would mount bodies on any dependable chassis.

1929

The standard National-Kissel limousine funeral coach was this Model 8-74-B. This car could also be supplied as a six-cylinder car, which was referred to as a Model 6-54-B and sold for less money. Both cars could be had as ambulances or as side loading hearses. A larger model was the deluxe funeral coach called a Model 8-90-B and it, too, was offered as an ambulance.

National-Kissel introduced its first town car funeral coach this year with this stylish vehicle. This car was powered by a 126 horsepower straight eight White Eagle engine and was mounted on a wheelbase of 162 inches. This style coach had been borrowed from premium models of passenger cars and was enjoying considerable popularity among the more affluent funeral directors.

This National-Kissel Model 8-90-B DeLuxe ambulance was also powered by the 126 horsepower White Eagle eight-cylinder engine. It featured a straight limousine style with curved top side windows and a leaded glass in the rear quarter windows. Note the wire wheels, fender mounted mirrors, and the three-tone paint finish.

Knightstown Funeral Car Co., not to be mistaken for Silver Knightstown, offered coaches with Galahad bodies and small refinements that made the products stand out among the smaller makers. This attractive limousine coach is mounted on a Packard chassis and features the straight metal back style with the front part of the roof having a fabric covering.

This was the new Gardner straight eight funeral coach model 187. The body was hand made and built in the St. Louis Coffin Company's own plant under rigid supervision. The body was Udylite processed which was a new feature of the Gardner construction, with all exposed parts of the chassis being chromium plated. Headlamps were cone shaped and stanchion mounted. The Gardner funeral car was powered by a Gardner Lycoming Yellow Jacket L-head vertical eight in-line engine. The Gardner Lycoming engine was especially adapted to funeral car or ambulance work and gave smoothness and remarkable performance at low speed with quick acceleration and noiseless operation at high speed.

Arthur Donnelly of St. Louis, operated these two Gardner coaches. The car in the foreground is a Gardner service car of yet another style.

R.G. Sipe Undertaking Co. of Trinidad, Colo., owned and operated this Gardner service car. Mr. Sipe is seen standing in front of his car after delivery. The Gardner was built by the St. Louis Coffin Co. and the chassis was supplied by the Gardner Motor Company of the same city. These cars had a straight eight engine and a four-speed transmission.

1930

It seemed that there was no other manufacturer who could utterly turn its back on styling as did Cunningham. This coach, designed and built by Cunningham for Dickinson-Streeter Co. of Springfield, Mass., features a return to the carved sided coach and combined this with the new town car look. This example features massive carved draperies and eight carved columns, wire wheels and massive headlights. In its own way, this is a truly beautiful car and a Cunningham through and through.

Another unusual Cunningham was this glass sided hearse with eight carved columns and heavy mohair draperies. Once again the car features the town car style, but the body is reminiscent of carriages that appeared in the early twenties. The main reason for the return to the carved style coaches was to satisfy a clamoring for funeral cars that looked like funeral vehicles, not large passenger cars. There was a demand for coaches with the traditional symbolism attached to the old horse-drawn coaches. With this demand, the carved car was entering a new era and one on the most interesting in its long heritage.

The stock market carsh on Black Friday of October, 1929, found the manufacturers ready to announce their 1930 models and with a substantial volume of orders on their books. The immediate cancellations were suprisingly light and 1929 ended as a very satisfactory year for this industry. 1930, however, saw critical reductions in sales that were to haunt the field for the next few years. Manpower in all of the plants was reduced and foremen went back to bench and assembly line work. Office personnel often had to go on the road to sell, while wages and salaries were slashed. The makers eliminated all unessential expenses. Although some of the plants continued through the depression and even survived, some did not or were damaged to a point that they could not survive for long after the depression ended.

Entering this financially tight year, Sayers & Scovill announced yet more new models with the Riverside town car funeral coach and the Worthington town car ambulance. Town car funeral coaches and ambulances were becoming quite popular and now almost every maker offered a town car model.

Realizing that the depression would severely restrict the buying habits of the funeral directors and the ambulance operators, the St. Louis Coffin Co. of St. Louis, offered the Gardner 6 funeral coach and combination car for $1,750, F.O.B. St. Louis. The St. Louis Coffin Co. had been building bodies for hearses and mounting them on Gardner chassis for quite a few years. One of its larger offerings for this year was the Gardner model 187 that featured a straight eight-cylinder engine and four speeds forward. These coaches were claimed to deliver smoother performance, and have a finer appearance than any funeral coach of the day. The models 177 and 187 were new for 1930 and sold for $2,950.

Henney, continuing to completely assemble their cars, offered this model funeral coach as their Nu-3-Way. This car had a full length, movable casket table of Eureka patent. This side loading system eliminated the necessity of sending pallbearers and the casket into the street. This attractive coach has curved top side windows and a stylish visor over the windshield. Note the appearance of height between the runningboards and the bottom of the side windows.

Shown in the showroom of one of their agencies, this Henney ambulance was equipped with a radio, patented reclining patient cot, bassinet, hot and cold water, a lavatory, inhalator, first aid and wreck equipment, an electric fan and a heater. Note the side opening doors hinged on the center pillar and the siren on the headlamp stanchion.

1930

Both Meteor and A.J. Miller had as much business as they needed throughout the 1920's and at Miller, an eye was cast toward the production of taxi cabs. But that is about all they did. Instead they set about perfecting their various ambulances and hearses. A special thrift model was introduced by Miller this year and was called the Benjamin Franklin line. These cars were offered on Ford and Chevrolet chassis initially. Cadillac became a standard chassis for the coach makers in this year, as it was producing a chassis especially engineered for funeral coach and ambulance work. Other chassis employed at this time included Nash, Packard, Chrysler, Buick, Dodge, Reo, Studebaker and Pierce Arrow, among others.

The A.J. Miller Company of Bellefontaine, Ohio, offered an expanded line this year. This model, called the Jefferson 30 Ambulance, was mounted on a Packard chassis. Miller claimed that depreciation for Packard ambulances and funeral coaches was less than on any other make and that because Packard styles remained constant year after year the coach would look new longer.

Offered on the Nash Six chassis was this A.J. Miller Mount Vernon 30. This car was at the lower end of the Miller price scale and sold for $1,950. The same body could be mounted on a Nash Twin Ignition Six chassis as either an ambulance or a hearse. In this form the car sold for something less than $2,700. The body was of straight limousine styling with a metal back and a flat flower tray above the casket compartment. Note the Gordon tire cover on the side mount.

Called the Westminster 30, this Miller-Cadillac ambulance with its leaded rear quarter window glass is quite a nice-looking vehicle. The Westminster 30 featured a cast aluminum upper body construction.

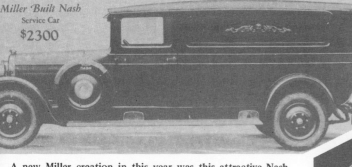

A new Miller creation in this year was this attractive Nash service car. With the small amount of carving on the rear body sides, this coach was a nice-looking and utilitarian vehicle for the mortician. It sold for $2,300.

Available in two varieties, the Miller Packard funeral coach was designated either the Jefferson or the Chief. Called the "boss of the road" the Miller-Packard Chief sold for $4,600 and had modern limousine styling. A lower priced version of the Packard funeral coach, called the Madison, sold for $3,475.

Cadillac had, for many years, made a chassis of unquestioned integrity and of sturdy construction with splendid performance. In this year Miller made Cadillac one of the standard offerings. This Miller-Cadillac, Model 341, sold for $4,200. It is a superb example of the mating of the Cadillac chassis and the Miller limousine body.

Continuing to build their own completely assembled cars, Sayers & Scovill introduced a new model to the trade in this year with the Riverside town car funeral coach. Seen from the front, the Riverside features a heavily chromed radiator shell of S&S design and large chrome plated headlight pods with the lamps being stanchion mounted. Note the S&S insignia between the lamps.

"True leadership," was the head of the introductory advertisement for this magnificent S&S Riverside funeral coach. The S&S Riverside town car marked the development of the finest in modern rolling stock. It had a classic style that was achieved through harmonizing the radiator, fenders, body and trimmings. A new Continental straight eight engine that developed 115 horsepower was utilized, while the body was the last word in luxurious coach craftsmanship. The interior was finished like a drawing room, with finely finished walnut panels, cove mouldings, and rich damask draperies and over draperies. Four candelabra sconces decorated the sides of the rear compartment and were specially designed for this car. The casket table was available in a choice of either solid walnut or covered with carpeting. The walls and ceilings were covered with mohair to match the outside color scheme. The S&S Riverside heralded a new era in rolling stock for the funeral director.

The town car had come to the ambulance with the new S&S Worthington. This aristocratic coach featured styling borrowed from the Riverside and the Signed Sculpture and combined it in an ambulance with all of the usual S&S ambulance features. The inset picture shows the roof of the driver's compartment closed, and highlights the metropolitan look of the coach. This special cover was quickly and easily adjusted and sheltered the driver in the case of inclement weather. Notice the ornate and intricate leading of the rear quarter windows, the rounding of the side window tops, and the carriage lamp on the "B" pillar.

The S&S Fairview ambulance was a very attractive vehicle and came complete with all of the most modern life-saving essentials of the day. Styling was in line with the continuing S&S Washington funeral coach with the addition of leaded glass in the rear quarter windows.

Continuing in production despite the depression and lagging sales, the beautiful S&S Signed Sculpture funeral coach was a car with all of the dignity and symbolism that any funeral car ever had. This car instantly proclaimed its identity and was one of the most interesting cars of the era. This coach sparked a return to the carved type coach and a more formal town car style.

At the lower end of the price scale, S&S announced the new Melrose hearse. They proved that an inexpensive funeral coach did not have to look cheap or makeshift. The Melrose thrift coach was an S&S car for those who thought they couldn't afford them. It was also ideally suited for use as a combination car.

1930

Superior Body of Lima, Ohio, continued the production of a complete line of coaches for Studebaker. This Superior-Studebaker Buckingham funeral coach was mounted on a 156-inch wheelbase and was powered by a 115 horsepower straight eight-cylinder Studebaker engine. Studebaker advertised that this car was for the funeral director who insisted upon the best. It was priced at $3,975 at the factory.

Purchased and used by Ed. Bond & Condon Co. of Atlanta, this Superior-Studebaker ambulance featured wire wheels, a front mounted siren and spotlights. Note the AAA plaque mounted on the headlamp stanchion.

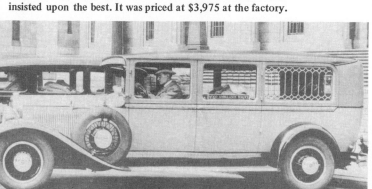

Called the Providence, this Superior-Studebaker ambulance came with a complete array of equipment. It was mounted on a 156-inch wheelbase chassis and powered by a 115 horsepower eight-cylinder engine. This car sold complete for $4,375 at the factory.

Wearing a body most likely built by Procter Keefe of Detroit, this Model A Ford funeral coach is a very nice-looking vehicle. Note the unusual wheels and the flower trays fitted to the interior body sides.

Built for the United States Navy was this White Model 602 ambulance. This car was built to the U.S. government specifications and was equipped with fire extinguishers mounted on the runningboards, a roof-mounted carry-all rack and hooks fitted to the rear body sides.

Another Model A, possibly built by Procter Keefe, is this service car. Note the side mount with cover, the plain rear body sides with a small carriage lamp and the wheels.

1930

Seen here mounted on a Packard chassis is another Eureka three-way funeral coach. The now famous Eureka three-way table was patented under the basic Heise patent number 1721391 and other patents were pending at this time. This picture gives one a good idea of the interior of the car and a look at the operation of the table.

This interesting photo was taken in Flxible's plant in Loudonville, Ohio when this car was new. Notice the incomplete hearse and ambulance bodies in the background. This Flxible Buick funeral coach was built in two versions — the Buick chassis Series 50 of which 91 were built this year and the Series 47 that had 173 Flxible professional car bodies constructed on it. This Series 47 funeral car has smooth rounded lines and a two-tone paint finish.

Photographed in front of the Park Avenue Baptist Church in Mansfield, Ohio, is this attractive Flxible Buick Series 60 funeral coach. In this year, Flxible built a total of 213 coaches, all on the Buick series 60 chassis. Of interest on this car are the chrome trim rings on the side mounts, the wire wheels, and the heavy velour draperies.

The American Legion often bought ambulances for their localities and they purchased this Flxible Buick Series 47 for the town of East Greenwhich, R.I. Flxible's approach to the rear quarter windows is interesting. These are much more intricately patterned leaded glass than some of the others. It also incorporates a beveled glass frame area. Flxible built a total of 264 professional cars this year.

The United States Public Health Service took delivery of a fleet of Flxible Buick ambulances in this year. The stand-up "Marine Hospital" sign is unusual.

1930

Gardner claimed that their large eight cylinder funeral coach and ambulance line were ten year cars. The eight-cylinder Gardner was powered by Lycoming Yellow Jacket L-Head vertical eight-in-line engine. This engine drove the car through a four speed gearbox and delivered smoother performance, a longer life and greater economy than any other coach offered – this is what Gardner claimed. This car is the 1930 Model 187 hearse.

Struggling to stay alive in bad times, the Gardner funeral coach offered some lower priced models. Gardner funeral coaches, built by the St. Louis Coffin Co. offered this six-cylinder car at a reduced price of only $1,750. Last year this car carried a price of $2,950, thus the car was reduced in price by a full $1,200. Gardner advertised that shrewd buyers would take advantage of a demoralized market and that everyone knew there was a big over production of funeral motor cars. Gardner went on to state that they, too, were caught in the crash with too much six-cylinder material, and they had to cash in on this and sell off these cars. This particular car was delivered to A.N. Wallace of Price, Utah, who drove it home from St. Louis in October of 1929.

A luxurious product of the General Motors Truck and Coach Co. was this funeral limousine and pallbearers' car. Basically a Chevrolet, this car was said to have been comfortable, easy riding, and comparable to the finest private limousine in every respect. The interior appointments of this car included every desirable refinement and the non-shatterable glass in the windows offered maximum protection for driver and passenger alike.

"Stamina" was the headline for a Knightstown advertisement. Knightstown dubbed the bodies of their products as "Galahad" and they received a fair amount of owner mail that proved these cars were very soundly constructed. This is a Galahad Knightstown funeral coach.

Seen from the front, the Meteor Model 666 six-cylinder funeral coach was vaguely similar to a Model A Ford, with the deep narrow radiator and full crown fenders. Bumpers were standard on this model.

One of the most attractive cars that the Owen Brothers of Lima, Ohio, ever built was this Model 958 ambulance on a Cord chassis. Note the overall lowness of the body and the two-tone highlights, the leaded glass in the rear windows and the small carriage lamp. Owen offered cars in this year on any suitable chassis but specialized in Cadillac, Packard, and Studebaker. These cars ranged in price from $3,450 to $3,750 when mounted on a Cadillac chassis. The Packard models included town car ambulances.

1931

The Detroit Department of Public Safety of the police department operated this Mack ambulance in this year. The body was most likely built by Proctor Keefe a prominent Detroit body maker. Note the siren and the pull-down shades.

This attractive A.J. Miller service car was built on an extended Chevrolet chassis and has decorative panels set into the rear body sides. The car is wearing thin line whitewall tires and wire wheels.

Meeting the demands of the funeral director with a dependable, economical service car, Gardner offered this six-cylinder vehicle for $1,500. This Model 301 service car combined three essentials in being sturdy, economical, and dignified. The cycus leaf and nameplate on the rear body sides were the only decoration on the car other than the two small carriage lamps. There is some speculation as to whether Gardner was utilizing a Chevrolet chassis in this year or perhaps Chevrolet body pieces.

The death of the Kissel left the National Casket Co. to find another chassis to mount a body on and call its own. At this point, National Casket made an arrangement with the Reo Motor Co. to supply chassis and with Henney to supply the bodies. Advertised as National Motor Equipment with bodies by Henney and Chassis by Reo, the cars were guaranteed by National Casket. National was stressing that prospective customers should compare the vehicles offered for quality, performance, appearance and genuine dollar-for-dollar value. The National range in this year consisted of three series, A, B, and C, in six models.

S & S was still marketing its Signed Sculpture model under a new name, the Masterpiece, and at a lower price. The company also introduced a new limousine hearse to the ever expanding line. The Clifton was a new, more modern looking limousine hearse with the special S & S chassis that was designed especially for funeral work. It was powered by a big, powerful, straight eight engine coupled with a new syncro-mesh transmission. In this year prices spanned from $2,925 to $6,950.

A.J. Miller continued to offer its bodies on Nash chassis. This coach is a Miller-Nash Mount Vernon funeral coach, mounted on a Nash Eight chassis. It listed as standard equipment a spare tire and wheel, Gordon tire cover, bronze nameplates, windshield sign, silk or waterproof curtains and spring rollers, rear vision mirror, automatic windshield wiper, dome light and velvet cords across the rear side doors. Fully equipped this car sold for $2,100.

The St. Louis Coffin Company continued to offer their line of Gardner professional cars. They now were concentrating on a line of smaller, lower priced vehicles that would deliver performance, dependability, and economy in an age of depression. This new Gardner light hearse, Model 201, had the Gardner four speed transmission, large extra heavy tires, special length rear springs, bumpers front and rear, a six-cylinder engine, one pair of name plates, a plated radiator shell, aluminum step plates and a flower tray — all for only $1,750.

1931

S & S offered a comprehensive range of ambulances utilizing both the larger chassis of the Riverside and Majestic and the smaller chassis of the Clifton. This ambulance utilized the Clifton chassis and was designated the Fairview. The unusually low center of gravity gave the S & S ambulances remarkable handling characteristics. The larger ambulances, called Kenwood, were powered by a Continental straight eight engine that produced 118 horsepower. Note the long, narrow rear quarter window that has been carried over from the Washington series cars. Leaded glass in rear quarter windows of ambulances was still popular in this year.

The S & S Riverside hearse of town car style was another S & S coach of rare dignified beauty. Note the flowing beauty of its styling and its sweeping curves. The Riverside was refined for its second year of production with an artistic new radiator, massive new headlamps, chrome plated disc type wheels, metal tire covers with chrome highlight band, two chrome plated horns, hood louvers instead of shutters, and a new symbolic radiator cap called the "Spirit of Grace." In this year S & S cars ranged in price from $2,925 to $6,950.

With the "Spirit of Grace" radiator ornament leading, this was yet another new S & S coach for the year. Called the Knickerbocker, this limousine style coach was the make's leading seller for many years. This coach featured all of the styling items described for the S & S Riverside town car. S & S continued to assemble their own complete cars, utilizing a Continental Red Seal straight eight cylinder engine.

To make S & S quality within the reach of almost every funeral director looking for a new coach, S & S introduced its new, lower-priced Clifton hearse. The Clifton shared many of the styling features with the larger, more expensive S & S coaches but was offered at a lower price. This car was also powered by a big Continental straight eight cylinder engine and was coupled with a new synchromesh transmission.

Capitalizing on the unique beauty of the Signed Sculpture, S&S announced a new model with a carving of the Angel of Memory in place of the rear quarter windows. Called the "Majestic," this new S&S coach commanded attention wherever it was seen. One example of a Majestic was parked in downtown Cincinnati, Ohio, for a few minutes by a company official and bystanders and pedestrians immediately crowded around the car admiring its beauty. The Majestic had the Angel of Memory in cast bronze relief in a smaller version of that seen on the famed Signed Sculpture.

Continuing through this year was the famed Signed Sculpture funeral town car hearse. It was now renamed the Masterpiece, but the car bore all of the original Signed Sculpture styling and features. This was the last year in which this car was to be marketed.

1931

Cunningham introduced a new line of funeral coaches in this year under the name Cathedral. This Cathedral town car model features wire wheels, and a highly carved rear body. Drapery carving was again in style and some of the makers were highlighting its comeback in a big way.

Shown here is the unique interior of the Cunningham Cathedral funeral coach. Finished in all hand-carved hardwood, the Cathedral interiors were styled to reflect a church-like atmosphere with panels shaped like stained glass church windows. Note the casket rollers set into the floor of this distinctive interior. This is a striking example of Cunningham craftsmanship.

Cunningham was another firm that held fast to the principal of assembling their own complete cars. This town car is another example of the Cunningham Cathedral and again combines the carved column style with the large plate glass windows.

Another example of Cunningham craftsmanship in yet another style of Cathedral hearse. This car owned by C.A. Gould of Wakefield, Mass., features disc type wheels and a fully enclosed carved drapery style body. Note the massive appearance of this car and how Cunningham made the rear body side styling blend into the roof line.

This Cunningham Cathedral carved side funeral coach was delivered to Voth & Anderson of Milwaukee. It has yet another style apart from the Cunningham Cathedrals seen in this chapter. Note the elaborate carvings that include columns and draperies, and how these traditional hearse ornaments have been blended with a more modern style coach.

Yet another Cunningham approach to the Cathedral features a combination of carved columns and glass windows on the rear body sides. Note the overly large tassels on the draperies that hang within the rear compartment windows and the fringe at the bottoms of these draperies.

1931

The aristocrat of service cars was this Cunningham town car service car with blanked off rear quarters and a two-tone paint finish. The wire wheels really help to set this coach off.

One of the more interesting Cunningham designs was this low deck combination service-flower car. Although not designed to operate as a modern flower car, it could carry flowers in the rear compartment or it could be used for first calls or general funeral chores. Note the landau bows on the fabric covered, fully padded roof.

Designated the Henney Model 42, this Nu-3-Way funeral coach was the top of the Henney range of offerings in the hearse field in this year. Note the rather high body styling and the wide opening of the center hinged side doors. The patented Eureka table would also glide through the rear door. Henney was still assembling their own chassis.

The Henney Model 44 was this firm's most prestigious ambulance. It came completely equipped, including bassinet in the rear compartment. It used wicker trim on the lower part of the rear compartment windows.

"A beautiful ambulance" was how the trade journals of the era described this Silver Knightstown town car ambulance. The car was finished in White with dark blue waist highlights. Note the leaded glass rear quarter window, chromed sidemount cover and the wire wheels. The inset photo shows the attractive frontal ensemble of this Cadillac chassis coach and the siren and warning lights that it used.

One of the lowest priced three-way coaches in the Henney line this year was this vehicle, designated Model 10. While retaining all of the traditional Henney features, this car was mounted on a smaller chassis and sold at a considerably lower price than the Model 42. Note the location of the flower trays. Henney also offered low priced models designated as Models 30 and 20 that rode on successively lengthened wheelbases and had better trim and features. With this array of models, Henney offered a complete price span and a coach for every budget.

All Flxible funeral coach models were available in ambulance form like this model with leaded and frosted glass in the rear quarter windows. Flxible built on two Buick chassis this year. It built a total of 181 cars on the model 8-870 chassis and 1 car on the model 9-90 chassis. The dimensions for the 1931 Flxible funeral coach and ambulance measured inside length 93 inches, side door openings 36 inches wide, rear door opening 39-3/8 inches wide. The complete interior was available in either mohair or leather upholstery. Flxible prices started at $2,695.

Flxible built only one ambulance in this year on the Buick Series 90 chassis. Note the chrome side mount cover, leaded glass rear quarter windows, bumper mounted siren, and the central glass partition within the driver's cab. This was a truly dignified looking ambulance but unfortunately did not enjoy wide popularity.

Flxible funeral coaches and ambulances were all built with chassis by Buick. The Buick special service chassis featured a wheelbase of 155 inches, a one piece frame, and a straight eight engine that developed 104 horsepower and could pull from 4 to 60 miles per hour in high gear. Bumpers in front and bumperettes in the rear, air-seasoned hardwood body frame and a duco finish in any choice of colors were additional features of Flxible coaches.

Continuing to build coaches on the Buick chassis only, the Flxible Company built a total of 182 cars this year. They were built on two different Buick series chassis — the Series 47 being the most popular of the two with 181 bodies being mounted on it. The Buick Series 90 chassis was of hardly any consequence to Flxible this year as they built only one coach on this chassis. Here we see the Flxible limousine funeral car for 1931. Note the flower trays in the rear compartment and the owner's nameplate in the side window.

One of the most beautiful cars built in this year by Eureka was this model mounted on a Pierce Arrow chassis. Eureka expressed new beauty and dignity with this car. Note the two-tone paint finish and the whitewall tires on this car. The crucifix is mounted in the rear quarter windows. This car had pull down shade type draperies.

EUREKA
BURIAL COACHES

Eureka built this Buick chassied town car style funeral coach this year and delivered it to the Didesch Funeral Home in Dubuque, Iowa. This body could be fitted to any suitable chassis and featured a metal back with landau bows. The flower trays are mounted on the interior body sides above the casket table. This coach is also a side loading model. Eureka was the first maker to be able to offer a three-way town car model to the trade.

1931

With a chassis by Reo and a body by Henney, the combination was completely guaranteed by the National Casket Company. This view of the National funeral coach with the Nu-3-Way feature shows the interior of this car and the casket table in operation. The wire wheels on this car add to its attractive appearance.

Wearing a Henney body, the new National funeral cars were also marketed as Nu-3-Way funeral coaches. These cars literally took the country by storm and some funeral directors claimed that this new car was the most distinguished hearse they had ever owned. Notice the overall low look imparted by the car and the gracefully curved fenders and window channels.

With the death of Kissel, the National Casket Co. of Boston lost its supplier of funeral coaches and sought another maker. It struck up a deal with the Reo company and another with Henney and began to market funeral coaches and ambulances again. Reo built the chassis and Henney supplied the bodies for the new National funeral cars. Note the modern styling of this coach.

Mounted on the medium priced Auburn chassis, this Knightstown body features its own version of the popular 3-way coach. Note how Knightstown has made the body lines flow with those of the Auburn chassis. The draperies in this car were of heavy purple mohair.

This Knightstown funeral coach body rides on the beautiful Cord chassis. Knightstown Galahad ambulances and funeral coaches achieved good blend of style and dignity at a price that was considerably lower than competitive coach offerings. They offered a complete line of Galahad ambulances, funeral cars in either rear loading or 3-way versions, and service cars.

This Chrysler chassis is wearing a Knightstown ambulance body. Knightstown called their bodies Galahad and would mount new 1931 style bodies on older lengthened chassis. Note the etched pattern on the rear quarter windows, along with the pillar mounted spotlights and the center warning flasher mounted on the front bumper.

1932

In Canada the Ingersoll, Ontario, based firm of Mitchell built this attractive funeral coach on a McLaughlin Buick chassis. This particular coach was delivered to C.L. Eeoy Funeral Home of Tavistock, Ontario, this year. Of special interest on this car are the etched designs on the side windows that form a frame within the glass area itself.

Attempts at streamlining funeral coaches and ambulances were beginning to show in this year. The products of the A.J. Miller Company appeared with windshields sloped slightly, and the window surrounds were more rounded. Some firms added V-ed windshields to the models in an attempt at streamlining.

Sayers & Scovill continued to assemble its own chassis using Continental engines until this year when a switch was made from the older power plant to a large Buick six-cylinder engine. All of the other components of this chassis remained more or less the same as before. Borg & Beck clutches, axles from the Timpkin Company, and transmissions from either Detroit Transmission Co. or Brown & Lipe were coupled to the new Buick power plant.

Even though the makers made attempts at presenting new models with new innovations, 1932 proved to be the poorest auto production year since 1918, a fact that caused the deaths of several more of the smaller makers of ambulances and funeral coaches. With the exception of some S & S models and a few Cunningham models, the limousine and the town car limousine style dominated the market.

Another McLaughlin Buick funeral coach built by Mitchell Hearse Company of Ingersoll, Ontario, was this example that was delivered to the Stephenson & Son Funeral Home in Alisa Craig, Ontario. This car also features the unusual side window etchings seen on the other Mitchell example. Most of these Mitchell photos were taken in and around Ingersoll with this one being taken at the city cemetery in front of the mausoleum.

Eureka continued to offer the famous three way casket table in their hearse models. This Eureka funeral coach is shown mounted on a Cadillac chassis. Note the sloped windshield and the crucifix in the rear quarter window. Cadillac was now supplying hundreds of chassis to the body makers every year and of all types; LaSalle, V-8, V-12 and V-16 chassis, were all to see hearse and ambulance bodies mounted on them.

This Eureka three way funeral coach is mounted on a Buick chassis and is quite similar to that seen earlier on a Cadillac chassis.

1932

Continuing to assemble their own chassis utilizing the Continental engine, S & S again offered one of the most extensive model lines available at the upper-medium price range. All body stampings were made by S & S including the attractive frontal design of this Masterpiece.

Convenience was the key word for the S & S Knickerbocker for 1932. The modern convenience was found in the new S & S extension table that glided smoothly through the rear door, and the large church truck storage compartment. These were the key features of the new Knickerbocker. S & S also stressed eye-appeal for this coach. Gently streamlined from bumper to bumper, this car wore a special dignity in the world of limousine coaches. Notice how the front door is quite wide.

With the exception of the carving on the rear quarters, the S & S Majestic funeral coach looked very similar to the Knickerbocker. This photo shows the bronze panel of the Majestic in detail. This panel was set into the place of the window of limousine models and was a miniature of that used on the Signed Sculpture coach that was no longer offered. The S & S Majestic was billed as "America's most distinctive funeral car."

The clear-cut beauty of the S & S Kenwood ambulance created a buzz of comment at the Funeral Directors convention this year. The front compartment styling of this coach closely resembled the metropolitan type of town car in a closed style. Among its features were an 118-horsepower straight eight cylinder engine, cushioned power impulse, syncromesh transmission, silent second gear, automatic ride control, and an unusually long wheelbase. The leaded glass rear quarter windows continued in popularity for ambulance use through this year.

S & S were masters of interior appointment, as this picture indicates. This is the rear compartment of the 1932 S & S Majestic and features cathedral panels on the front partition and the rear doors, casket bumper and dome light set into walnut medallions, and candelabra sconces placed on walnut panels mounted on each side wall and set off by plush drapes. Note the patented S & S extension casket table.

The S & S Wellington represented one of the greatest funeral car values ever offered. It was also the lowest priced S & S coach offered and still retained all of the traditional S & S quality. Priced at only $2,995, the Wellington was available in either a straight end-loading type or as a three way coach at additional cost. It was powered by the 118 horsepower eight-cylinder engine and rode on a wheelbase of 154 inches. The chassis featured such items as duodraulic shock control, synchro-mesh transmission with silent second gear, a super trussed and sound-proofed body, shatter-proof glass all around, a streamlined sloping windshield, and finger tip steering.

1932

The top of the Superior-Studebaker range was this Westminster, priced at $2,685. This coach featured six wire wheels as standard equipment, and was highlighted by Studebaker's ovaloid headlamps and twin chrome plated horns that flanked the sloping grill. Bumpers were extra cost options. Drapes were heavy brocade and deep blue in color.

Pierce-Arrow, now sold by Studebaker, was also to have special coach bodies mounted on their chassis by Superior Body of Lima, Ohio. The new Pierce-Superior models were unveiled at the National Funeral Directors Convention in Milwaukee in September, 1931. This Pierce-Superior limousine funeral coach was powered by the 125 horsepower Pierce-Arrow eight-cylinder engine. It featured free wheeling, silent synchromesh gearbox, ride control, and many other Pierce-Arrow features. The magnificent Superior body was mounted on a 160-inch wheelbase chassis, and this large wheelbase coupled with the patrician frontal treatment, graceful cowl, sloping windshield and wide sweeping fenders, gave this coach a dignified and impressive appearance. The interior was appointed with rich mohair that covered the sidewalls, seats and the casket table. This casket table could also be ordered in genuine walnut. Bracketed candle type lighting fixtures and flower trays made of genuine walnut finished off the luxurious interior. Note the elaborate draperies.

This Superior Studebaker funeral car features styling that made Superior one of the most popular medium priced coaches on the market. Mounted on a Studebaker chassis, this car was one of the company's most popular offerings.

Superior also offered their new Pierce-Arrow models in town car versions. Both limousine and town car models were available in ambulance or funeral coach styles, and with standard end loading or fitted with Eureka side loading equipment. This Pierce-Superior town car is seen in ambulance form and rides on the 160-inch pleasure car chassis with a 125 horsepower eight-cylinder engine.

Superior also entered into building of coaches on chassis other than Studebaker or Pierce-Arrow with this especially built town car funeral coach on a Cadillac chassis. This was one of the most attractive Superior coaches built in this era and was driven from Superior's Lima plant to its home by owner A.E. Kingsley.

1932

At the lower end of the Superior-Studebaker range was the Arlington funeral car that was available in two versions. The Arlington 6 was priced at $2,187 and was powered by an 80 horsepower engine. The Arlington 8 was priced at $2,325 and was powered by the 101 horse engine used in the Samaritan. The Arlington series cars rode on a wheelbase of 147 inches. For $2,100 a service car called the Elmwood 8 was offered.

Reflecting quality service, this Cunningham carved funeral car was built for Joseph Flavo of Rochester, N.Y. It featured the same frontal ensemble shared by the style 315-A and 333-A limousine funeral coaches. Notice the intricate carving work on the rear body sides and the metal spare cover. Disc wheels are also seen on this car, although they were no longer as popular as they had been through the late twenties.

Continuing to assemble their own cars, Cunningham offered this ambulance, Style 318-A in their line. Very little has been done to streamline this coach. It retained the leaded glass pattern in the rear quarter windows.

This fleet of three Cunningham side loading three-way funeral coaches was owned by Zabriske & Scott Inc. of Paterson, N.J. This firm wrote to Cunningham with such comments as, "When we think of motor equipment we invariably think of Cunningham." Zabriske & Scott had ordered 10 Cunningham cars over the past few years and claimed that they were "the best in service and equipment at all times." Cunningham records show that this funeral home had been doing business with them for over 30 years. As did other makers, Cunningham was utilizing the Eureka three-way casket table for their cars.

This Cunningham style 315-A limousine funeral coach is a three-way unit. Cunningham also called this type of coach a Nu-3-Way and offered this feature on all of their hearse models. Cunningham also offered a service car this year.

Cunningham began to mount their bodies on the manufactured chassis of other makers, and this Cunningham-Cadillac funeral coach is a prime example of this type of Cunningham adaptation. In this model, the windshield pillars have been slanted a little. The natural gracefulness of the body lines added a touch of streamlining.

1932

Designated a Model 10, this Henney limousine funeral coach now featured an electrically operated three way casket table. Henney was now building cars on chassis other than their own, and offering a complete line of models. This particular coach is mounted on the Henney assembled chassis and features heavily curved side window frames and a high waist line. The nameplate is mounted by the kick plate.

The National Model C was this small Reo Flying Cloud 6 chassied car. The L-head six developed 85 horsepower and the complete car rode on a wheelbase of 152 inches. In this model line there were only two types of cars offered, an end loading hearse as shown, or a service car.

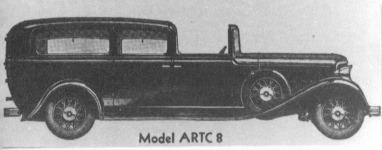

The National model ARTC8 was a funeral coach version of the ambulance. It too, featured town car styling and was mounted on the Royale eight-cylinder chassis. This car was supplied with shades instead of draperies in the rear compartment. These town car National Reo funeral coaches and ambulances were among the most attractive cars offered this year.

National Casket Co., the Reo Motor Co., and the Henney Co. continued to build a line of high quality funeral coaches and ambulances. This model AATC8 was mounted on the Reo Royale 8 chassis and was fitted as a town car ambulance complete with leaded glass in the rear quarter windows and pull-down shades.

This National hearse was called a Model B and was mounted on the Reo Flying Cloud chassis powered by an eight-cylinder engine that developed 125 horsepower and rode on a wheelbase of 159 inches.

This service car was also included in the National range and was built on a shorter version of the Royale eight chassis. Designated model CS8 this car featured a four door type body that looked like a panel delivery. Reo called this chassis the Flying Cloud 8.

Designated as National Model A, this coach was mounted on the Reo Royale 8 chassis that was powered by a 125 horsepower L-head straight eight engine. This car and the models AATC8 and ARTC8 rode on a wheelbase of 162 inches. This car was available as a rear loading hearse, a side loading or three-way model, or as an ambulance.

1932

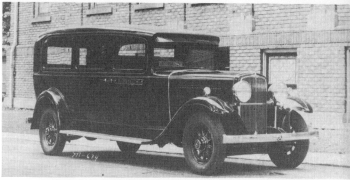

Miller continued to offer a line of funeral cars and ambulances on the Nash chassis. Notice the retention of the cadet visor over the windshield and the simplicity of the lines.

Gardner, entering their last year of funeral car production, offered an advanced model for 1932 with this 85 horsepower V-8 coach. It offered free wheeling, ride control, synchro-mesh gear box, and a body with a frame of seasoned oak construction. The interior was lined with Packard blue mohair and the flower trays were finished in walnut. This car looks as if Gardner was buying some of its body panels from Studebaker. It sold for $2,485. F.O.B. St. Louis.

The part of the business that Flxible is most famous for, and most remembered for, is their busses. For many years Flxible used the Buick chassis, extended, for their comprehensive bus range. This Buick bus was built by Flxible in 1932. Flxible continues to build some of the finest busses on the market today, although they have ceased to produce funeral cars and ambulances.

Flxible built a total of 171 funeral coaches and ambulances in this year, mostly on the Buick chassis. Buick chassied cars were built on two different frames as two separate series. The Series 80 was the most popular, and 160 coaches were built on this chassis. The Buick Series 60 chassis had 8 coaches built on it. This example is of the 80 series, powered by a 118-horsepower engine. It featured such items as synchro-mesh transmission and ride regulator shock absorbers that made riding smoother and more silent.

This Buick chassis wears a funeral coach body built by Knightstown Hearse Company of Knightstown, Ind. Note the elaborate draperies and the flowing lines of this car. The Buick long wheelbase or special hearse and ambulance chassis was aptly suited to this type of car.

In addition to the regular line of Buick funeral coaches and ambulances, Flxible built three cars on the Cadillac V-8 chassis. This Flxible-Cadillac features smooth graceful lines with chrome highlights on the full wheel covers and the ring around the side mount.

1933

Joining the switch to the manufactured chassis, Meteor discontinued the assembly of its own chassis and took up that of Buick. As the depression was still on, the A.J. Miller Company began to offer another lighter-weight, lower priced coach on an Oldsmobile chassis. Miller, which had pioneered streamlining in 1932, added a new touch to the coaches in this year. It replaced the "tucked under" style rear end with a more modern "beaver tail" type. This new rear end style could also have a small compartment placed in it and turned out to be just the right size to accommodate a church truck.

The limousine style that had come into prominence in the early 1920's had never fully captured the market. Carved cars retained about 10% of the sales with the limousine taking the rest. With the introduction of the S & S Masterpiece and Signed Sculpture models the industry began to have second thoughts about abandoning this type of car completely. Within the profession, there were some grumblings about the undignified limousine style, signifying a change was needed if the manufacturers were to keep the customers happy.

Called the Chieftain line, the 1933 Eureka funeral coaches and ambulances were again offered with the three-way loading that they had originated many years earlier. This attractive Eureka Pierce-Arrow features smooth rounded rear lines and side mounts. The windshield is sloped and the windows have pull-down shades.

This ambulance, on a Lincoln chassis, was built by Eureka and offered the same lines that were found on the Chieftain line of funeral cars. Ambulances were still clinging to the leaded glass patterns in the rear quarter windows. Note that this car has a two tone paint finish and a complete set of wire wheels.

"Distinctive" and "impressive" were the words Eureka used to describe this Packard three-way funeral coach.

This Eureka offering was among the most stylish of the year. Mounted on a Cadillac chassis, this Eureka town car featured carved drapery panels on the rear body sides and the whole car was very well streamlined. Notice the beavertail rear, the curved rear portion of the roof, the slanted windshield and how the Eureka body blends so well with the lines of the Cadillac chassis. The whole design is highlighted with the addition of the wire wheels and the whitewall tires.

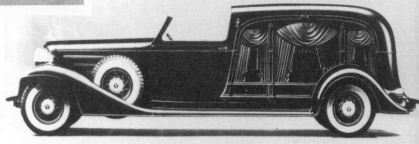

1933

This imposing Cunningham ambulance, built for W.H. Sardo & Co., Funeral Directors, of Washington, D.C., took an important part in the inaugural celebration for F.D. Roosevelt on March 4th that marked the beginning of the "new deal" for the nation. This ambulance was designated as the official police ambulance for the inaugural parade and festivities.

Quality first, last and always was the Cunningham motto in these years and this example of Cunningham craftsmanship was designated the 321-A town car funeral coach. This particular model was delivered to K.C. Urban of Buffalo and was especially designed for this firm. Among its features were French glass plates in the rear compartment, divided by carved columns. Heavy velour draperies hung behind these large windows. The body was mounted on a standard Cunningham chassis.

An interesting combination carved funeral coach town car is seen in the styling of this Cunningham offering. Built on their own chassis, this car features an elaborately carved rear body, drum type headlamps, and wire wheels.

Called the Style 333-A funeral coach, this Cunningham was built on their own chassis. Cunningham bodies began to acquire some streamlining as the sloped windshield of this coach indicates.

This Cunningham custom built creation was called a style 322-A service car and wore styling quite similar to that of the limousine hearse but with blanked off rear quarters. Around this era, even with stiff competition, Cunningham retained high quality and a reputation for building some of the finest funeral service vehicles and ambulances available. In many cases they were looked upon as the Rolls Royces of the professional car field.

1933

This A.J. Miller LaSalle ambulance shows the intricate leading work that was applied to the rear quarter windows of Miller ambulances. Cadillac and LaSalle had become a standard Miller chassis offering earlier, and the orders for these cars were rising with every year.

A.J. Miller offered town car style coaches for that added touch of dignity. This Miller Packard limousine town car funeral coach is a prime example of Miller's design in this vein. Notice the "veed" windshield and the new beavertail rear end. Other streamlining details are seen in the curved tops on the rear side windows and the graceful curves of the fender lines.

With a more pronounced beavertail rear treatment, this Miller-Hudson funeral coach also features a slanted windshield and rounded lines for the windows. Miller was the first maker to offer the slanted windshield and the beavertail rear in a search for a more modern and streamlined coach. Notice that this coach has draperies on the rear quarter windows only.

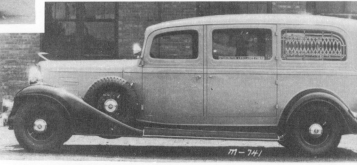

Offering coaches on designated chassis, the AJ. Miller Company of Bellefontaine, Ohio, built this ambulance body and mounted it on a LaSalle chassis. The car, painted gray overall, featured ornate leaded glass in the rear quarter windows, a bumper mounted siren, windshield pillar mounted spotlights, and side mounts. This car was built for the Stonington Ambulance Corps.

Without the modern beavertail rear design, this Miller Chevrolet funeral coach is still quite an attractive vehicle. The walnut flower trays are visible through the rear windows and the wide whitewall tires and wire wheels highlight the overall appearance of this coach. Of interest is the extremely raked windshield line.

This Miller-LaSalle funeral coach is a rather distinctive looking car due to the chrome wheels and radiator shell and the metal sidemount cover. Notice how the front windshield is slanted back a little in an effort to introduce streamlining these cars. The heavy velour draperies add to the overall luxury appearance of this vehicle.

1933

Entering a new and lower priced line of coaches to their offerings, Miller introduced the Benjamin Franklin series of low priced cars. This Miller Ford was the most popular of this new series and came complete with wire wheels, walnut flower trays, and a side flash of another color.

This Miller offering on a Nash chassis was a handsomely styled service car. Note the decorative panel on the rear body sides and the side mounts. Service cars were usually the lowest priced vehicles in any maker's range and varied in style and chassis.

This stately hearse was the S & S Knickerbocker for 1933. Although the Knickerbocker was priced below the Mannington, it was a high quality car throughout. The interior of the Knickerbocker was fitted with genuine walnut cathedral panels and all of the garnish panels were also of genuine walnut. The interior was appointed in high grade mohair and draperies were of heavy mohair or velvet to match the interior color.

An advertisement for the S & S Majestic was headlined "Turn out America's finest funeral car!.. came the order!" And S & S built and advertised the Majestic as just that. Once again the Majestic featured a miniature of the Bronze Signed Sculpture plaque in the rear quarters and was the top of the S & S line. In this year S & S abandoned the use of Continental engines and utilized the Buick eight cylinder engine.

The new Buick powered S & S cars were now more streamlined, with sloped windshields and curved top side windows. This S & S Mannington limousine funeral coach represented the most popular of the S & S limousine funeral cars and enjoyed a wide clientele typical of its own character and quality.

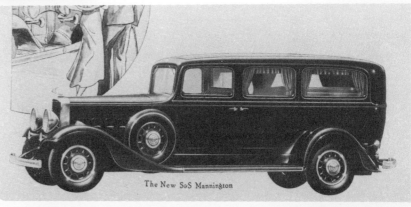

1933

Superior Body of Lima, Ohio, continued to build all of the bodies for the Studebaker comapny's funeral cars and ambulances on a wide variety of chassis lengths and types. This Arlington funeral coach was powered by a 101 horsepower straight eight engine and rode on a wheelbase of 154 inches. This particular model sold for $2,995 this year. It was also offered in a six-cylinder model and a service car called the Elmwood.

The big Superior Studebaker Samaritan ambulance was an imposing sight with its leaded glass rear quarter windows, ovaloid headlamps, wire wheels and side mounts. This car was also powered by the 101 horsepower eight and utilized the 154-inch wheelbase. Note the large siren mounted on the front fender.

Proctor-Keefe of Detroit built this interesting Dodge combination hearse-service car this year. The small rear side window and the carriage lamp add distinction to this little utilitarian vehicle. Proctor-Keefe built these cars to special order only and was not a major manufacturer in this field.

Flxible again offered a full line of professional cars on the Buick chassis. They enlarged the line by offering coaches on three series of Buick chassis and built a total of 222 cars this year. This Buick series 33-80 ambulance was one of 160 various coaches in this series. The series 33-60 only had one coach built on it, and the series 33-90 saw 49 Flxible bodies mounted on it. Flxible also built a grand total of 12 Pontiac chassied professional cars in this year.

In this year Meteor discontinued the assembly of their own chassis and began to use a Buick chassis. Meteor coaches still carried a body that was made totally by them and that included the frontal ensemble and the radiator shell. This distinguished looking Meteor town car was among the stately offerings this year.

The Stephenson Funeral Home of Alymer, Ontario, owned and operated this interesting coach. Mitchell mounted a 1933 body with three-way loading on the chassis of a 1927 McLaughlin Buick that the funeral home supplied and came up with this attractive creation. Note the flower trays mounted on the side walls and the etched pattern in the rear side windows.

1934

As a general business recovery began in 1934, the funeral coach and ambulance manufacturers found themselves in a strong and competitive climate.

Every year in October, the National Funeral Directors Assn. holds its national convention. As was the custom, all the coach manufacturers unveiled the new models at the 1934 convention, held in October 1933.

Mounted on a Hudson chassis, this Knightstown funeral coach wore the Galahad body. The Hudson was powered by a 108 horsepower eight-cylinder engine and looked very nice with the Knightstown body. Knightstown had a rather streamlined body. They specialized in the use of the Hudson chassis, although they claimed to have been able to make this same type of adaptation to any standard make of chassis.

Called the National Nu-3-Way electric funeral coach, Model AE-8, this car continued to use the Reo chassis and wore a body built by Henney. Naturally, when Henney began to use an electric side loading device, they built this feature into the bodies they supplied to National.

Superior of Lima, Ohio, stole part of this show with a new and completely streamlined funeral coach and ambulance line mounted on a Studebaker chassis. Another part of the new vehicle show was stolen by the James Cunningham Co. with its new streamlined "Cathedral" type town car style carved funeral coach. Featuring softly rounded lines and carved side panels, this car was painted a striking silver.

The introduction of the Signed Sculpture and Masterpiece cars several years earlier sparked a return to the carved style coaches. But by now people had become accustomed to the limousine style and the modern streamlined curves of their passenger cars, so the new carved coaches had to be streamlined and spohisticated. Added to this, was the fact that many funeral directors were demanding funeral coaches that looked like funeral vehicles and not like big pleasure cars. The A.J. Miller Company responded to this demand with a totally new type of carved funeral coach in the form of the new Miller "Art Model." This car, along with the earlier entries from S & S led the revival of the more ornate type of funeral vehicle. The Art Model featured large carved arches within which were several beveled plate glass windows. These were separated by carved columns and the cars always featured a small coach lamp just behind the front compartment windows.

This Superior Studebaker ambulance was the epitome of streamlined emergency vehicles this year. It was mounted on the Studebaker President chassis, and was fitted with an eight-cylinder engine that developed 103 horsepower and had cylinder heads and pistons that were made of aluminum.

Utilizing the Studebaker Dictator chassis, this Superior service car was powered by a 88 horsepower six-cylinder engine and used 5:50 x 17 inch tires. Note the unique design that Superior used on all of their service cars.

Another innovation to make its debut in this year was the electric side servicing funeral coach. A take-off on the patented Eureka 3-way table, the new electric table offered the funeral director push button ease together with added dignity in loading and unloading. This feature was offered on some models by Henney, and National coaches with Henney built bodies.

This year, a total of 1,885 funeral cars and ambulances were built in the U.S.

1934

Shown at the top of this photo is the sedan type version of the S & S Olympian that had the same features as those of the town car version shown earlier in this chapter. The lower car is the S & S Mannington funeral coach that had the same chassis and engine as the Olympian but wore a limousine style body.

S & S marketed two versions of car called the Olympian. The most exclusive model was this town car version that featured carved drapery panels on the rear body sides and windows set into these draped panels. These cars were powered by a Buick 8-cylinder engine and the bodies were hand crafted by the Sayers & Scovill craftsmen at the Cincinnati plant.

This is the S & S Kenwood ambulance that was called the "aristocrat of ambulances" in 1933. The Kenwood was powered by the Buick eight and that was matched to some Buick drive train components. These were mounted in the S & S built frame and the bodies were all hand built. Note leaded glass in the rear quarter windows was still popular.

With this year Henney began to use manufactured chassis from the major passenger car makers instead of assembling their own. This Henney Lincoln three-way funeral coach was said to have been the "world's finest funeral coach" by comparing it with the competition. The Henney-Lincoln was precision built to the highest mechanical excellence thru the unlimited resources of the Lincoln engineering laboratories, especially for the Henney Motor Co. The three way casket table was electrically operated and the car had formal interior appointments that reflected the traditional Henney craftsmanship.

At the lower end of the price scale Henney introduced the Progress series. This Henney Progress Straight 8 Oldsmobile was available in funeral coach or ambulance forms. It had among its refinements such items as X-K Girder frames, knee action wheels, center control steering, ride stabilizers, synchro-mesh transmission, and super hydraulic brakes. Interior finish of these cars featured newly styled appointments and oxyplate table hardware. This body style was called slipstream and was the latest development in streamling by Henney. The progress models were also available with the famous Henney three-way servicing and electrical assistance.

1934

Called the style 347-A limousine, this Cunningham was also mounted on the 148-inch wheelbase Oldsmobile chassis and powered by a 90 horsepower engine. This car had almost boxy lines then compared with some of the competition's limousine offerings, but the magic of the Cunningham name was still there. They claimed that this car was at a price that was within the reach of every progressive funeral director. The price was $2,850 plus taxes at Rochester, N.Y.

At the upper end of the Cunningham price scale was this carved hearse mounted on their own chassis. This car, delivered to Ingimire & Thompson Company of Rochester, N.Y., featured an elaborate carved drapery pattern on the rear body sides, a gold initial on the front doors, side mounts and carriage lamps for around $6,500.

This Cunningham Cathedral was a town car model and in this side view the soft rounded lines of the rear end are easily seen. Note the crucifix mounted in the center rear side window and the metal sidemount cover. Once again Cunningham used the slogan, "Quality, first, last and always."

Although 1934 was not what could be called an overwhelming year for the industry, production-wise, with only 1885 units built by all of the makers, it was a year in which some new models made their debut. One of the most interesting of these was the Cunningham Cathedral, built on a newly revised Cunningham chassis. This car was finished overall in silver and featured imposing styling with rounded lines and carved panels on the rear body sides that were fitted with small glass windows within the drapery folds.

In an effort to stay alive and remain competitive, Cunningham was forced to begin building coaches on chassis other than their own and those of luxury cars. They were put in a position of having to offer the trade lower priced, high quality vehicles in this time of depression. To do this they turned to Oldsmobile for chassis and began to build some rather conservative limousine type bodies on these chassis. This Cunningham style 353-A carved hearse was mounted on an Oldsmobile chassis with a wheelbase of 148 inches. It was powered by a 90 horsepower engine and priced at $3,350. "A special modest price embodying all of the same high quality workmanship as found in our higher priced vehicles — making possible maximum value at a minimum price," was the way Cunningham described this car.

In what was to become one of the most beautiful and unusual offerings for the year, Cunningham built and designed this magnificent town car Cathedral model on a Cadillac chassis. This car featured a massive rear body with a domed center. The sides of the rear compartment were covered with elaborate and intricate draped carvings. This car, an excellent example of Cunningham craftsmanship, was delivered to Math. Hermann & Son Funeral Home of St. Louis.

1934

Announcing the most revolutionary new design to hit the field in years, the A.J. Miller Co. of Bellefontaine, Ohio, introduced the new Art model to the field. This car featured an arch shaped panel on the rear body sides that featured three small beveled glass windows within its confines. Other ornamentation on the body sides included a coach lamp and a place for the owner's initial at the peak of the arch. Here is a Miller Art model on a Packard chassis.

This Nash-Miller features town car styling on a carved side coach. The intricate draped carvings and columns were all encolsed within a large arch. This was a totally new concept in carved car design and was called an Art Carved car. Note the small carriage lamps.

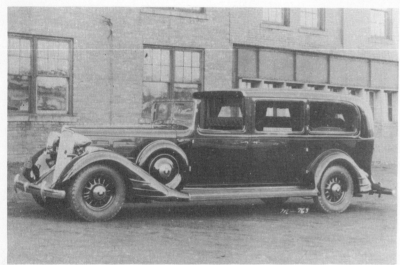

Built by A.J. Miller, this Nash town car funeral coach is a stylish vehicle indeed. This uniquely styled Nash was built for George J. Devine. This car was a three-way loader and the removable section of roof was easily placed behind the driver's seat when not in use.

Referred to in Miller records as the "Coulter job" this little Willys chassied funeral car was built as a child's hearse and was finished in white. The arch on the rear body sides quickly identifies this unusual car as being one of the new Art models that caused such a sensation within the trade. The side windows on this car were of French (beveled) plate glass and the heavy draperies were of a contrasting color. The edge of the arch was finished in gold for highlighting.

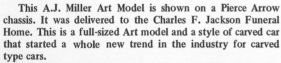

This A.J. Miller Art Model is shown on a Pierce Arrow chassis. It was delivered to the Charles F. Jackson Funeral Home. This is a full-sized Art model and a style of carved car that started a whole new trend in the industry for carved type cars.

1934

This attractive Miller-Ford service car was delivered to the Tom Bartlett Funeral Home. The chassis of this car was extended and it looked like a super long panel delivery truck with artistic stampings on the body sides.

This Miller funeral coach was built on the Chevrolet chassis for the Stapf funeral Home. A full length flower tray can be seen through the rear windows. The sloped windshield and the beavertail rear accentuate the modern lines of this attractive coach. Note the wire wheels and the metal sidemount cover with the Chevrolet emblem.

Shown on a Packard chassis is the full sized Miller ambulance for 1934. This body was identical to the funeral car body and could be supplied on almost any chassis.

This small Miller ambulance, built for the Hackensack (N.J.) Hospital, was mounted on a Reo chassis. It had the intricately patterned leaded glass in the rear quarter windows and the streamlining that was found on other Miller coaches.

Flxible built a grand total of 106 Buick chassied funeral cars and ambulances this year. They were constructed on three chassis; Series 90 of which 28 were built, Series 70, (seen above in ambulance form) which saw 60 professional car bodies mounted on it, and the Series 40, that had 18 bodies mounted on it. Note the leaded glass in the rear quarter window and the fact that Flxible still used the old style tucked-under rear end instead of the streamlined beavertail type.

1934

Eureka built this Chieftain funeral coach body, mounted on a Lincoln chassis, to come up with this magnificent vehicle. The Eureka body was very streamlined with a sloped windshield and a beavertail rear end. Eureka referred to this car as a distinctive symbol of distinctive service.

This Eureka Chieftain body was custom built and mounted on this 1931 Cadillac chassis. Eureka would custom build any body, ambulance, hearse or service car, to fit any chassis, new or used. With the notable exception of the beavertail rear end treatment, this car features all of the streamlining features of the Eureka models mounted on 1934 chassis.

This Canadian built Mitchell was owned and operated by the Evans Funeral Home of London, Ontario, and was built on a modified 1926 Cadillac chassis. The body was constructed by Mitchell Hearse Company of Ingersoll, Ontario, this year. It features etched patterns within the centers of the side windows, and smooth rounded lines.

This Canadian built Mitchell was owned by the Needham Funeral Home of Chatham, Ontario. It was built on the chassis of a 1931 Packard. The 1934 Mitchell body features special etched glass in the rear quarter windows to signify that this car was used for ambulance duties. The smooth, graceful lines of this Mitchell ambulance made this and other Mitchell cars trend setters in Canada and also strengthened their position against the American imported competition.

An interesting picture snapped as the cortege proceeded down the street with a 1934 Mitchell Buick in the lead, shows the funeral of a Toronto policeman accompanied by six other Toronto policemen walking beside the coach. The hearse was operated by the McCougal & Brown Funeral Home of Toronto. Note the interesting chromed wheels and small wheel discs. This car, properly identified, should be called a McLaughlin-Buick. It was built by General Motors of Canada in Oshawa, Ontario. Mitchell was located in Ingersoll, Ontario.

1935

[T]he classic era of the funeral coach and ambulance [bega]n back in 1933 and by 1935 all were well into the [era] of the truly classic vehicles in this field. With the [dep]ression over, the demand for new funeral coaches and [amb]ulances was at an all-time high. The manufacturers [were] ready with new models and innovations to please an [eage]r car-buying public.

[M]any manufacturers began to use the Oldsmobile as a [stan]dard chassis this year. Among these were Henney [with] its new lower priced Progress series and Sayers & [Sco]vill with its Arcadian funeral coach and Abington [amb]ulance. S & S also introduced several other new [mod]els in this year and among them was the Hamilton [fune]ral coach. While using a Buick engine, the remainder [of t]he chassis continued to be of the assembled variety, [but] this new S & S model marked a sweeping change in [desi]gn from former S & S models. It was also in this year [that] S & S introduced the year marks to eliminate the [ann]ual styling changes that automatically dated vehicles. [An]other new S & S line was the Corinthian carved coach [and] the Clayton ambulance.

[H]enney began to build coaches on the Pierce Arrow [cha]ssis. These cars were designated Henney Arrowlines [and] stressed utility plus beauty at the low price. [Res]trained streamlining and a roomier interior were [sev]eral features of this design, with a custom tailored [inte]rior that was the pride of the manufacturer. The [Hen]ney people were also expert at thinking up names for [new] features and their expertise was not wasted when it [cam]e to tacking a tag on the electric driving mechanism [for] the 3-way casket table. Called Electdraulic, the [pat]ented new innovation was made standard on some [mo]dels in the range. It was also in this year that Henney [beg]an to build cars on the Packard 120A chassis and went [into] the Packard fold in a big way. The new Packard [fun]eral coach was titled the Henney 800.

[S]uperior had not followed the crowd using the Eureka [cas]ket table to develop a three-way coach. Instead, the [eng]ineers set to work and designed a three-way coach of [the]ir own. Using a pull-out roller by each side door and a [num]ber of small rollers built into the coach floor, they [dev]eloped their own style of side-loading coach, thereby [elim]inating the need to purchase tables from an outside [sou]rce. The Flxible Company in Loudonville, Ohio, [wh]ich had concentrated on a line of coaches on the Buick [cha]ssis (and a few on Cadillac and LaSalle) also intro[du]ced a model on the Pontiac chassis. Flxible had a [uni]que way to give the coaches a better look from a [pro]portional aspect. As most coach bodies were necessar[ily] high and the hood lines low, Flxible added a piece to [eac]h side of the hood to raise it in proportion to the rest [of] the car. As far as Flxible and the Pontiac chassis were [co]ncerned, this was the first and last year of production [of] this make.

[T]he Shop of Siebert of Toledo, Ohio, had been in the [cus]tom coachwork field since 1853, and in this year [thr]ew their hat into the funeral coach and ambulance [are]na. Specializing on the popular Ford V-8 chassis, Siebert was to find a comfortable nitch for itself in the field. The Owen Brothers of Lima, Ohio, had been designers and builders of custom coaches since 1899. Although never a large organization, it built some very interesting coaches on manufactured chassis over the years. In 1935, it introduced the model 1080 custom built, streamlined funeral coach on the Packard chassis. This car featured carvings of a special design and was claimed to have been one of the most distinctive and dignified funeral coaches of the year. At the end of the year, a 36-year history of contribution and achievement ended with the death of this Lima based company.

Most all of the cars sold in this era were either through the mails or via a traveling representative of the company. In either case, the customer dealt directly with the manufacturer. With the increased volume in sales, the A.J. Miller Co. also increased the numbers of field representatives employed. This was done in an effort to cover the entire United States more thoroughly. A special sales force was created for Miller vehicles in Canada. The depression was, without doubt, definitely over for the professional car builders.

Hearse and ambulance production was up substantially this year, compared with the previous depression years, and a total of 2,623 such professional vehicles were built in the U.S.

Called the "Mobile Sanctuaire" this unique car was built by the Mitchell Hearse Co. of Ingersoll, Ontario, especially for Edwin W. Morris of Windsor, Ontario. It was said to have been the only one of its kind in North America and was finished in light gray with contrasting black running gear. The rear body sides featured massive carved drapery panels but these were not of carved wood. The entire body was made of aluminum. The interior of this car was finished in the gothic pattern complete with stained glass windows and lights that illuminated the casket compartment when the rear door was opened. This car was built in this year and mounted on a 1931 Packard chassis.

A masterpiece of the hearse builders art was seen in the A.J. Miller Art models. This Miller Art model is mounted on a Packard 120-A chassis and features the carved arch and columns with the top center of the arch featuring the owner's initial in gold. All of the windows in the rear compartment were of heavy beveled plate glass.

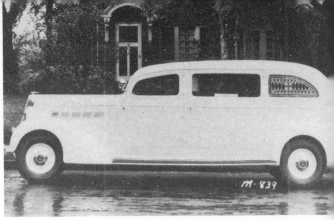

Mounted on a Packard 120-A chassis, this ambulance was especially constructed by Miller and was called the "Billups job." This design consisted of a special rear end and wider rear side doors that gave the car a lower overall appearance and added to the streamlining. Note the small rear quarter window with the filigree leading.

Miller referred to this coach as the "Newcomb carved, job" in their files. This is a fine example of one of Millers Art Carved models mounted on a Packard 120 chassis. Note the finely detailed carvings on the rear body sides and the size of the arch that encloses these draped carvings.

The Norwalk (Conn.) Hospital was the recipient of this Miller ambulance built on a Chrysler chassis. Instead of having ornately leaded rear quarter windows, this car has frosted glass in this area to insure privacy. Fender skirts and side mount add to the exclusive appearance of this Miller coach.

This car was a special job done by Miller and has a reconditioned horse-drawn hearse body mounted on the chassis of a Ford. Notice how Miller blended the boxy old coach body into the lines of the new Ford chassis. This type of work was common with some of the coach builders in earlier years, but this is most likely one of the last jobs of this type to be carried out by Miller.

The city of Lynn, Mass., ordered this special Miller-Chrysler ambulance for the police department. This ambulance was a stylish looking job with fender skirts and side mounts. The rather high body could carry two stretcher patients and attendants and the coach was equipped with special grab handles and spotlights.

1935

A.J. Miller would mount one of their bodies on any suitable chassis, and this limousine hearse body is on a 1935 Pontiac Eight chassis. Note the crucifix in the rear quarter window and the side mounts.

Another startling example of the fine work done by Cunningham was this Packard hearse they built for Drehmann-Harral of St. Louis. This beautiful coach features town car styling and a carved panel rear body side. The eight column rear compartment has carvings of heavy draperies between the columns and the center section is raised. Among discriminating funeral directors the trend was toward the carved panel hearse, and those that could afford them, toward Cunningham.

The Cunningham carved panel Model 365-A is seen here mounted on a Packard 120-A chassis. This car featured massive carved drapery panels set between eight carved wooden columns. This car also has a small coach lamp behind the front door and was finished in black.

Cunningham built this high looking Lincoln ambulance for one of their long-time customers, the Scully-Walton Ambulance service of New York and London. Finished in limousine styling, this car was painted in gray and black and had a three door configuration.

A combination of modern limousine styling with carved wooden drapery panels was seen in the Cunningham style 358-A model. This car is seen here mounted on a Packard 120-A chassis and sold for $2,985 F.O.B. Rochester. The car has a very modern looking heavertail rear end and smooth flowing limousine lines.

The Cunningham carved panel model 361-A is seen here on a Packard 120-A chassis and was built to order for Fred J. Lowe of Troy, N.Y. This car had a massive carved rear panel area with eight columns and draperies. Notice the highly mounted side mount and the rear-hinged front doors.

In daily use in China at this time was this unusual Chinese hearse mounted on a Ford chassis. From the cowl back, the car was a mass of carvings and detailed reliefs. The roof, shaped like a Buddhist temple, featured a carved dragon on each side. Much of the car was painted in gold, with the front half in black.

1935

The newest addition to the S & S line in this year was the S & S Hamilton funeral coach. This car was designed especially for the funeral director who attached more importance to beauty and refinement than price. The Hamilton replaced the Mannington in the new line and featured all of the high quality features that had made the Mannington so popular. All S & S cars were now powered by Buick engines and S & S styling was superb on all models.

Following in the model line of the new Hamilton, the latest addition to the S & S range of carved funeral cars in this year was the Corinthian. There were, at this time, more and more funeral directors that were beginning to see advantages in owning two hearses, and S & S appealed to these people by offering them the highest quality in carved side funeral cars for one type of funeral, and limousine styles for yet another style of service.

S&S OLYMPIAN FUNERAL CAR

Continuing with the styling of earlier S & S cars, the 1935 S & S Olympian carved hearse was popular because it revived the spirit of symbolism and heightened the atmosphere of reverence. This car was popular among funeral directors that prided themselves with the exclusiveness of their service and its beauty was of the classical type that expressed true dignity. This was the last year for this older styled model that was powered by a Buick engine.

Derived directly from the Hamilton was this stylish S & S ambulance called the Clayton. This car features leaded glass in the rear quarter windows and spotlights mounted on the A pillars. A two tone paint finish and side mounts completed the style picture for this Buick powered S & S ambulance.

THE S&S CLAYTON

Sharing the chassis and body characteristics of the Sayers-Arcadian, the Sayers-Ablington ambulance was moderately priced and completely equipped. It, too, was powered by the 100 horse Oldsmobile engine and rode on a wheelbase of 159 inches.

Mounted on a wheelbase of 159 inches, this Sayers-Arcadian funeral coach was powered by an Oldsmobile 100 horsepower engine and utilized Oldsmobile fenders and hood. The car featured such items as a chassis that was built by S & S exclusively for funeral coach or ambulance service, oversize axles, spindles, and bearings designed especially for funeral car work.

Superior built a wide range of Studebaker based coaches this year and also ventured into mounting their bodies on other chassis. This Superior Studebaker funeral car was called the Mission and was mounted on the Studebaker Commander 8 chassis. It sold for $2,065 and had a large casket compartment that was over 100 inches long. The side doors of this car were 54 inches wide, with a rear door loading height of only 27 inches. Superior side loading funeral coaches featured an improved and simplified Sidroll that saved 4 inches of inside height and eliminated the need for a casket table. This Sidroll feature was optional at extra cost on all Superior Funeral coaches. Other models in the Superior Studebaker range were: the Arlington on the Dictator 6 chassis at $1,850, the Westminster on the President 8 chassis at $2,490, the Samaritan ambulance on the President 8 chassis for $2,665, and the Elmhurst service car on the Commander chassis at $1,905.

This Superior funeral coach carries lines notably different from those seen on the Studebaker models. Mounted on a Packard 120-A chassis, this car has rounded window lines and a straight-through belt line. Superior would mount a body on any chassis but Studebaker was the prime customer for these cars.

Eureka also built this attractive three way loading funeral coach body on the chassis of a Pierce Arrow. Notice the graceful curve of the rear end with the beavertail and the smooth rounded curves of the roof. This Eureka custom built hearse was another so called "heir to unmatched beauty."

Sold and serviced by Studebaker dealers, the Superior Studebaker line of funeral service vehicles and ambulances were aimed at the middle and lower end of the upper price range. This distinguished looking Superior service car was called the Memorial, and was mounted on the Commander 8 chassis. This car sold for $1,905 and was complete with carriage lamp and wreath type work on the rear body sides.

Called an "Heir to Unmatched Beauty," this Eureka ambulance was mounted on a Reo chassis and had up-to-the-minute streamlining and graceful lines. Note the pattern in the leaded rear quarter windows and the fact that this car could also be loaded from the side.

This stylish Cadillac wears a carved-side body by Erueka that features one of the largest carved panels ever offered. The massive arch that enclosed the carved drapery area was to become a symbol of Eureka carved funeral cars. These were the largest panels offered, and were intricately carved and detailed. Note the small carriage lamp on the rear body sides and the nameplate mounted on the running board.

1935

Offering low priced, economical coaches on the Ford chassis, the Shop of Siebert of Toledo, Ohio, offered a full line of cars this year. Selling for $1,370, this Siebert Ford funeral coach was the bread and butter car of the line and outsold all other models that Siebert offered this year.

The second most popular body style from the Siebert coach makers was this Ford ambulance that went for $1,470. The two-tone paint finish was an optional extra. Note the leaded glass in the rear quarter windows. Wire wheels were considered standard equipment.

The least expensive and least popular model in the Siebert range was this Ford service car. This vehicle shared all of the body components with the limousine funeral car and ambulance but had the window openings covered over. This car sold for $1,300.

This strikingly handsome Henney Packard limousine funeral car was mounted on the Packard 120-A chassis. It featured modern limousine styling with a beavertail rear end. Note the interesting draperies that this car uses.

Called the Henney Arrowline, this attractive Pierce Arrow funeral car wore a body built by Henney of Freeport, Ill. The car featured the Elecdraulic three-way servicing that Henney had become famous for, and a styling approach that gave this car a long, low appearance. The finest appointments were utilized for the interior of these Henney Pierce Arrow cars and they were second to none in luxury and convenience.

The Campbell Funeral Home of Chatham, Ontario, was among the progressive and fortunate establishments able to afford one of the fine Henney Arrowline funeral cars. This three-way Pierce Arrow funeral car is seen here in front of the Campbell funeral home in all its black shining glory. Note the beavertail rear end styling and the high waist line.

Built on the Oldsmobile chassis, this Henney Progress funeral car was one of the lowest priced offerings from the Henney Motor Co. of Freeport, Ill. Once again this car utilizes the fringe draperies that line the tops of the rear windows. The flower trays can be seen through the rear quarter window. Note the extent to which the rear door opens.

Flxible revealed its big news for 1935 with the all steel top that pre-dated this type of construction by any automaker by one full year. This car was built on a Buick Series 60 chassis and was one of 71 such vehicles to ride on this chassis. Finished in two-tone gray with black fenders, this vehicle was delivered to the Emery Funeral Home and there it was used as an ambulance. Note the windshield pillar mounted spotlights and the windshield "AMBULANCE" sign.

With a hood line that has been raised a considerable amount, this was the Flxible Pontiac funeral car for 1935. The company built a total of 78 such Pontiac chassis cars in this year with the remainder and the bulk of the production being devoted to Buick chassis. Although the Flxible styling that this car wore was pleasing, the orders were not outstanding.

The U.S. Government, always a stickler for high quality in anything they bought, purchased a fleet of Flxible ambulances in this year for the Navy Department's medical corps. This example of the Flxible Navy ambulance features a windshield pillar mounted spotlight, sidemounts, leaded glass quarter windows and an overall finish of Navy gray.

Appearing for the last time was a professional car offering from the Owen Brothers of Lima, Ohio. This Style 1080 carved panel hearse was mounted on a Packard 120-A chassis, but Owen advertised that this body could be mounted on any suitable chassis. Owen retired from the funeral car and ambulance field after many years of offering high quality and many times unusual vehicles.

The Flxible Company of Loudonville, Ohio, built a total of 96 professional cars on the Buick chassis this year and they all featured the revolutionary all-steel top. They were built on the Buick Series 90 (like the one shown) which saw 13 Flxible bodies mounted on it, the Buick Series 60 chassis which had 71 mounted on it, and the Series 40 that saw 12 Flxible professional car bodies on it. Note the smooth graceful lines of this limousine funeral car and the modern rear styling.

A special offering was this rare service car from Proctor-Keefe of Detroit. This car featured ornamental chrome wreath and cycus leaf details on the rear body sides along with a small coach lamp. Offerings from Proctor-Keefe were to special order only. This car is mounted on a Dodge chassis.

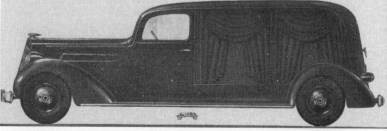

1936

Cunningham said that their new hearses and ambulances were the best built cars since they began in 1838. "Cunningham, the best in hearses, limousines and service cars for 98 years," was the slogan. This was also their last slogan, as Cunningham discontinued the construction of these high quality cars at the end of the model year. This hand carved Cunningham style 374-A carved hearse was mounted on a Cadillac V-75 chassis and featured carved drapery panels between eight columns with windows inset in these areas.

In an attempt to increase their sales and survive hard times, the Auburn Motor Car Co. of Auburn, Ind., introduced a funeral coach and ambulance. Utilizing an X-type frame and a 115 horsepower engine built by Lycoming, this car sold for $1,850 in funeral coach form. This was the only year that the company ever offered these vehicles due to extremely low sales.

Knightstown built this attractive funeral coach on the Hudson chassis. It continued to utilize a top with a center section of fabric. This Knightstown body is mounted on the Hudson Eight chassis and is complete with side mounts and bumpers.

After a rather brisk sales year in 1935, the manuf turers were ready with more new offerings for the selling season.

Sayers & Scovill finally made the switch to all-Bu mechanical components, installing them within their o long wheelbase, specially built chassis frame. This cluded the famous Buick knee action front end, Bu rear axle and, of course, Buick transmissions and engin They also began to use the standard Buick front and r fenders in this year.

Superior began the production of a full line of Pont chassis funeral service vehicles and introduced a n multi-purpose funeral coach. In its standard form it w a limousine style hearse that could be loaded from eitl side or the rear, a three-way car. Within a few minutes pair of carved aluminum panels could be attached to t sides of the car to convert it to a carved-side style fune coach. It could also be used as an ambulance. This was mounted on a Studebaker chassis this year, with t innovation to be applied to other Superior coaches succeeding years.

All-steel stamped sub-floor panels superseded the ol sill structure in Miller coaches of this year. This was important advance toward all-steel funeral coach a ambulance bodies.

The Auburn Automobile Co. of Auburn, Ind., mad venture into the funeral coach field in this year with attractive limousine coach for $1,895. Meanwhile, t Siebert people were concentrating on their line of lc priced, Ford chassis funeral service vehicles. "Relie yourself of the load of costly equipment, its burdenso upkeep and big depreciation. Enjoy the practical ecc omy as being practiced by hundreds in the selection the Siebert Aristocrat with its streamlined, mode though dignified beauty. The luxurious appointmen economical operation and low initial investment a making it the favorite of America's funeral directors The Siebert Ford line was available through local Fo dealerships and, therefore, sales and service facilties we already quite adequate and well established.

One of the true pioneers in the funeral coach a ambulance field folded operations with the end of t model year. James Cunningham, Sons and Co., Rochester, N.Y., had always been considered the "Ro -Royce of the funeral car field" and had been establish in 1838. Although they did not always follow the trend they built some truly beautiful coaches over the year Sometimes eccentrically standing firm with coaches th resembled rebuilt horse-drawn coaches mounted on motorized chassis, the firm had a long list of repe customers that swore by Cunningham.

Again, total production figures for all funeral cars a ambulances produced in the U.S. were higher than previous years, with 3,247 such vehicles having be built.

1936

Showing the styling that started the whole trend back toward more ornate and carved funeral cars, this A.J. Miller Art model is mounted on a Hudson chassis. Notice the arch that has windows within its confines. The rear door is almost invisible except for the handle seen at the front of the door. This car was a side loading coach that also utilized the Eureka patented three way casket table. The nameplate in the rear center window reads "Hawkins."

When carved panels were desired along with the Art model's arch feature, the end product was called an Art Carved model. This Art Carved Miller funeral car is seen mounted on a Packard 120-A chassis and features metal stamped carved panels, side loading and whitewall tires. The nameplate mounted on the running board reads "Truscott Brothers."

Miller built coaches to appeal to all pockets, and offered this ambulance body on the Ford chassis. The car is finished in gray overall and has frosted glass in the rear quarter windows. This particular unit was delivered to the Long Beach, Cal., Hospital.

This Miller Packard limousine ambulance is all decked out in its winter wear that includes a radiator blind. Notice that this car has leaded glass in the rear quarter windows and the same type of siren/warning light on the front bumper bracket. In addition this car has a large fog light on the bumper and a spotlight mounted on the A pillar. The center sliding glass division can be easily seen. The nameplate reads "Jack Jones."

Miller also offered a complete line of limousine body styles as seen here on a Packard. In these years there wasn't a maker that wouldn't finish a car in white, but none would guarantee this color not to fade. Notice the warning light/siren combination mounted on the bumper brace, and the spotlight. The rear quarter windows were of frosted glass.

The Miller Art model body was applied to many kinds of chassis. Here it is on a Chrysler. This car has fender skirts and some of the details are more clearly visible. The nameplate in this car signifies that this car was delivered to "Higgins" Funeral Home.

1936

The Progress series in the Henney line also included this ambulance on the Oldsmobile chassis. This car was called the Henney Progress Model 764 and had frosted glass on the lower half of the rear quarter windows and blinds that could be pulled down. Notice the two-tone paint finish. The Henney Motor Co. was located in Freeport, Ill.

Built on the standard pleasure car chassis that has been extended by Miller, this Chevrolet-Miller limousine funeral car appealed to the funeral director that could not afford cars mounted on the more expensive chassis. Although mounted on a Chevrolet chassis, this coach had all of the high quality features standard on all Miller cars. The flower trays are easily seen through the rear quarter windows. The nameplate reads "Whiteley."

Henney continued the production of the Progress line of coaches built on Oldsmobile chassis. This Henney Progress features the Henney blind-type draperies and the Nu-3-Way loading table assisted with an electric motor. Note the graceful sweeping curve of the rear end and the side mounts on this coach.

This Henney Packard Nu-3-Way funeral coach, loaded with a body, is returning to the Towers Funeral Home in St. Thomas, Ontario, with all of the drapes closed. The car is mounted on a Packard 120-BA chassis and there were 750 such Packard chassied Henney coaches built this year. This car, still in original condition with the original owner, was recently sold to Walter M.P. McCall of Windsor, Ontario, who plans to add it to his collection of older funeral coaches.

Henney continued with the production of their distinctive Henney Arrowline funeral coaches on the Pierce Arrow chassis. Here we see a row of these Arrowline funeral cars that belonged to the Funeral Auto Co. of Louisville, Ky. This company was one of the city's major hearse rental firms. Henney called the Arrowline coaches the "world's finest funeral cars" and all of these cars were equipped with Henney Electdraulic casket tables and three way servicing.

Here we see a close-up of the famous Henney Arrowline funeral car. Among the cars features were ball bearing casket table casters, an all steel casket table mount, and Electdraulic Nu-3-Way side servicing. Henney now claimed to have been the home of the genuine side servicing funeral coach although they were using the three way table under license from Eureka.

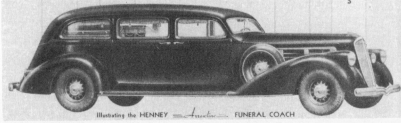

1936

S & S made the complete switch in this year to all-Buick running gear and installed them within their own frame. These components included knee action front end, Buick rear axles, and of course Buick engines and transmissions. They also began to use Buick front and rear fenders this year. Shown here is the S & S Alexandria carved type funeral coach. This model was completely new for 1936 and featured modern styling while maintaining a distinguished and symbolic appearance.

Another new S & S introduction in this year was the Romanesque carved funeral coach. This car was more elaborate than the Corinthian but used the same chassis and power train. Once again the drive train was Buick mounted within a frame built by S & S especially for this type of vehicle. The front and rear fenders were all Buick. The hood and radiator were built to fit the otherwise Buick front end ensemble but were made to S & S design.

With leaded glass in the rear quarter windows, a bumper mounted siren and warning light, spotlight, and two-tone paint finish, this was the S & S Brighton ambulance. Like all S & S radiator models, this car utilized Buick components in both the body and drive train. The S & S radiator treatment is more clearly seen in this picture.

Another new model in the S & S line-up was the Claremont funeral car. Commonly called the Sayers Claremont, this car utilized Buick chassis and frontal ensemble with a refined S & S grill and hood.

This is the large S & S Corinthian carved funeral coach finished in gray for Groves & Co. Notice the unusual S & S grill with special chrome ornamentation at the top of the radiator. The carved drapery panels were quite elaborate and enclosed within an arch. This model was also referred to as a Style 3600.

Proctor-Keefe of Detroit continued to build a limited number of service cars on special order. This one, mounted on a Dodge chassis, features wreath insignia, carriage lamps and a nameplate that reads Young & Son. Notice the streamlined rear lines and the overhang of the body.

1936

This picture was taken in front of the Ingersoll, Ontario, mausoleum when this Mitchell Oldsmobile funeral coach was built. This Canadian company was one of the major suppliers of coaches to the Candaian trade. It features styling very similar to that found on models built in the United States.

Canada and the Canadian funeral car makers generally followed the styles set by the makers south of the border. This Mitchell Packard 120-A features a distinctive type of carved panel that shows drapery and a center panel where the rear door hinges with a wreath and eternal flame draped. Mitchell was still using carved wooden panels while some of the U.S. makers had begun to use stamped or cast metal ones.

Featuring the largest carved area of any funeral coach, the Eureka Air-Flow carved type hearse is shown here mounted on a Cadillac chassis. The panel depicts heavy draperies and thin columns with a vast blank area. This car did not have any rear side doors. It features small coach lamps. It was built on the long wheelbase Cadillac sedan chassis.

The Eureka three-way funeral car, finished in white and mounted on a Cadillac chassis. All Eureka limousine models were called Chieftain Series cars.

Eureka built this attractive Buick Chieftain limousine funeral coach which featured the Eureka three-way casket table and streamlined flowing lines.

Built by Eureka and mounted on a Cadillac chassis, this ambulance is with special rear quarter window frosting details that include a maltese cross and a frame around it. This car was delivered to the Atlantic City (N.J.) Hospital. It is finished in white overall.

1936

Called a Model 606, this Meteor was mounted on the Cadillac 60 chassis and sold for $2,302.27 at Piqua. The price included such standard items as safety glass, front and rear bumpers, "funeral coach" nameplate for the windshield, side window nameplates, windshield wipers, electric clock, removable flower trays, and Fisher no-draft ventilation. Side servicing was optional and the standard model had chrome plated bier pins and casket table hardware. The body was also available on the Cadillac 70 and the LaSalle.

Wearing Superior hub caps and radiator identification badges, this Superior funeral coach was built on a Studebaker chassis. It was the remaining link that the Superior Body Co. maintained with Studebaker. This car has rear side doors that were significantly wider than those found on the competition. The flower tray in the rear quarter windows can be seen between the draperies.

This new Superior built funeral coach embodied all of the latest developments in the field and entered at a price substantially lower than some of the competition. In this year, Superior made Pontiac a regular offering within the product line. This car was extremely attractive, with its Superior streamlined body and large rear side doors with oval side door windows.

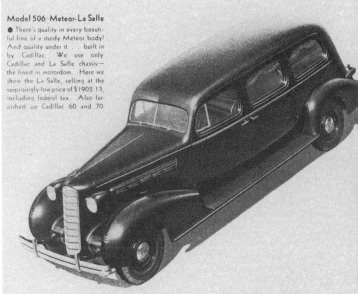

Designated as a Model 506, this Meteor limousine funeral coach was mounted on a LaSalle chassis and featured a solid steel top and Meteor craftsmanship. This unit sold for $1,902 F.O.B. Piqua, and this price included federal tax. This body could also be supplied on the Cadillac 60 or 70 chassis.

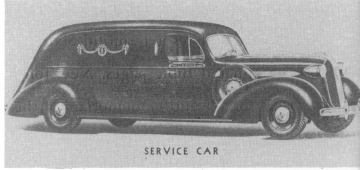

Offered on both the Pontiac and the Studebaker chassis was this interesting service car from Superior. This model was called the Linwood and had small coach lamps on the rear side door and the same service car ornamentation seen on Superior cars earlier.

This Superior ambulance wears only Superior identification plaques and hub caps although it is mounted on a Studebaker chassis. The frosted glass rear quarter windows contain a sunburst pattern with a cross in the center of the rays. This same body was offered by Superior on the Pontiac chassis, but in this variation the car carried Pontiac name plates.

1936

Building 206 professional cars this year, Flxible offered both limousine funeral coaches and ambulances on the Buick chassis. This Series 60 ambulance featured a solid steel roof. In fact all Flxible coaches had a solid steel roof, as Flxible was the first maker in this field to build cars with this type of roof. This gray finished ambulance has a bumper mounted siren, spotlight, and a unique pattern in the leaded glass rear quarter windows.

One of the least expensive models in the Flxible Buick model line-up for 1936 was this attractive service car that features a solid steel top and a small golden wreath on the rear quarters. The partition wall between the front and rear compartment is easily visible in this view of the Series 40 service car.

With differing body styling and features, this Eureka Buick ambulance has leaded glass in the rear quarter windows, a two-tone paint finish and a red warning light on the roof. Dual windshield pillar mounted spotlights and side mounts complete the attractive appearance. This body could be mounted on any chassis.

Flxible continued to build coaches mounted on the Buick chassis. In this year they again offered these cars on three series chassis. The Series 40 (seen above in limousine funeral coach form) was the firm's best seller with 105 units manufactured. The Series 90 chassis had 23 bodies mounted on it, and the Series 60 saw 64 professional car bodies being mounted on it. Flxible, of Loudonville, Ohio, specialized in Buick coaches although they did build 14 cars on the Pontiac model 40 chassis this year.

Siebert continued to produce a diversified line of professional cars on the Ford chassis. The range included the funeral coach (shown) at $1,420, the ambulance for $1,455, the combination hearse and ambulance at $1,470 and a service car for $1,350. They advertised for the funeral director to relieve himself of the costly burden of expensive equipment. The Shop of Siebert was located in Toledo, Ohio, and had been involved in coachbuilding of one type or another since 1853.

This unique service car wears a body of unknown origins. The photo was supplied by John Conde, the historian at American Motors, who could give us no further information on it. Mounted on a Hudson Six chassis, this car features an oddly shaped rear window and smooth, unusual body lines. The nameplate reads "Tulare Funeral Home," and the car has little coach lamps on the rear body sides.

1937

At the National Funeral Directors Convention held in the fall of 1936, at Louisville, the new 1937 professional cars were previewed, with some rather startling innovations among them.

The transportation of flowers to the cemetery had for quite a number of years plagued the funeral director, who many times had to hire several open phaetons to carry the floral tributes. Later the hearses had built-in arch-type flower trays in the casket compartment just above the casket. Later still, smaller flower trays were hung along the side walls of the coach so that the flowers could be seen through the windows. In this year the Henney Motor Co. tackled the problem and introduced a coach specifically designed to transport flowers. Built on a Packard commercial chassis, this car had a simulated convertible-coupe roof over the driver's compartment and a long, low open deck at the rear, fitted with a series of movable trays to accommodate the various types of floral tributes. When not in use as a flower car, this vehicle could quickly double as a useful first-call or service car. It had a drop-style rear door (tail gate) equipped with casket rollers and a full-length compartment that was large enough to carry a casket or a cot. An immediate success, other manufacturers were soon to add flower cars to their model line-ups. Another Henney pioneered feature to debut this year was the Leveldraulic, or vehicle leveling hydraulic system. The leveldraulic, patented and used under the original trademark by competing makers for several years, was a manually operated electric-motor-powered device that leveled the car from side to side after it had been loaded. It was also used to level the vehicle while it was standing at a curb loading, eliminating much exertion by the pall bearers loading or unloading the coach.

New for 1937 was the Flxible Challenger service car built on a GMC panel truck chassis. Featuring elaborate scroll work on the body sides and a rather plush interior for a truck, this was another Flxible model that only saw one year.

The 1937 A.J. Miller models were available on various chassis, among which Oldsmobile had become a standard offering. But it was the Miller-LaSalle that attracted the most attention. This model was reported to have "out-shone and out-sold the field." All previous production and sales records were broken at Miller for the 1937 model year.

In this year Sayers & Scovill made the complete switch to the Buick long wheelbase commercial chassis for all models. While continuing to use the date marks on the grilles, the cars were much more easily detectable as Buick chassied coaches. In addition to the Romanesque carved coach introduced by S & S last year, the company launched two new carved type funeral coaches in this year. These were the Athenian and the Byzantine. S & S also demonstrated a unique flair with names.

For the fourth year in succession, production of funeral cars and ambulances in the U.S. rose substantially, as 3,745 vehicles rolled out of various plants.

The S & S line was now mounted on the Buick chassis with an extended Buick frame. The radiator grille and hood louvers were modified, and they placed date marks over the grille to make it harder to distinguish the annual model changes of the cars. These date marks were always the same, especially the chrome ornamentation at the top of the grille. This was the S & S Romanesque carved type funeral coach that was quite popular in these years.

Another S & S carved model was the Byzantine that was built on the Buick Limited chassis with the longer wheelbase. This car, although generally similar to the Romanesque, was a larger, more ornate and more expensive carved car, and one of the top offerings from this eminent Cincinnati coachbuilder.

S & S called this new Manchester limousine funeral car matchless in the more moderate price class. The car was a stately looking limousine funeral coach with large windows and heavy drapery. The date marks on the grille partly eradicate the Buick heritage. These cars also wear S & S hub caps.

A companion to the Manchester was the S & S Springdale ambulance that was available with rear compartment air conditioning. The car, wearing date marks on the grille and S & S hub caps, features special rear quarter window treatment and spotlight on the A pillar.

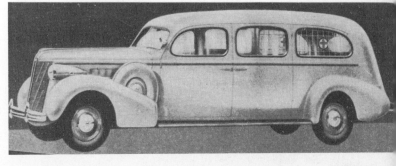

1937

In a move that was to affect the whole future of the funeral coach, Henney introduced this first flower car that was designed from the ground up as a flower car. Unfortunately this rather poor photo is all that could be found of this interesting vehicle. These cars had a convertible type top with a low rear deck that was complete with rails to keep the floral tributes from sliding off the stainless steel deck. The car was mounted on the Packard chassis only and was quite popular for leading processions. It also had a large compartment under the flower deck that could be used for carrying equipment or for first calls.

Henney called their 1937 Progress series "bigger and better at no extra cost." The Henney Progress was still mounted on the Oldsmobile chassis and this was the last year in which Henney would use any chassis other than Packard. In the fall of the year, Packard, annoyed at the success of Cadillac and LaSalle in this market, handed down an ultimatum that any maker wishing to use Packard chassis could not use any other chassis than Packard. Henney was the only maker to accept this directive and concentrated on the Packard vehicles. This last of the Henney Progress Oldsmobiles once again featured the Henney Nu-3-Way loading system with Electdraulic operation being offered at extra cost.

Some funeral homes used all Henney and Packard equippage, as did the Heinen & Son Funeral Home shown here with their fleet. The Henney Packard limousine funeral car is about a 1935. They were one of the recipients of the stylish Packard Henney flower cars just introduced. The other two cars are Packard livery limousines that were built by Packard especially for the funeral director.

Designated as Henney Packard Model 864, this ambulance was mounted on the Packard 120-BA chassis. This year, Henney built 1800 professional coaches of all types: hearses, ambulances, service cars, flower cars and combinations. One of Henney's direct factory distributors was an old hearse manufacturer, A. Geissel & Sons of Philadelphia.

Continuing to wear the largest carved panels in the industry, Eureka offered this carved side-way funeral car on the Cadillac chassis. This coach was engineered and styled by custom craftsmen and was supposedly unrivalled for beauty and quality.

The same coach, shown open, illustrates the plush interior appointments and the workings of the famous Eureka casket table. The car was available with the rear loading feature only if desired, and cars fitted with the straight rear loading feature had a partition between the front and rear compartments.

Eureka built this ambulance and mounted the body on a LaSalle chassis. The car contained every feature for rendering perfect service. All Eureka coaches were sold on a direct-to-buyer basis at attractive prices. There were no dealers at this point of coach development. Eureka also mounted coaches on Buick chassis or any other suitable frame. Notice the paint scheme of this car, the twin spotlights, and the roof mounted siren/warning lamp. A frosted glass pattern was applied to the rear quarter windows.

1937

The A.J. Miller Art Model was always an attractive coach, and this version mounted on a Cadillac chassis upholds this reputation. Styling of the Art model had changed very little since its introduction in 1934, but it continued in its popularity.

The Art Model was not only available as a funeral coach, but also as an ambulance, as seen here with this Art mounted on a Chrysler chassis. The center window in the rear compartment contains frosted glass with a pattern etched on it. The center of the arch at the top had the owner's initial T. This car was owned by the Trevey Funeral Home.

This Miller limousine funeral car was placed in direct competition with the Pontiac chassied offering from Superior. The car doesn't have the smooth, graceful lines that adorned the Superior Pontiacs, but rather had a high utilitarian look. The nameplate in the rear window read "Updike." Miller didn't always put drapes in the center rear side windows.

Mounted on the LaSalle chassis, this Miller ambulance was of more orthodox limousine configuration. This car has leaded glass in the rear quarter windows and side mounts. Finished in white overall, this car was a very handsome vehicle in its day and the ultimate in emergency care. LaSalle and Cadillac had become very popular chassis for this type of vehicle and their popularity was increasing each year.

Standing in front of one of his creations, John Little of Ingersoll, Ontario, poses proudly. This photo was taken outside Mr. Little's Shell service station where he hand-built custom funeral cars and ambulances that were to become so popular all over Canada. This coach is seen mounted on a LaSalle chassis.

On the lower end of the Miller price scale was this Chevrolet limousine funeral car. Miller still had not adopted an all-steel top for its cars. A glass partition was fitted between the front and rear compartments.

The Superior Pontiac Guardian ambulance was mounted on the Pontiac 6 chassis and received the same flowing lines that characterized the whole Superior line. Notice the clean sweep of the rear end lines and how the body blends with the Pontiac chassis, cowl, fenders and styling. Notice, too, the leaded glass pattern in the rear quarter windows.

The Superior Pontiac Oakridge and other Superior three-way models now featured the more popular type of limousine styling, with sidemounts and formal draperies all around. Notice the unique type of casket roller that Superior used for their side loading models. This car looks like some of the more expensive offerings from other makers.

Superior was now concentrating its advertising dollars toward the promotion of Pontiac chassied cars. This Oakridge three-way funeral coach was the middle offering in the Superior Pontiac line, and was of conventional limousine styling complete with sidemounts. The car was mounted on the Pontiac 6 chassis.

Mounted on the Pontiac 8 chassis and extended like all Superior Pontiac models, this was the top of the Superior Pontiac range. It features mildly restyled frontal treatment with parking lights mounted on the front fenders. Also offered in the Pontiac 8 Superior line were Benevolent ambulance and the Woodlawn service car. The bumper name plate had been changed from the Pontiac Indian insignia to a Superior emblem, and the hub caps all carry the Superior nameplate instead of Pontiac.

At the lower end of the Superior Pontiac offerings for this year was the Rosehill service car, mounted on the Pontiac 6 chassis. The new Rosehill had ornate service car wreaths and other decorative touches on the rear body sides, and a carriage lamp mounted on the rear side doors. The style of the door-mounted nameplates were new for 1937.

Although Superior was now more oriented toward the General Motors chassis, they did not completely sever their ties with Studebaker, as is evidenced by this strikingly attractive Superior Studebaker carved type hearse on the Studebaker chassis. This coach was mounted on one of the Studebaker 6-cylinder chassis, and could have been a dual purpose car for use as either a hearse or a service car. The carvings on this car were stamped metal. Some screw heads are visible on the carved panel. The whole panel was removable to convert this coach to a limousine style car.

Superior did not abandon the Studebaker chassis altogether as is evidenced by this attractive Superior-Studebaker limousine funeral car with full velour draperies and side mounts. This car was a side loader and white wall tires were optional.

Utilizing one of the most attractive Buick chassis of the era, Flxible offered their coaches in two series, the 40 and the 60. This extremely attractive ambulance was one of only 69 coaches mounted on the Buick 60 chassis by Flxible. Notice the duo-tone paint job, the fog lights on the bumper, spotlights and raised hood line. To complete the picture the car wears whitewall tires.

Introduced for the first time in this year, Flxible entered the carved funeral coach field. This model called the Classic A, featured two small arches on the body sides within which were carved draperies. The frame around these areas and the space between them were filled with elaborate ornamentation. This coach was a radical departure from the other styles of carved cars and opened a whole new era in carved car styling. This Classic A is mounted on the Flxible Buick chassis, Series 40, and was but one of 237 cars built on this chassis this year.

This unusual Flxible service car could double as a hearse in an emergency situation. The car was of the usual Flxible limousine body style with the rear compartment window areas filled in with carved panels. Altogether Flxible built 306 vehicles in this year and they were all on the Buick chassis.

Also mounted on the Series 40 chassis was this attractive Flxible service car that was complete with wreath decoration on the rear body sides and duo-tone paint finish.

The Shop of Siebert also offered a line of Ford DeLuxe coaches in this year. They were offered as a funeral coach for $1,495, a combination for $1,520, an ambulance at $1,500, and a service car for only $1,350. Seen above is the Siebert Ford DeLuxe funeral coach that was complete with fender skirts. National Hearse & Ambulance Co. of Toledo were the exclusive distributors for all Siebert vehicles.

The Shop of Siebert in Toledo, Ohio, was continuing to turn out interesting coaches on the Ford chassis. This Ford funeral coach was the standard model in the Siebert line, and sold for $1,470. Notice the drapery.

1937

Mounted on a Cadillac Series 60 chassis, this Meteor ambulance was delivered to the American Ambulance Company. The car features a duo-tone gray finish, side mounts and fender mounted parking lamps. This photograph was taken in front of Meteor's plant in Piqua, Ohio.

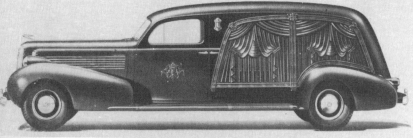

Meteor built this attractive ambulance on a Cadillac chassis. This car is complete with roof mounted siren/warning lamp unit, spotlight, and leaded glass in the rear quarter windows.

Meteor's limousine funeral cars were as attractive as any made, and this one mounted on a LaSalle chassis is no exception. The full formal draperies and the light color all make the car look larger than it is. Notice the flower trays in the rear compartment.

Meteor was also offering carved side funeral coaches in side loading models. This attractive coach also features a very large area covered with carved drapery panels. Meteor also included a carved initial panel on the front doors and carriage lamps on this model. This particular car is mounted on the LaSalle chassis but the style was offered on both Buick and Cadillac.

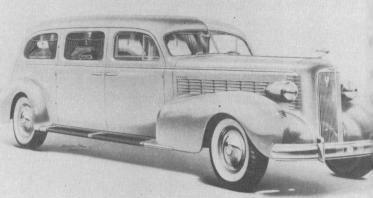

Seen here are three mystery flower cars all mounted on LaSalle chassis. These cars were found in a Cadillac advertisement of this year and there was no identification as to body maker. Styling is such to hint at their being Meteors. It is proof positive that the flower car was becoming very popular with more than one maker.

Canada's major coach builder, Mitchell of Ingersoll, Ontario, turned out this interesting hearse this year. Although the belt line seems rather high, the car was mounted on the Packard eight chassis and was complete with sidemounts and formal draperies.

This attractive Packard 120-CA ambulance was built by Knightstown and features whitewall tires, a bumper mounted siren, spotlights and is finished overall in medium gray. Notice the interesting ambulance pattern that they used in the rear quarter windows. Small coach lamps adorned the sides between the doors.

1938

In the fall of 1937, the Packard Motor Co. served notice to the manufacturers that they must discontinue use of all chassis other than Packard. During the twenties, the coaches that used the Packard chassis enjoyed a wide popularity, but with the entrance of Cadillac and La Salle into the business of supplying chassis to the ambulance and hearse field, Packard sales had greatly diminished. This rather heavy-handed ultimatum was flatly refused by every manufacturer in the field with the exception of Henney. While Henney was content to build only on Packard, the other makers developed a strong line of equipment based on General Motors chassis, eventually standardizing on Buick, Cadillac and Oldsmobile. Out of this came a rather interesting situation with Henney in Canada. Brantford Coach and Body Limited of Brantford, Ontario, had, several years earlier, become the official Henney distributors for Canada. When Henney accepted the Packard "offer," it obviously did not apply to the firm's Canadian operations. Brantford continued to mount Henney bodies on other makes of chassis for Canadian consumption and marketed them under the name Brantford-Henney. Meanwhile, Henney developed the first air-conditioned ambulances equipped with a genuine mechanical refrigeration system, such as was utilized for the air-conditioning of railroad cars and passenger busses. The first such equipped vehicle was delivered to the Kreidler Funeral Home in McAllen, Texas, with another going to L.T. Christian Inc. of Richmond, Va.

Superior Body Co. of Lima, Ohio, again scooped the funeral coach and ambulance industry with the introduction of the first all-steel, welded construction coaches. This was closely followed by Miller and Meteor cars with the same feature. Superior also expanded their multi-purpose coach to include the Pontiac line.

Another of the innovations that was to spark a whole new wave of vehicles throughout the industry was introduced in this year by Sayers & Scovill. With the advent of the new Victoria, the industry was sent scattering within the next two years to imitate again another new style. The Victoria was a throwback to the old leather back or landau style of the early twenties. But this time the rear quarters were completely blanked off and ornamented with a pair of large landau irons. The roof of the Victoria was padded to add to the luxurious appearance. S & S introduced yet another carved type coach in 1938 with the Damascus. S & S had switched to the Buick chassis and body components in 1937, but in 1938 they switched all of their models to the Cadillac manufactured chassis, using both the Cadillac and the La Salle styles.

Of the carved coaches of this era, the Eureka had the largest carved area on the sides of the body. This carving was enclosed in a large arch on the rear body sides and was in the form of massive draperies.

For the first time in four years, the production of ambulances and funeral cars in the U.S. took a drop. And, it was a substantial drop, with only 2,529 such vehicles leaving the various shops.

Henney, now building coaches on the Packard chassis only, offered this stately looking funeral car on the Packard 1601-A chassis. In 1938 Henney built a total of 1000 Packard chassied coaches of all types. This car is a three-way loader, with Electdraulic on the table. The Henney Packard enjoyed a great amount of popularity in these years.

Henney offered this incomparable three way funeral car with Elecdraulic assisted casket table in this year. This car is of the limousine style and this photo clearly shows the magnificent interior used on these cars.

Offering mechanical air conditioning for the first time, Henney introduced their 1938 line of ambulance models. This Henney Packard ambulance was also mounted on the 1601-A chassis and had leaded glass in the rear compartment quarter windows. The sliding glass partition is clearly visible.

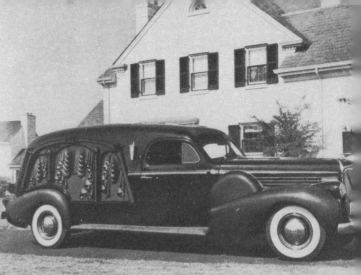

S & S of Cincinnati was the maker carrying the big card for 1938 with the introduction of a complete new line of models, a new styling innovation that was to sweep the industry, and settling on a new chassis line. This car called the S & S Victoria, was the newest and most impressive car ever to be announced by any maker. Its impact is still being felt 35 years later. The styling of this car was called a landau. The top was covered at the rear with padded fabric and the rear quarter panels were covered with a large landau bow and a small carriage lamp. Beginning with this year's production, S & S mounted their cars on either Cadillac or LaSalle chassis exclusively. Notice the date marks on this LaSalle grille and the S & S hub caps.

This is the S & S Damascus funeral coach of more conservative styling and fully enclosed. Having this coach at the head of a procession added new dignity to this solemn rite. The Damascus properly symbolized the old values and traditions, and this was the interpretation of the carving work on the body sides.

Called the Woodlawn, this was yet another new model in the S & S line up. Mounted on a LaSalle chassis, this car was of straight limousine styling. Notice the date marks on the radiator and grill and the S & S hub caps. Other models of S & S limousine funeral cars were known as the Emerson and were available on either the Cadillac or LaSalle chassis.

The ornate carving work of the Damascus was carried throughout the car and is to be seen even on the rear door. Carved drapery panels were set into an area below the rear door window and fancy carved type trim was applied to the area above the rear window. Notice the S & S insignia on the rear bumper in place of a LaSalle emblem.

Called the Damascus Town Car, this S & S model featured carved panel sides with a town car style. Once again this coach is mounted on a LaSalle chassis and there are date marks and S & S hub caps applied to this car. The glass partition was etched with a sunburst pattern and the coach was a rear loader only.

This dignified ambulance was called the S & S Lakewood and was complete with the S & S air conditioning system that represented a triumph in S & S engineering. This picture clearly shows the S & S date marks on the grille. This car is equipped with spotlights, sidemounts, fog lights, and plain rear quarter windows. Notice the S & S nameplate in the center of the bumper in place of a LaSalle ornament.

1938

This A.J. Miller Cadillac town car ambulance was built for Claude A. Lord and featured Miller styling that was prominent on their limousine coaches. It had ornate leaded glass pattern in the rear quarter windows and small carriage lamps on the sides.

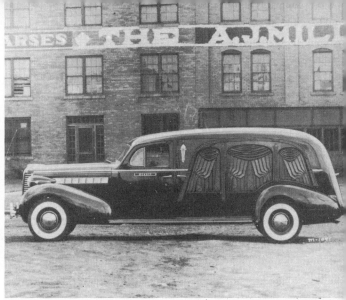

Continuing with their successful Art models, this is the Miller Art Carved model for 1938, mounted on a Buick chassis. The draped carvings within the arch were stamped steel, but they retained the look of hand carved wood. Complete with whitewalls, this car carries a Morris nameplate.

This Miller service car was also mounted on the Chrysler Royal chassis. The completed unit was delivered to the Cavanagh Funeral Home. Looking very much like a Chrysler panel delivery truck, this car had only three doors and a decorative panel on the rear body sides.

Mounted on a Chrysler chassis, this Miller ambulance had frosted glass on the lower portions of the rear quarter windows. The warning light on the roof is unusual and the car also has a combination siren and warning light on the fender. A pillar mounted spotlight and nameplates for the owner complete the package.

This Meteor LaSalle ambulance styling was exactly like that of the limousine funeral cars. The ambulance had the frosted glass rear quarter windows and interior details that differed from the funeral car version but otherwise they were identical. This car is very appealing with its white finish and whitewall tires.

Meteor offered this attractive carved side funeral coach on the LaSalle chassis. Once again the unit was complete with a carved type decoration on the front doors, massive carved drapery details on the rear body sides, a coach lamp and whitewall tires. The rear side door is practically invisible except for the handle. It was hinged in the center of the carved panel area.

1938

Flxible entered this carved coach to the trade and the firm began to mount bodies on Cadillac and LaSalle chassis as well as Buick. This carved funeral car was one on 20 Cadillac chassied coaches built by Flxible this year.

Flxible continued the unique Classic A model in two versions — with windows in the arches or with carved panels in these areas. This Classic A is mounted on a Cadillac 60 Special chassis. This year Flxible mounted 20 professional car bodies on Cadillac chassis, 81 on LaSalle chassis and 209 on the Buick 40 chassis. An additional 45 cars were built on the Buick chassis.

Called the Flxible Challenger, this service car was mounted on a Chevrolet truck chassis and carried special decorative ornamentation on the rear body sides. A carriage lamp was also mounted on the side to add distinction. This year Flxible built a grand total of 373 professional cars of all types, of which 18 were Challengers.

This illustration of a Eureka Buick shows clearly just how high these coaches were. This Chieftain Buick three way limousine funeral car was built to high standards of quality and engineering. Eureka also offered their cars on Cadillac and LaSalle chassis.

Carrying the largest area of carved panel of any coach in the industry, this Eureka carved three-way funeral coach was mounted on a Cadillac chassis. Notice the little ornament on the front door and the carriage lamp on the rear door. Rock Falls, Ill., was the home of Eureka, which had been in this field since 1887.

This strikingly handsome ambulance on the LaSalle chassis was built by Eureka. It featured shades over the rear quarter windows and a spotlight on the windshield pillar. The beauty of this Eureka LaSalle ambulance was just a reflection of its high quality and refined appointments.

1938

Derham, of Rosemont, Pennsylvania, built very few professional cars and when they did they were bound to be innovative and unusual. This Derham ambulance was mounted on a Lincoln Zephyr chassis for the Chester Hospital of Chester. Their cars were built on special order only and this feature makes this car all the more rare and interesting.

Setting another first in the industry, was Superior Body of Lima, Ohio, with the introduction of the first all-steel welded body construction. Previously the bodies of all makes of funeral cars and ambulances were of wood framing but now Superior had taken a step forward and used welded steel for their body frames. This Superior carved coach on the Cadillac Series 60 Speical chassis was a distinctive looking car with massive carved drapery panels that were stamped steel, coach lamps, and drapes on the central partition window and the rear window. The doors for the rear compartment were hinged in the center of the carved area.

Superior continued to produce a strong line of vehicles on the Pontiac chassis, and this limousine funeral car is a fine example of Superior styling and coachwork. This car was mounted on the Pontiac 8 chassis and enjoyed a considerable popularity.

Called the Elmwood, this Superior Pontiac had carved panels on the rear body sides that were easily removable. Removing these panels would quickly convert the coach into a limousine style car. Lawson & Son nameplate is on the running board.

This Superior Limousine style funeral coach is also seen on a Cadillac 60 Special chassis. Notice the extremely wide rear side door and the side mounts. This was a side servicing coach with the unique Superior casket roller system.

Another limousine style funeral car from Superior, on a LaSalle chassis. The rear side door width is even more visible in this shot, taken in front of the Lima Fire Hall. White wall tires and side mounts really set this coach off.

1938

This year, in addition to the demand for Bender ambulances and funeral cars at home, Bender Body Co. received a great rush of orders for ambulances from foreign countries. These orders were for air conditioned ambulances and came from South Africa, Columbia, South America, Ceylon, Canada, and Great Britain. Here a fleet of Bender Studebaker ambulances is ready to be shipped overseas.

When Superior withdrew from using Studebaker chassis and went to General Motors, Studebaker made arrangements with the Bender Body Co. of Cleveland to build coaches on the Studebaker chassis. Of course when they (Studebaker) switched makers, the traditional Studebaker professional car names went along. Shown here is the distinctive Bender Studebaker Westminster carved type funeral coach. Note the carved area with carved draperies around the edges and a cross in the center. Step plates and fender skirts were standard on this coach, as was the Bender nameplate on the hood.

The sedan type ambulance and funeral car had all but disappeared from the United States, but in Canada the Mitchell Co. of Ingersoll, Ontario, built this sedan type Chrysler hearse. The car was a side loading coach only and the rear door or trunk lid was practically unused. There were some of these cars built as three way coaches.

In Canada the Mitchell Company of Ingersoll, Ontario, was still turning out high grade funeral coaches and ambulances on almost any chassis. This car has the original "FUNERAL COACH" sign in the windshield that was standard equipment for all coaches of this era. The car is mounted on the Packard 1601-A chassis.

The hearse and ambulance carftsmen of Weedall Nicholson of Halifax, England, built this interesting and attractive coach this year and mounted it on a 1929 Buick chassis. Note the etching in the heavy French plate glass window in the rear compartment, the coach lamp, and the flower rack on the roof. This car was one of hundreds of Buick chassis exported to the UK in this era, and has the British regulation fender mounted parking lamps.

This 1928 Rolls Royce Phantom I chassis wears a 1938 Woodall Nicholson body. This Halifax firm was renowned throughout England for high quality work. Note the etched glass in the rear side windows, the coach lamp, and the flower rack on the roof. Styling is typical of what was popular in the UK at the time and the chassis undeniably the finest.

1939

The decade was wrapped up with the introduction of some rather special models from several makers. Certainly the most attractive emanated from Superior. In its last town car creation, Superior built one of the most attractive versions of this style ever seen. This coach, mounted on a 1939 Cadillac chassis, was first displayed at the Michigan State Funeral Directors Convention and was indicative of the type of opportunities afforded funeral directors to personalize their funeral cars. This unit was designed exclusively by Superior craftsmen and used a Cadillac Fleetwood town car front end. The exterior was finished in black with all exterior mouldings, windshield frame, side mount tire brackets, radiator grill, hood ornament, coats of arms and name plates, all gold plated even to the license plate frames. Coats of arms with small crest were on the front doors evolving a new idea in dignified distinction. The casket compartment was fitted with the Superior side-loading table and was trimmed in turquoise mohair, with hand-trimmed draperies in green velvet with silver fringe. This car was especially built for Ralph J. Balbirnie of Muskegon, Mich.

New carved models were introduced by both Miller and Meteor and were quite similar in configuration. Miller's new carved coaches were called Cathedral models and featured three panels shaped like gothic church windows on the rear body sides. The Meteor version was called the Gothic Carved and featured the same type of church window shaped carvings but they were enclosed within a large arch. The difference was that the Meteor version's window insets contained stained glass with removable panels (with gothic-filigree centers) to facilitate washing.

Although Henney never entered the carved car set, the company introduced a new version of the standard limousine to counter the carved cars sold by the competition. The Henney "Formal Limousine" had large decorative metal panels in the center of the side doors with the same treatment duplicated on the rear door and with woodwork on the inside of the car.

With the opening of the 1939 World's Fair in New York City, Flxible announced that its ambulances had been exclusively chosen for the official ambulance service at the Fair. The coaches used at the Fair were of a particularly distinctive type. They counted among their standard features Bomgardner cots, special rubber mattresses, walnut medicine cabinets, two attendant seats with spring cushions, hot water heater, a Federal siren with flashing light, a spot light, and six-ply tires.

Total U.S. production of funeral cars and ambulances went up slightly this year, but only to 2,702 for the total industry.

Fully open and showing the interior details is the large Eureka carved side LaSalle chassis funeral coach. The Eureka casket table is swung fully open.

The trend setting S & S Victoria continued in popularity and is seen here mounted on a LaSalle chassis. The whole roof of the Victoria was now fabric covered and padded for added distinction. The landau bows and carriage lamps were retained. Notice the unique S & S date marks on the grille and the S & S insignia in place of the LaSalle emblems.

Available in either side servicing or rear servicing models, the S & S Masterpiece was mounted on either Cadillac or LaSalle chassis. The name Masterpiece had been resurrected from earlier S & S models for use on this new carved coach. Once again notice the S & S date marks on the front end, the S & S logos on the hub caps and hood louvers. The carved panels were now of cast aluminum and looked as detailed as the hand carved wooden ones.

1939

Supplied to the Cincinnati General Hospital, this S & S was among the very first high headroom ambulances ever designed. Mounted on a Cadillac chassis and fitted with almost every conceivable type of warning light, siren, and fog light, this car was finished in red with gold lettering and a gold crest on the driver's doors. This vehicle could easily accommodate four injured patients.

S&S offered the three way side servicing coach as an option on all of its models in this year. The casket tables were hand fashioned in a little adjoining shop at the S&S Cincinnati plant. Here we see this excellent example of the woodworker's art complete with inset rubber casket rollers and bier pins.

The S & S Parkdale ambulance was one of the standard S & S offerings in the ambulance line. This car came with a completely equipped patient compartment and pull down shades in the rear door windows. The leaded glass was retained in the rear quarter windows. This car is fitted with a bumper mounted siren, and spotlights. The S & S frontal treatment on this Cadillac chassied ambulance is easily seen here.

This S & S Stratford ambulance was especially designed to meet the requirements of Phillips-Robinson & Co. Funeral Home. Besides being equipped with every known device and facility for saving lives, this ambulance was equipped with special red warning lights mounted in front, and on top. The car also has a uniquely patterned rear quarter window and a set of whitewall tires.

Introducing their own flower car this year, S & S came out with this design. Mounted on a LaSalle chassis, complete with all of the S & S identification marks, this new flower car was designed almost identical to that announced in 1937 by Henney. The low, sweeping flower deck has a railing to keep flowers from falling out, and the body is of coupe style instead of a mock convertible like the Henney.

The dramatic and luxurious S & S Damascus carved type funeral coach was continued in the S & S production schedule through this year. This was the very top of the S & S carved car line, with beautiful genuine walnut paneled interiors and etched partition windows. This Damascus is mounted on the Cadillac chassis and contains a crest on the front doors for the owner's initial and nameplate.

1939

Here is the last town car funeral coach that Superior Body Company of Lima, Ohio, was to build. The town car was a copy of the same style seen on passenger cars of the premium class and they were becoming extremely expensive since the all-steel body had come into its own. This exceptional vehicle was finished in deep black with all of the exterior mouldings, windshield frame, side mount tire brackets, radiator grille, hood ornaments, coats of arms, nameplates and even the license plate frames finished in gold plate. The coats of arms, with small cast nameplates, are mounted on the front and rear doors and were a new idea in dignified distinction. The casket compartment was fitted with the new Superior three-way casket table and was trimmed in turquoise mohair with hand tailored draperies in green velvet with silver fringe. The front part of the car was that of a Cadillac Fleetwood town car, while the chassis was that of a Cadillac commercial car. The car was designed by Superior's design department and was built for Ralph J. Balbirnie of Muskegon, Mich. It was shown at the Michigan State Funeral Directors Convention, where it was the center of attention.

This Superior Cadillac limousine funeral car was devoid of the gold plate and fancy trimmings that the special Town Car coach had, but was in every way a fine example of Superior craftsmanship. This coach was delivered to the Century Burial Assn. and featured side mounts and end loading.

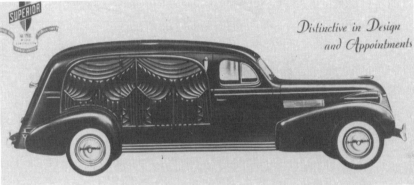

Offering their coaches on both Cadillac and LaSalle chassis, Superior offered this attractive carved model called the Tarrytown. Mounted on a Cadillac chassis, this car makes a very impressive sight. Notice the width of the rear side door that is hinged just forward of the rear fender. Carved drapery details were stamped in metal and small coach lamps were standard.

Called the most distinguished carved funeral car ever presented to the profession, the new Meteor carved funeral car was made available on the Cadillac, LaSalle or Buick chassis. This was the biggest styling stride ever taken by Meteor. Shown on a LaSalle chassis, it featured carved panels of early English gothic, inspired by the world's most beautiful cathedral windows. Within this ornate filigree were stained glass windows. The carved panels that covered these could be removed to facilitate washing.

This Superior Cadillac ambulance is seen with unusual roof lighting, color scheme, siren, spotlights and side mounts. Other features of this Superior coach include frosted glass in the rear quarter windows and a nameplate that reads "Drake Braethware Company."

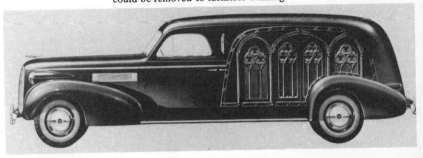

Meteor also used the Buick chassis as one of the standard offerings. Here are two of the 1939 Meteor Buick offerings, the limousine funeral car and the ambulance. The Meteor Buick hearse was designated as a model B41-520 while the ambulance carried the designation B41-521.

1939

In front of the Meteor Motor Car Co. office is this Meteor Cadillac ambulance. The coach with roof warning light pods, spotlight, and whitewall tires, is a very handsome vehicle.

Bender Body Company of Cleveland and Elyria, Ohio, branched out to build coaches on chassis other than Studebaker this year. This model, called the Washington, was their top offering in the carved coach line. Seen here mounted on a Cadillac chassis, the car features three cathedral shaped areas on each side and a huge coach lamp that was almost as high as the body of the coach itself. A special nameplate was mounted on a carved ornament on the front doors. Bender called this the most aristocratic and distinctive funeral car in America.

This beautiful Meteor flower car, mounted on the LaSalle chassis, was one of the striking styling achievements of the year. The car could be used as a flower car, service car or as a hearse. There was ample room under the flower deck to carry a casket. The side doors, when opened, were wide enough to allow loading of chairs or other articles. The car was of graceful coupe styling with draperies hung in the rear quarter windows of the driver's compartment.

This Bender Cadillac combination coach features a sliding casket extension table that was removable for rear loading. When removed, it revealed an ambulance floor. This car was also available with removable carved side panels similar to those used by Superior. The peaked roof section over the windshield was a Bender trademark. This body was available from Bender mounted on Cadillac, LaSalle or Buick chassis.

Another Bender ambulance is seen here on the LaSalle chassis. Notice the illuminated ambulance sign over the windshield and the warning lamps mounted in roof pods. In this view, the Bender peak over the windshield is easily seen. Bender used this area to mount a special air intake for the ventilation system.

The Bender Studebaker Samaritan ambulance was mounted on the same chassis as the Bender Westminster hearse. These cars shared the Studebaker President 8 chassis while the Arlington used a Studebaker 6 chassis. Note the frosted glass in the rear window quarters, high mounted sidemounts, spotlight, and the Federal combination warning light siren on the roof. Fog lights are seen mounted on the front bumper.

Called the new mode in funeral car styling and design, this Bender Studebaker Westminster funeral coach features a peak over the windshield and sidemounts. Notice the wide rear doors and the whitewall tires. Other models in the Bender Studebaker range were the Arlington funeral coach, Elmwood service car and the Elmhurst service car.

1939

This interesting looking Henney ambulance was completely equipped with sidemounts, windshield ambulance sign, and a Federal combination siren warning light. Finished in white, it is quite attractive.

Henney, building coaches on the Packard chassis only, had never gone in for the elaborately carved styles that were offered by all of the other makers, sticking instead to a line of solid limousine style funeral cars and ambulances. This year they introduced a model that was as close to a carved type car as they ever came. Called the Henney Packard Formal Limousine, this car had decorative plaques mounted in the center of the side doors and on the rear door. These plaques also held the owner's nameplate. This particular model, in original condition, is owned by Walter M.P. McCall of Windsor, Ontario. His name appears in the nameplate area on the side doors.

Henney claimed that their Packards had custom built quality at production prices and that this Henney Packard Nu-3-Way, Leveldraulic, Elecdraulic, funeral coach was the choice of particular funeral directors. The Henney patented Leveldraulic system kept the coach level throughout the loading and unloading procedure and while the car was moving with a load, no matter what the pavement angle or condition. The Henney patented Elecdraulic system was an electrical assistance for the operation of the three-way casket table. This year Henney built a total of 1,200 coaches of all types on the Packard 1701-A and 1703-A chassis.

This interesting Cadillac chassied coach is the product of an interesting international concern. Henney had their cars assembled and distributed in Canada by the Brantford Coach Co. of Brantford, Ontario. While the agreement with Packard to mount their bodies on Packard chassis only was binding in the U.S., it was obviously not in Canada where this Henney Formal Limousine body is seen mounted on a Cadillac chassis. Cars derived from this mating were called Brantford-Henney because they were Henney bodies assembled and mounted by Brantford Coach. This car was owned by the Thorpe Brothers Funeral Home of Brantford, Ontario.

This Brantofrd LaSalle funeral coach maintained the high traditions set by earlier Brantford models. The Henney-type Nu-3-Way table is swung out and the interior details are clearly visible.

Designated the Brantford Model 3692, this was a body built by Brantford and mounted on chassis by them in Canada. This Brantford Cadillac 3-way side servicing funeral coach was a beautiful example of the fine work done by this Canadian company.

1939

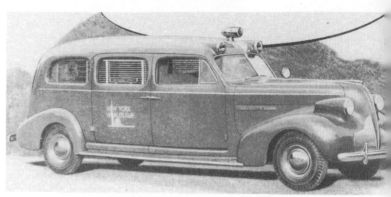

Loudonville's Flxible Co. was honored this year to be picked to supply the official ambulances for the New York World's Fair. They supplied the Fair with a fleet of blue and orange (the official Fair colors) Buick chassied vehicles. Many of these were built to identical specifications while others differed somewhat. This particular vehicle features warning lights in special roof pods, a Federal combination roof siren and warning light, spotlight, venetian blinds in the rear compartment windows, and special Fair insignias painted on the body sides. In this photo, the raised hood line is quite obvious and the headlamps were put on pedestals to keep them in proportion with the overall design. This coach was mounted on the Flxible Buick Series 40 chassis, on which a total of 145 were built.

This photo was shot at the New York World's Fair, with the fair trappings in the background. This car varies from other Fair ambulances in exterior details only. It has the roof mounted Federal combination siren and warning light, plus some other minor details. In this year Flxible built on two Buick chassis — the Series 40 on which 145 were built, and the Series 60 that saw 33 bodies mounted on it. Total Flxible production for the year was 275 vehicles, of which 22 were on the Cadillac chassis, 73 on LaSalle, 178 on Buick and only 2 on Chevrolet.

Flxible continued to build bodies for Buick, Cadillac and LaSalle. This limousine funeral car is one of 73 mounted on the LaSalle chassis. Notice the raised hood line and the striking effect that the whitewall tires lend this car.

Called the Flxible Sterling Service car, this coach was mounted on the Buick chassis only. The car, all new for 1939, wore distinctive styling that automatically marked it as a service car. The unusual ornamentation of the rear body sides and the small coach lamps highlighted the styling. The car was a very utilitarian vehicle that was also aptly suited to duties as an emergency hearse.

Called America's finest and most beautiful funeral cars, the Eureka limousine coaches were now mounted on either Cadillac or LaSalle. The drapes run along the very top of the side windows only. The flower trays are visible through the rear quarter window.

Siebert introduced a new service car in this year called the Aristocrat. Like all Siebert professional models, this was mounted on the Ford chassis. Notice that the rear quarter window has been replaced with a decorative panel and that the nameplate is mounted in the center window. Siebert offered a complete line of models; hearse, combination, ambulance, and service cars, all with relatively the same body styling.

1939

A.J. Miller was now building their cars on a strong line of General Motors chassis. The large LaSalle and Cadillac chassis were the most popular. Here one of the Miller Art Carved funeral coaches is mounted on the LaSalle chassis. This four door, five including the rear door, body featured a choice of either straight rear loading or three-way servicing. They continued the massive and ornate draped carvings within the arch of this model and the rear side door was hinged in the center of the carved arch. Coach lamps were standard on this model.

This Miller Art Model was continued virtually the same as the original version introduced in 1934. The arch contains three beveled glass windows and two columns. All three windows have drapes in them. This car is also mounted on the LaSalle chassis. Other chassis offered were Buick, Oldsmobile and an occasional Chrysler.

Another new carved type model introduced by Miller this year was the Miller Cathedral. This car features three panels on the rear body sides that depicted gothic windows in shape, and contained smaller stamped metal cathedral shapes. On either side of the car were two torches depicting the eternal flame of memory. The Cathedral model seen here is mounted on a LaSalle chassis.

Called the Miller flower car and utility vehicle, this new Miller model is shown here loaded with flowers on the rear deck. This model was also of coupe styling with a low rear deck. The tonneau seen on most flower cars contained a cover for the stainless steel flower deck. Under the flower deck there was room enough for a casket or a load of chairs or other necessary items. These cars were often used as service cars, then doubled as both flower car and hearse during the funeral service.

One of the last funeral coaches built by Woodall Nicholson before the outbreak of World War II was this interesting vehicle mounted on a Rolls-Royce Phantom III chassis. Streamlined for a tradtional styled coach, this car featured the typical large glass area in the casket compartment, and a roof mounted flower rack. Notice the clean flowing lines and the traditional British razor edge styling. The Rolls-Royce chassis was the best that was available, and when combined with the high quality craftsmanship of Halifax based Woodall Nicholson, the finsihed product was superb.

The International Harvester Company of Chicago showed this car, which was a conversion of their panel delivery truck. Called a combination funeral coach and ambulance, this car utilized the standard 125-inch International wheelbase of the ½-ton panel truck. Available equipment included venetian blinds, curtains and rollers, double faced drapes, lever type cot fasteners, a removable casket table, chrome plated attendant seats, medicine cabinets, spotlight, Federal siren, red emergency lights, heater, radio and air conditioner. The rear compartment floor was made of ship-lap lumber with an optional linoleum covering. The body featured four neatly installed safety glass side windows and two side pillar coach lamps. The interior was trimmed in imitation leather, mohair or velour, depending upon the customer's preference.

1940

One of the finest looking ambulances produced this year was the Miller LaSalle with roof warning light pods and cream finish. The pattern with the owner's initial in the rear quarter window was etched into the glass.

The Miller Art carved models were alive and thriving in 1940 with one example being seen here on a Buick chassis. Very little change had been made in the Art models since they were announced in 1934. This car was built for export to Quebec, and is adroned with the roof mounted crucifix that was the vogue in this Canadian province.

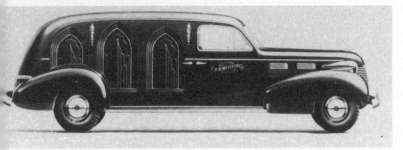

A.J. Miller of Bellefontaine, Ohio, continued to offer the Cathedral carved models on either the Oldsmobile, Cadillac (shown) or LaSalle chassis. These models carried a stylish nameplate on the front doors and torches between the cathedral window panels on the rear body sides. They were available as end loaders, straight side loaders, or three-way coaches.

Styling had made a complete about face with regard to the carved side car. Almost every maker offered this style in some variation and there was now a wide variety of carved type cars from which to choose. But now the carved panels were made of metal instead of hand worked wood. Some of these panels were cast aluminum while others were of steel stamped with the carved indentations made by huge presses. Superior offered a carved style coach that bore some similarity to the Cathedral and Gothic models introduced earlier by Miller and Meteor. Called the Tarrytown, this car featured simple, modern church window shaped inserts on the rear body sides. These were not filled with ornate filigree or drapery work but used a succeedingly smaller repetitive church window shape. Sayers & Scovill introduced another new carved model to the line in this year with the Majestic funeral coach. Mounted on the LaSalle chassis, it featured a "modern carved panel motif" that reflected the look of heavy draperies within an arch. The Victoria landau model was continued on the Cadillac and LaSalle chassis as both ambulance and hearse models.

The limousine style was greatly streamlined by many makers. Superior and Bender gave these styles lines not dissimilar to those being found on passenger cars of the era. Sloping backs, fender skirts and full body length character lines were now commonplace on funeral coaches and ambulances.

In 1940, Henney dropped the 800 series and adopted a model designation system that was to remain with them until 1954. This new system was based on model years. Henney also made big news with the introduction of what they called a "veritable singing chapel on wheels." This was a music box hidden under the hood and was said to "further aid the funeral directors in their efforts to conduct a beautifully-complete funeral service." Posters were provided dealers depicting musical sounds emanating from the vehicle parked next to a memorial tower. Any of the 1940 Henney Packard models could be equipped with this feature. An interesting vehicle Henney began to produce in 1939 was continued through this year and was a new innovation on the landau style. The model 898 three-way funeral coach was built along the lines of a custom landaulet passenger car, with landaulet construction resembling a fabric-covered folding top even down to the dummy bow joints. The side windows had a lowered or "chopped" appearance and both front and rear doors closed on the same post. This car almost resembled 4-door convertibles of the era and appealed to the truly style-conscious funeral director.

Again, production of professional vehicles in the U.S. was up, with 3,118 such cars being produced by the various shops in the country.

The landau funeral car introduced in 1938 by S & S with the Victoria was beginning to have an impact on the other makers. This year the A.J. Miller company made this body style available on their coaches. Their version of this type of car is shown here mounted on a Buick chassis. Note the small landau irons and the coach lamps. The roof was fabric covered just short of the windshield. This car is a three-way loader. It has fog lights mounted on the front bumper.

1940

Smith & Smith Funeral Home of Newark owned and operated this handsome Miller LaSalle service car. Although rather plain, this car was set off by its whitewall tires and light color.

This was the year that Miller made Oldsmobile one of their standard chassis. The Miller Oldsmobile, in rear loading limousine form, had draperies in the rear quarter windows only. Another model in the limousine's body style was the Miller Combination car that was suitable for use as either a hearse or an ambulance and could be switched over to either of its forms in a matter of minutes. Combinations were popular with funeral directors who could not afford two separate vehicles.

Owned by the Swedesboro (N.J.) Community Hospital, this Miller Chrysler ambulance features built in roof warning light pods, a Federal combination light-siren, and frosted glass in the rear quarter windows. This ambulance, like all Miller emergency vehicles, was completely equipped with all the facilities available for saving lives.

Some Miller bodies were mounted on Chrysler chassis, as seen here with a three-way limousine style funeral coach. The Chrysler chassis was sufficiently long so extending the frame was not necessary.

Bender claimed that for more than a score of years the finest in coach building had come from the Bender Co. In 1940 this firm, producing coaches on the Studebaker chassis and others, installed a new addition to their plant that was six times larger than the older part of the factory. They claimed that their facilities were second to none and that the craftsmanship and quality of their cars was unmatched. Unfortunately, there are no known Bender Studebakers to check this by. The car shown here is the 1940 Bender Studebaker Westminister funeral coach.

Bender of Cleveland continued to build coaches on the Studebaker chassis, and completely redesigned the line for 1940. The lines of the cars were smoother, and gone was the unusual air intake that was mounted in a windshield peak. The 1940 Bender Studebakers offered the funeral director distinction at a moderate price, and all of the service facilities of local Studebaker dealers. Note the full formal draperies and the fender skirts on this Westminster.

1940

Originally designed by Henney, the modern style flower car was designated a Model 4096 this year. It is seen here loaded with flowers. These vehicles were very versatile and were also quite dignified when seen at the head of a procession. Their popularity was the strongest in the metropolitan areas where funeral homes operated large fleets.

This view of the Model 4098 Henney Packard Landaulet shows the low overall height of the car and its extremely attractive lines. Note the seeming length of the roof line and the absence of a rear door. This car was a side loading vehicle only.

The most distinctive of the Henney Packard funeral coaches was this Model 4098 landaulet funeral coach. This car looked like a four door convertible with a chop job, and was one of the most beautiful of the Henney offerings. In this year Henney built a total of 1,200 coaches of all types on the Packard 1801-A and 1803-A chassis.

Henney and Brantford of Canada collaborated on this attractive ambulance owned by the James H. Sutton Funeral Home. This car features a white finish with a Federal combination siren and warning light, dual spotlights, and side mounts with the owner's name imprinted in gold.

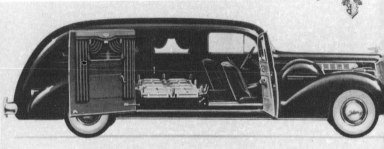

Henney said that this coach was worthy of two great names, Packard and Henney. This is the Henney Packard limousine funeral coach with Leveldraulic, Electraulic, and three-way loading. This view of the car shows the plush interior appointments that were standard on all Henney vehicles. This model was designated as Model 4092.

This view of the Model 4098 shows the interior appointments of this unique vehicle. The Model 4098 was the most attractive vehicle built by Henney this year and the most expensive.

Siebert continued to build a strong line of vehicles mounted on the Ford chassis, and this year had made Mercury part of the product line also. This is the Siebert Aristocrat funeral coach. Other models included the service car, combination car, ambulance, and municipal ambulance.

1940

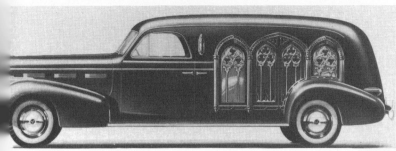

The top offering from Meteor was the Carved Gothic with ornate filigree in the side windows. There was glass behind the filligree and the ornamentation was easily removable for cleaning. This was an end-load coach and it is seen here mounted on a LaSalle chassis.

Meteor offered a nice looking flower car in this year, shown here on a LaSalle chassis. The car has draperies in the rear side windows and wide access doors to the rear compartment. The interior of the rear compartment and the top of the flower deck were finished in stainless steel.

Meteor also announced a version of the landau funeral car this year, but this car did not have the fabric covered roof. The landau bows and small coach lamps on the rear quarters were similar to those found on other landaus. This car has fog lamps on the bumper and a spotlight on the windshield pillar.

Meteor added a styling item to the cars this year that was to become a company trademark. The unusual shape of the front door window made a roof for a coach lamp and heightened the dignified look. This limousine funeral coach, mounted on a Cadillac chassis, was a standard Meteor offering. Note the formal draperies and the flower trays in the rear compartment.

When the rear tailgate door of the flower car was opened it revealed rollers on its interior side that matched rollers on the casket compartment floor. This car could also be used as a hearse, used to carry the body of the deceased from the place of death to the funeral home, or as a utility vehicle. The small hydraulic mechanism in the front left corner was used to raise or lower the flower deck in a step arrangement for the upright placement of baskets of floral tributes.

With a duo-tone paint finish, dual spotlights, fog lights mounted on the bumper and a red warning light, this Meteor ambulance was photographed in front of the Meteor plant in Piqua, Ohio. This car is also equipped with a radio, dual built-in roof warning light pods, and frosted glass in the rear quarter window. Again, the Meteor front door coach lamp trademark is part of this Cadillac's design.

198   1940

This was the Meteor LaSalle service car for 1940. The addition of the small coach lamps and the side decoration helped distinguish this vehicle as a funeral service car.

Dramatizing the custom character that was so widely sought by funeral directors in their motor equipment, the Sayers & Scovill Co. (S & S) of Cincinnati announced the new Majestic carved type funeral coach this year, once again drawing out an old name and reviving it on a new series. The Majestic was regarded as the authentic interpretation of the carved car design and brought back the traditional symbolism to the funeral cortege. The carved drapery panel was updated and made to look more modern and flowing. There was a special mounting on the front door for the owner's initial. The S & S radiator and grille date marks are highly visible on this LaSalle Majestic. This car was available as a three-way car, a straight end loader, or a side loader, on either the Cadillac or LaSalle chassis.

On the lower end of the Meteor price scale, the company produced a line of Chevrolet professional cars that included this service car. The Meteor Chevrolet funeral coach, combination and ambulance all shared this body with the full three window styling. The side windows of the service car were blanked off and had decorative panels in their place.

Available in either three way, end loading, or side loading versions, the 1940 S & S Victoria was an impressive looking vehicle and the forerunner of the modern landau hearse. Note the sweeping curve of the rear window and the grace of the landau bow. The roof was fully padded and covered with fabric. Two coach lamps adorned the rear body quarters to add distinction. S & S continued to place date marks on the radiator top and the grille. Hubcaps bear the S & S emblem. The Victoria is seen here on a Cadillac chassis.

Still bearing LaSalle grille, radiator, and hubcaps, this S & S service car was very dignified, with a unique decorative plaque on the rear body sides upon which the owner's name could be placed.

The impressive S & S Victoria was also available in ambulance form, and was called the S & S Victoria Parkway. This model was available on either the Cadillac or the LaSalle chassis, as seen here. Notice the leaded glass in the rear windows and the same roof treatment as on the hearse. A combination siren and warning light is nestled between the fender mounted headlight and the hood, and a spotlight is mounted on the windshield pillar.

1940

Available on either the Cadillac or the LaSalle chassis, Superior offered the carved Tarrytown Model. The styling of this carved car consisted of three stamped metal cathedral type window shapes on the rear body sides, divided by columns.

Superior Body of Lima, Ohio, offered their limousine funeral coach on Cadillac, LaSalle and Pontiac chassis this year. This Pontiac funeral coach, with body by Superior, is a stylish number indeed. The belt line of the new Superior body curved down at the tail end of the car to add a new and sweeping dimension to the car's lines. The coaches still featured the widest side doors in the industry.

The Superior Pontiac service car was a handsome little utilitarian vehicle that was complete with coach lamps and decorative ornamentation on the rear body sides. This car was a three door model only and was mounted on the Pontiac 6 chassis. The Superior Pontiac coaches enjoyed a considerable amount of popularity and were at the less expensive end of Superior pricing.

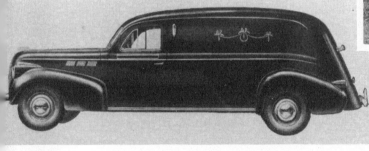

Building a total of 454 Buick chassied coaches, Flxible had one of their best years ever. In addition to these coaches, they built 88 on the LaSalle and Cadillac chassis for a grand total production of 532 coaches. This Series 60 ambulance was purchased by the American Legion Post 15 of East Greenwich, R.I. Among this car's features are: dual spotlights, bumper mounted fog lights, roof mounted warning light, rear warning lights, a radio, and blinds on the side windows.

This year, Flxible built on four different Buick chassis. The Series 60 had 160 Flxible bodies mounted on it, while the Series 70 Buick chassis only had 1 built on it. The Buick Series 50 chassis had 2 bodies mounted on it, and the most popular, the Series 40, saw 291 Flxible professional car bodies mounted on it. This Series 40 Flxible Buick limousine funeral coach was purchased by the Weinwright Funeral Home. Notice that the company was still raising the hood lines of these cars.

Flxible also built a total of 80 funeral coaches and ambulances on the LaSalle chassis this year, and eight on the Cadillac chassis. This Flxible Classic-A LaSalle hearse went to the Cornell-Dibble Co. and featured carved drapery sections in the arched areas with ornate carving dividing the dual arches. To complete the side styling, carving work was applied to the area underneath the arches. This five-door coach was a straight end loader.

By now the landau style had been firmly entrenched along with the limousine body type and was quite a popular configuration used in both funeral coach and ambulance models. Flxible entered the landau segment of the market this year and also announced its first flower cars. The Flxible model line consisted of limousine ambulances and funeral cars, the new landau and flower cars, service cars and an innovation in the carved car style called the Classic A. Flxible had introduced the double A and Classic A model line in 1937 and continued to market these attractive carved type cars through this year. The A models featured two arches above the belt line on the rear body sides within which were windows of beveled plate glass. Variations on this style coach featured modestly carved trim within these arches in the form of drapery.

The A.J. Miller Company scored another industry first with the introduction of Tu/Level attendant seating in ambulances and Duplex floors for combination models. Tu/Level seating utilized what was otherwise dead space under the front end of the rear compartment floor and made an entry way and chair-type leg room for seating within this space. Although patented by Miller, this feature became widely copied throughout the industry and is with us today. Along with Tu/Level seating, Miller originated the reversible floor sections which made the true combination funeral coach and ambulance a self-contained unit.

The announcement of a new addition to the Siebert line of extended chassis Ford funeral service and emergency vehicles was made for this model year. Known as the triple-purpose Aristocrat, the newly designed Ford landau served as either a service car, ambulance or funeral coach. The Siebert-Fords of this era were larger and longer than previous models with bodies of modern design and attractive styling. In addition to offering coaches on the Ford chassis, Siebert branched out and began to build on the Mercury chassis also. In fact, the Mercury Aristocrat was declared a vast improvement over the Ford models.

With the economy at a fairly good level, the production of funeral cars and ambulances took a substantial up-swing, and a total of 3,801 such cars were built in the U.S. this year.

Building on the Cadillac chassis only, S & S offered two lines called the Superline and the DeLuxe. The 1941 S & S Superline Victoria was designated as the most luxurious funeral coach ever offered by Sayers & Scovill of Cincinnati. The car was totally redesigned for the last time for many years to come. Longer flowing body lines characterized the Victorias. The rear quarters as well as the entire roof were covered with heavy leather grained material called "dreadnaught" and developed for this usage. This Victoria was the Superline model and was offered in either side servicing or rear servicing styles. The DeLuxe models did not have fender skirts, and lacked some of the luxury highlights of the Superlines. Note the graceful curve of the rear side windows and the S & S date marks on the hood.

Called the Superline Macadonian, this was the S & S carved funeral coach. Standard specification included coach lamps on the rear side doors and a crest on the front doors with the owner's initial.

The S & S Statesman was the limousine offering from this Cincinnati firm, and was available in either Superline or DeLuxe models and as either a side servicing or rear loading coach. Shown is the Superline Statesman as a rear loading funeral car.

The S & S DeLuxe service car was derived from the Statesman limousine hearse, but the rear side windows were replaced with carved panels of stamped metal. These panels depicted draperies and the complete car was finished in any color desired by the customer. This is the DeLuxe service car with mounting brackets on the doors for the owner's nameplate. The S & S date marks on the hood are easily visible, as are the special S & S crest at the front of the hood.

1941

The deceiving roof line of this Henney Packard Model 4196 flower car would lead one to believe that it was a convertible. But like the Landaulet funeral coach, the top was fixed and covered with padded fabric.

With very pleasing lines, this utilitarian Henney Packard service car had the new Henney wreath and bars on the rear body sides, sidemounts, windshield nameplate, and chrome fixtures for door nameplates. The rear compartment of the service car was very plain and utilitarian, and there was ample room for rough boxes, chairs, baskets and anything else the funeral director would need.

Offered for the first time was this interesting Henney Packard custom built airport limousine. With modern fast back styling and eight doors, this car would accommodate up to 12 people. One of these interesting vehicles was operated by United Airlines.

Henney's most beautiful offering was the Model 4198 Landaulet Funeral Coach on the Packard 1901-A and 1903-A chassis. The low lines of this car were highlighted in the narrow side windows and the long low fabric covered padded top that looked like it was convertible. The landau bows were painted black to match the roof, although they were not functional. A small coach lamp was placed on the door pillar and the nameplate was on the front doors. Henney built a total of 1,200 cars this year.

Designated the Henney Packard Model 4194 ambulance, this car is shown with all of its standard equipment. Spotlight, sidemounts, windshield sign, and ambulance rear quarter window flashers were all part of the standard specification. This car was finished in green.

Designated the Model 4192, this was the Henney Packard side servicing funeral car. Finished in black, this car had a beautiful maroon mohair interior and a walnut casket table.

This is the Henney combination funeral car and ambulance. The car was quickly converted from its duties as a hearse, as it is seen here, to that of an ambulance.

1941

Superior continued to offer the carved Tarrytown funeral coach with regal styling in the classic gothic design. The coach lamps had been changed and the body was streamlined similar to the rest of the Superior range.

The 1941 Superior Cadillac ambulance, like all Superior Cadillac models, was available on either the Cadillac Series 75 or Series 62 chassis. The unique Superior body styling offered more interior room than some of the higher priced competition and was complete with all of the life saving essentials of the day. Frosted glass in the rear quarter windows and shades on the rear door windows were standard equipment, while the roof lights and siren as well as the two-tone paint finish were optional at extra cost.

Finished in duo-tone, this Superior Cadillac Landau was a very attractive coach indeed. They have given the coach a more modern fastback look simply by painting the roof to follow the body lines. Two-tone paint finish was optional at extra cost, but worth while in appearance.

For the funeral director seeking dignified simplicity, Superior offered the Briarcliff limousine funeral coach on the Cadillac chassis. The Superior styling approach for these cars placed them among the most modern and stylish vehicles in the field. Note the swelled roof line and the sweeping blet line and the unusually large windows sandwiched in between.

The buyer of a 1941 Superior Pontiac chassis funeral car or ambulance had a choice of two lines — one with the Pontiac 8 and the other powered by the Pontiac 6-cylinder engine. Superior called the Pontiac a symmetry in styling that was within the reach of the thriftiest budget. Seen here is the Superior Pontiac limousine funeral coach on the 8-cylinder chassis.

Also offered in either 6 or 8-cylinder models was the Superior ambulance. These cars were priced well below the Cadillac versions and were designed to fit the smaller funeral home's budget. Once again the roof lights, siren and two-tone paint finish were optional at extra cost. Note the frosted glass in the rear quarter windows and the pull-down shades in the rear door windows.

Shown on the Pontiac 6-cylinder chassis is the 1941 Superior Pontiac service car called the Elmwood. This five door coach had decorative panels on the rear body sides, but followed the general theme of Superior Pontiac styling for the year.

1941

Superior continued to offer the Pontiac chassis line of coaches this year, and they were totally restyled similar to the Cadillac models. Superior Pontiac Landau funeral coach is finished in two-tone paint with landau bows of gleaming chrome. The Pontiac was also totally restyled this year.

A.J. Miller, now building on Cadillac and Oldsmobile only, offered this limousine funeral car on either chassis. The C-pillar was slanted forward, streamlining the car's appearance and adding a new touch to Miller styling. This model, on a Cadillac chassis, has drapes in the rear quarter windows only.

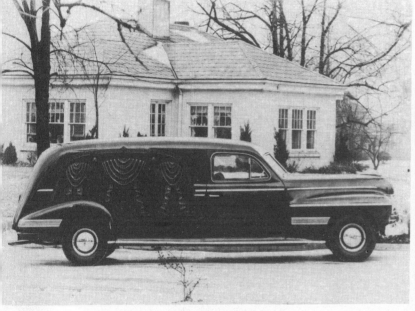

Called the Art Carved model, this version is mounted on an Oldsmobile chassis. The Art Carved model continued to feature heavy drapery details on the rear body sides. The Oldsmobile chassis was a very attractive unit this year.

Miller continued production of the Cathedral carved coach on either the Cadillac or the Oldsmobile chassis. This view of the carved panel section of a Miller Cadillac Cathedral shows the details that were in the carved sections, the torches and the interesting door nameplate that Miller used. The rear side door is hinged just ahead of the center carved panel.

Owned by the George Casey Funeral Home, this attractive Miller flower car is shown with its tonneau cover snapped over the stainless steel flower deck. This Miller body is on a Cadillac chassis. Miller made the rear deck flow with the coupe cab styling. This body style was not offered on the Oldsmobile chassis.

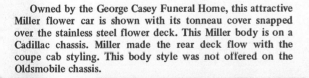

Far removed from the delivery truck look was the Miller service car on either the Cadillac or Oldsmobile chassis. This model on an Oldsmobile chassis, was equipped with leather seat covering, a rear floor covered in linoleum, chrome skids and a roller assembly at the rear door to facilitate loading of caskets. Miller offered bodies on either the Olds 6 or 8-cylinder chassis.

This Flxible Cadillac was featured at the National Funeral Directors Convention, finished in white. The interior was finished in high grade mohair in the front and stainless steel in the rear. Wide side doors permitted access to the rear compartment and the rear door was equipped with casket rollers. This car was referred to as a Model C61-FC.

Flxible of Loudonville, Ohio, began to build flower cars and mounted them on Buick and Cadillac chassis. This was the only year in which this company ever ventured into the manufacture of this type of vehicle, and they built only three coaches of this type. This one, mounted on the Buick Series 70 chassis, was one of two mounted on the Buick chassis. It was delivered to the J.J. O'Donnell Funeral Home. These cars had a low curved rear deck for the floral tributes.

One of the 100 Cadillac Flxible coaches built this year was this ambulance of limousine styling that was supplied to Carl J. Ballweg of Syracuse, N.Y. This coach was finished in white with blue lettering and fitted with a Federal combination light siren on the roof and warning lights in roof pods. This year Flxible turned out a total of 503 professional cars on Buick and Cadillac chassis.

The landau funeral coach was a new body style offering for Flxible customers this year. An unusual rear window and door treatment set the Flxible landau apart from the competition. The landau bows were painted the same color as the body.

This year, Flxible built Buick professional cars on the Buick Series 50 and 70 chassis. They built 230 on the Series 50 chassis and an additional 170 on the Series 70 chassis. This Series 70 Classic A funeral car has windows within the arches and a decorative plaque on the front doors. Once again the hood has been raised with an insert.

This strikingly handsome Flxible Cadillac Landau was a new body style this year. Note the shape of the rear window and the large landau bows on the rear quarters. The Federal combination siren warning light and the warning lights built into roof pods were optional, as was the spotlight. Notice that the hood line of the Flxible coaches was still raised, and that they placed the Flxible nameplate on the sides of the hood. This year, Flxible built 100 professional cars on the Cadillac chassis.

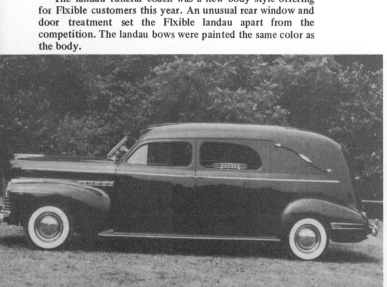

1941

The Meteor limousine coach could be purchased as a three-way car or a straight end loader, as a straight funeral car, a combination, or as an ambulance. Seen here in its hearse form, the car reflects a dignity and a sense of refinement of the new styling and the new Cadillac chassis. Note the door mounted coach lamp and the shape of the front window. The flower trays can be seen through the rear quarter window.

Complete with an ornamental wreath and cycus leaves on the rear body sides, this was the Meteor service car for 1941. Once again there are the unusually shaped Meteor door windows.

The ornate and expensive Meteor carved Gothic continued as the premium offering. The rear side doors were hinged at the center of the middle gothic panel and had a coach lamp on the upper front side.

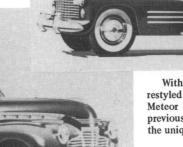

With the new Cadillac chassis, most companies totally restyled their bodies and Meteor was no exception. This new Meteor Landau funeral coach had smoother lines than the previous landau offerings. Meteor continued to incorporate the unique door mounted coach lamp trademark.

Meteor also continued to offer a line of less expensive vehicles mounted on the Chevrolet chassis. Seen here is the Meteor Chevrolet combination hearse ambulance. These cars were also offered as straight funeral coach or as an ambulance and were aimed at the funeral director who could not afford the more lavish coaches.

The least expensive Meteor offering this year was the Meteor Chevrolet service car. Mounted on an extended Chevrolet chassis, it had decorative panels set into the window frames to blank off the rear compartment. All of the Meteor Chevrolet funeral service cars and ambulances were built on an extended Chevrolet Special DeLuxe chassis.

1941

This unusual Cadillac hearse was recently retired by its second owner. It was one of the most admired coaches in Canada for years. Built in the shop of John Little of Ingersoll, Ontario, this coach was hand fashioned from a Cadillac coupe and delivered to the Needham Funeral Home of London, Ontario. All of the rear body side carvings were hand made of real wood, and they covered blue stained glass windows. Several years later the original owners sold the car to the Dodson & Habkirk Funeral Home of Kingsville, Ontario, which used it for many years.

This view of the unusual John Little Cadillac carved funeral coach shows the hand carved details of the rear body sides. Made from a Cadillac coupe, this car was finished in gray with a black crinkle-type top. The glass behind the filigree was stained blue and the interior was finished in high quality maroon mohair. The car was a side servicing unit and was retired in 1958 to a barn and later moved into a field outside Kingsville, Ontario, where it began to rot away.

This interestingly painted Knightstown Cadillac combination features long straight lines and pleasing styling. This car was built by Knightstown, not to be confused with Silver Knightstown, and they were both located in Knightstown, Ind. The Cadillac chassis, with few exceptions, had become the standard funeral coach and ambulance mount.

With unusually high body lines, this chrome plated ambulance was built by Silver Knightstown Body Co. of Knightstown, Ind. Notice the rear quarter window vents, the chrome rocker panel and stone shield, the roof-pod mounted warning lights, spotlights and the Cadillac installed fog lights below the headlights. The rear quarter windows feature an etched sunburst pattern with a cross at the center.

Another creation of the crafted hand of John Little of Ingersoll, Ontario, was this ambulance built for the John Labatt Brewery Limited of Toronto. This car was used as an employee ambulance and was fitted with a fog light, radio, roof mounted siren, and special ambulance decals on the rear quarter window. This car was built from a Cadillac four-door sedan. Mr. Little built most of these cars completely by hand in his own garage in Ingersoll. He had formerly been employed by Mitchell Hearse and Body Limited of Ingersoll.

Called the triple-purpose Aristocrat, Siebert announced a new landau model on the Ford and Mercury chassis. This car could be used as a hearse, a service car, or as an ambulance, and could be changed into proper attire for any of these functions in a matter of minutes. The Siebert Triple-Purpose Aristocrat Landau, on the Ford truck chassis, was also offered as a straight hearse or ambulance. Other models offered were a limousine style funeral car and ambulance or limousine triple purpose coach.

1942

The new 1942 models were announced in the fall of 1941 at the National Funeral Directors Convention in St. Louis. Among these new models was the most revolutionary funeral coach design to be introduced in many years.

The Flxible "Innovation" attracted wide interest at the convention display and reflected a totally new concept in funeral coach design. Featuring a long low deck, like that of a flower car, with coupe-type styling, this car was mounted on a Cadillac chassis and was to be offered on a Buick chassis also. The rear deck was smooth and rounded, not intended to carry flowers, and featured carved rear side panels made of genuine walnut with a natural finish. The interior was exceptionally roomy with a rear door that was a full 37 inches wide. The Casket compartment was available in only two trim choices. One was full fabric trim of either mohair or sumatra, while the other was a combination of either mohair or sumatra with carved walnut panels. The interiors were set off by six semi-indirect side panel lights that operated automatically with the opening of the rear door.

The Sayers & Scovill trademark had become synonymous with high grade funeral coaches and ambulances. In this year two young men who had grown up with this firm acquired the controlling interest in the company. Charles A. Eisenhardt and Willard C. Hess bought the company and brought their sons into the business to follow in their footsteps. The company name was changed to Hess & Eisenhardt Co., but they continued to utilize the famous S & S trademark for their funeral coaches and ambulances. They also did some custom and semicustom coach work on passenger car chassis, but the bulk of their work was devoted to the construction of high-grade funeral coaches and ambulances.

The Henney stylists triumphed in this year with a flower car that looked very much like a convertible. Easily the best looking vehicle in the business, it was yet another offering from the style-conscious Freeport, Ill., firm. While Henney was the only firm building cars on Packard chassis, the other large firms had switched to Cadillac and, with the death of the LaSalle, had added other General Motors chassis to the standard line. Superior continued to offer the Pontiac chassis cars and Miller had added Oldsmobile to its line as a standard model. This was a year in which the last of the true carved-side cars were to be seen. After the war, lack of materials and new chassis designs would eliminate this type car altogether.

When the bombs fell on Pearl Harbor, the manufacturers were busy filling large orders for new cars only, for production of all cars was to be banned in the early months of the year. Some of the firms began to switch over to wartime production, while others maintained a parts warehouse for their products to keep them on the roads for the duration of the war.

Obviously, with World War II cutting heavily into the manufacture of all private vehicles, the production of ambulances and funeral cars was down substantially as the builders found it impossible to obtain chassis or materials. Thus, only 1,070 such vehicles were built in the U.S. this year.

The most interesting new coach line for 1942 was introduced by Flxible and called the Innovation. With a low rear deck that resembled that of a flower car, the Innovation was a design triumph for Flxible. The low rear deck was smooth and rounded, but not intended to carry flowers. The car was finished in coupe styling. The interior was exceptionally roomy with a rear door that was a full 37 inches wide and 30½ inches high. The floor of the rear compartment was carpeted and had eight rubber casket rollers plus a set of large rollers on the rear door sill. The rear side panels were finished in genuine walnut carvings in natural finish. The design shown on the rear body sides was but one of the style choices available. The casket compartment interior trim was made available in two choices, of full fabric trims in either mohair or sumatra with carved walnut panels. Each interior was exceptionally rich and set off by six semi-indirect side panel lights that operated automatically with the operation of the rear compartment door.

The Flxible Innovation was truly a novel funeral coach design, but unfortunately, there were only three such vehicles ever constructed. Two were mounted on the Cadillac chassis. The second version is shown here in downtown Loudonville, Ohio. The other was mounted on a Buick chassis and reliable sources report that this car did not have the rear body side panels. The Innovation was a three-door coach and was strictly a rear loading hearse, with no rear side doors to spoil the lines.

Flxible continued to produce Buick chassis funeral cars as their main product this year, and the styling of the coaches was refined along with the new Buick chassis. This Flxible Buick Landau funeral car was mounted on the Buick Series 70 chassis of which a total of 151 cars were built. The company also built 32 cars on the Series 50 Buick chassis.

Flxible built a grand total of 183 professional cars on the Buick chassis in this year, not including one Buick Innovation. The stylish 1942 Flxible Buick Series 70 service car is finished in white. This car had the stylized chrome wreath and chrome strips on the rear body sides. Flxible was still raising the hood lines of the professional cars. The Flxible nameplate is seen on the lower portion of the hood.

There were very few of the new 1942 Flxible cars produced before the war broke out and all vehicle production was stopped. This Series 50 Flxible Buick ambulance was one of only 32 cars of this series built before production was halted. This view shows the interesting rear treatment of the Flxible Buick models and how the famous Buick Fireball taillights were retained for use on professional cars. Note the rear marker lights on the roof, and the large "back-up" light mounted on the left side of the rear bumper.

A grand total of 26 Cadillac chassis funeral coaches and ambulances were built in this short production year by the Flxible Co. Two of these cars were the stylish Innovations. This Flxible Cadillac ambulance features futuristic styling with dual warning lights in roof pods, fender skirts, metallic paint, and, uniquely styled rear quarter window decals with the red cross enclosed within a wreath. The grand total of Flxible professional cars produced this year amounted to 209.

The Henney Packard Model 4290 side servicing funeral coach was a very attractive hearse. With dual side mounts and formal draperies, this car was finished in black with a maroon mohair interior.

The attractive Henney Packard Model 4298 Landaulet funeral coach continued to represent the top of the Henney range of funeral cars. The styling of the car changed very little with the notable exception of the Packard front ensemble. This car featured a very low fabric covered landau roof with a pair of coach lamps on the door pillars. This car had dual side mounts and was one of 300 professional cars built in this short production year by Henney on the Packard 2001-A and 2003-A chassis.

Impeccable interior trim was a feature of the Henney Packard Model 4292 side servicing funeral coach. The casket table is drawn fully out. The draperies are of the formal type.

Henney continued to offer a very distinguished looking service car. This car, finished in deep black overall, featured a metal wreath on the extreme rear quarters of the rear body sides. This car also has side mounts and whitewall tires. Henney was the only maker to use the Packard chassis, and the slogan, "the body is Henney, the chassis by Packard, the combination exclusive."

The Henney Packard landau funeral coach was one of 300 professional cars built this year by the "world's largest hearse and ambulance maker." Note how this car follows the styling of the Model 4298 landaulet, but with a regular uncovered landau roof.

This Henney Packard limousine style ambulance was one of the many offerings in this short production year. Note the side mounts and the roof mounted warning lights in built-in pods. The decals used on the rear quarter windows were exclusively styled for Henney.

The Henney Packard Model 4291 dual purpose funeral car and ambulance was a combination that was quickly and easily convertible from a hearse to an ambulance in a matter of minutes. The combination was a very popular body type with the funeral director who could not afford two separate vehicles.

Loaded with floral tributes, this was the 1942 Henney Packard Model 4296 flower car. This car retained the imitation convertible top that was fabric covered, similar to the Landaulet funeral coach. The low rear deck was finished in stainless steel and the flower deck was hydraulically operated for the display of flower baskets. A tonneau was supplied to cover the flower deck in inclement weather and when not in use. These cars made very impressive lead cars in funeral corteges, and were very popular in the metropolitan areas.

Almost every maker in the field now offered a flower car of original styling but following the coupe-cab, low rear deck pattern. This interesting variation was built on the now popular Cadillac chassis by Eureka. Note the Eureka approach to the rear deck styling and the placement of the nameplates.

1942

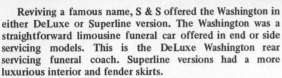

Reviving a famous name, S & S offered the Washington in either DeLuxe or Superline version. The Washington was a straightforward limousine funeral car offered in end or side servicing models. This is the DeLuxe Washington rear servicing funeral coach. Superline versions had a more luxurious interior and fender skirts.

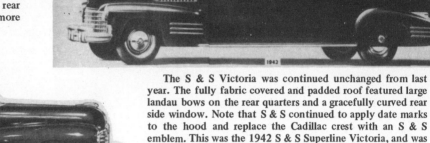

The S & S Victoria was continued unchanged from last year. The fully fabric covered and padded roof featured large landau bows on the rear quarters and a gracefully curved rear side window. Note that S & S continued to apply date marks to the hood and replace the Cadillac crest with an S & S emblem. This was the 1942 S & S Superline Victoria, and was offered in either straight end-loading or as a three-way car. It was also available in a lower priced DeLuxe version without the fender skirts and with less luxurious interior trim.

The Superior Pontiac Landau was but one model in the extensive Superior line. Like the Cadillac limousine, this Superior wore all-new body styling for 1942 that was devoid of the sweeping belt line of the earlier models. Notice the pleasing effect of the two-tone paint finish and the extremely large rear side doors of this side servicing funeral coach. The Pontiac line was also offered in limousine style, as well as an ambulance. The rear fender skirts and the two-tone paint job were optional at extra cost.

The Superior Body Co. of Lima, Ohio continued to offer coaches on either the Cadillac or the Pontiac chassis. All Superior bodies were restyled for 1942 in a more straight line type of appearance. This Superior Cadillac funeral coach was finished in limousine styling, but was also offered in a Landau. This same styling was seen on the Cadillac ambulance offered by Superior. Superior now offered an electric side loading casket table as optional equipment.

Superior offered a service car on both the Cadillac and the Pontiac chassis. Here is the Superior Cadillac service car complete with coach lamps and decorative rear body side mouldings. This car was of five-door style, and was considered the workhorse of the line.

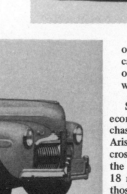

This flower car was the product of the Meteor Motor Co. of Piqua, Ohio, and reflects yet another approach to flower car styling. This coach has small, non-functional landau bows on the rear sides of the coupe roof and a slanted door window.

Siebert of Toledo continued to offer a strong line of economy professional cars on either the Ford or the Mercury chassis. Here we see the popular and economical Siebert Aristocrat ambulance on the Ford chassis. Note the winged cross in the rear quarter windows and the pull-down shades in the other rear compartment windows. The company offered 18 models on the Ford or Mercury chassis, and asked that those interested in purchasing a car place orders as soon as possible due to a national emergency and a shortage of materials.

1942

Shown at the 1942 National Funeral Directors Convention in October, 1941, is this beautiful new Miller flower car. Finished in two-tone with small landau bows on the stylish roof line, this car was one of the main attractions of the Miller display.

A.J. Miller continued to offer the Cathedral model on either the Cadillac or the Oldsmobile chassis. The body was completely redesigned and featured the traditional six gothic filigree-type stamped steel church window shapes and two stylized torches. The car was offered in either side servicing or end servicing models. The door-mounted nameplates were optional. This car is seen on the Cadillac chassis.

The A.J. Miller landau was offered on either the Cadillac or the Oldsmobile chassis and was completely restyled for the model year. This car had small landau bows and carriage lamps on the rear quarters and a rakishly curved rear door window. The top could be finished in decking of a dull paint for the heightened landau effect.

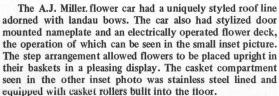

Rare as a professional car was the 1942 Oldsmobile. Miller was the only maker listing the Oldsmobile chassis as a standard offering. Seen here is the A.J. Miller limousine funeral car on the Olds chassis in a rear loading version. The bodies for Miller funeral cars were totally restyled for 1942 and enclosed the new pontoon fenders that stretched into the door panels. Note that this coach had draperies in the rear quarter windows only.

The A.J. Miller flower car had a uniquely styled roof line adorned with landau bows. The car also had stylized door mounted nameplate and an electrically operated flower deck, the operation of which can be seen in the small inset picture. The step arrangement allowed flowers to be placed upright in their baskets in a pleasing display. The casket compartment seen in the other inset photo was stainless steel lined and equipped with casket rollers built into the floor.

Carrying the same silhouette as the Miller limousine and landau models, the Miller service car was also offered on either the Cadillac or Oldsmobile chassis. The newly designed chrome plated decorative plaque on the rear quarters breaks the plain panel and was shaped to add to the lines of the car. The rear quarter and the top were of stamped steel and could be finished in dull or high gloss at the option of the buyer.

With yet another approach to flower car styling, the Knightstown Body Co. of Knightstown, Ind., offered this vehicle. Note the high sill around the flower deck and the tonneau at the rear of the car.

1943 - 1945

Superior supplied the Medical Department of the U.S. Army with a number of civilian type ambulances for home bases. This Cadillac chassis vehicle is one such example of these coaches.

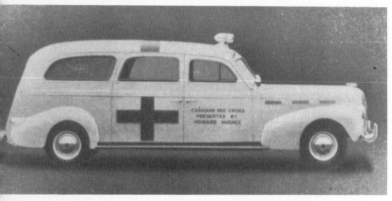

In the war years, a number of city or metropolitan type ambulances were quickly converted for war-time uses. When the war broke out in 1939, the British Commonwealth nations were all committed to the fighting. In that year, American millionaire Howard Hughes donated this Miller ambulance to the Canadian Red Cross. Note the inscription of the front door, the large red cross insignia on the rear doors and roof, and the siren and spotlights.

With the government ban on the production of automobiles, the funeral coach and ambulance manufacturers found themselves out of business. There were divided opinions among the makers about future policies. Some favored closing up the plants, while others believed that they should work only at automobile repair and reconditioning. Some felt that they should seek government contracts and keep their people working.

Those that did seek government contracts found that conversion to war work was quite costly and most manufacturers that chose this route showed a financial loss for their 1942 operations.

Superior began to produce military vehicles, including both field type military ambulances and mobile work shops, and rocket launchers to meet the country's wartime needs. This contribution earned Superior the Army-Navy "E" pennant for excellence. Eureka sought and won government contracts for busses, ambulances, turret baskets for tanks and other war ware. They purchased used auto transport trucks and converted them into 100-passenger busses to be used in ordinance plants.

Miller gained government contracts and most of its war work was in the aircraft field. Miller's largest single job was a sub-contract from Curtiss Wright for nacelles for cargo planes. They also produced hulls for fire boats, camera mounts, gas tanks, fire walls, engine cowlings, assembly fixtures, dies, air ducts and exhaust vents. Hess & Eisenhardt (S & S) built trailers with special suspensions, truck cabs, cargo bodies for trucks and wood parts for the standard army truck. Flxible's war efforts won them several awards, among them the Army-Navy "E," the Guidon Award, The Minute Man flag and an approved quality rating from the Army Air Force. They produced fighter and bomber parts, control gondolas for blimps, and components for tanks and ships.

Although Henney did not get into war production like some of the other manufacturers, it did offer some very interesting civilian defense ambulances styled along the lines of van trucks. These interesting vehicles were mounted on Packard chassis and were built in number of about 3,500. The government did allow some vehicles to be built during the war but these were almost all bought by the government and pressed into wartime service. A few metropolitan type ambulances were built with no chrome trim at all. Called "blackout" models these cars were painted a dull color overall and were built in very limited numbers.

Although the war had ceased the production of all professional cars, the Flxible Co. of Loudonville, Ohio, produced a few special blackout ambulances for some high priority customers. This vehicle was built for the AFL-CIO and donated to the Trenton, N.J., Defense Council. Note that with the exception of the bumpers and siren, all of the parts that would have regularly been chrome plated are finished dull in compliance with regulations at the time. Only buyers with a very high priority could receive the limited numbers of these cars that were allowed to be produced by the Office of Price Administration. The OPA allowed a trickle flow of these cars after June, 1943.

1943 - 1945

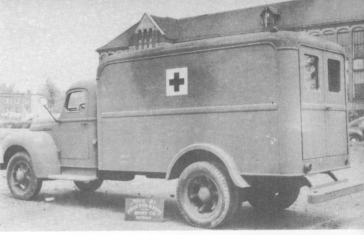

Henney did not convert its plant facilities over to war time producion, but did build some special ambulances for use by the Civilian Defense organization. This special DC ambulance featured an ice cream truck appearance, with a box-like rear compartment. This particular unit was built for the City of Springbrook Civilian Defense Unit Number 12. It was designed and engineered for use in this type of emergency work. Henney Packard ambulances served all over the world with a remarkable service record that was to be recalled after the war in Packard sales catalogues.

Mounted on an Interantional 2-ton 1942 chassis, this ambulance of military specification was built by Proctor-Keefe Body Co. of Detroit. Note the box-type ambulance body with the prominent Red Cross emblems. Blackout taillights and double-opening rear doors were an integral part of this design.

A rear view of the Superior ambulances supplied to the U.S. Army Medical Department reveals special rear side window treatment and the 1941 Superior body styling. Note the high lines of this car. The overall coach was painted an olive gloss.

A large front line ambulance with a box type rear compartment was this model built by Superior Body of Lima, Ohio. This vehicle was operated by the U.S. Army Medical Service and these vehicles were used extensively on the world's battlefields. Note the large Red Cross emblems, the folding rear step, and the ladder on the rear end of this vehicle.

This interesting little vehicle was not an ambulance. However, it was built by Superior as part of the war effort. This was a 1½ ton 4X4 Ordinance maintenance vehicle mounted on a GMC chassis. It was powered by a 6-cylinder 100 horsepower engine and was a derivative of the short wheelbase school bus built by Superior before the war. There were only 78 of these vehicles produced before they were replaced by a larger model also built by Superior.

Serving somewhere in England during the Blitz was this Miller LaSalle ambulance that was fitted with blackout lights, roof pod lights, and combination warning light — siren. The American style metropolitan ambulances were very popular in Britain in this period, and many served with distinction. This car also carries British army vehicle numbers and license plates.

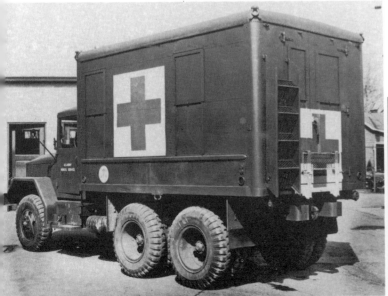

1946

When peace came, the reconversion to funeral coach and ambulance production was costly for those that had been employed in the war service. The losses in this late 1945 reconversion wiped out almost all profits gained during the war. Material shortages were especially acute in this post-war period. This was especially felt at A.J. Miller because the company had been employed in the airplane field and consequently used aluminum. As a result of this, Miller had been dropped from the customer lists of the steel companies. To carry out war contracts the coach makers made extensive purchases of machinery like drop hammers and presses. And in clearing their plants for war work, all of the tools and dies had been dismantled. For a new model run, it was necessary to engineer and design complete new bodies for their cars thus necessitating all new tools and dies.

By May of 1946, most of the large makers were in full production and deliveries of new cars had begun. Production was hampered by material shortages and customers were clamoring for more cars than all of the makers combined could possibly produce. O.P.A. regulations caused further complications and the costs of the new cars varied. Coaches had to be sold "subject to price at the time of delivery."

Henney was not fully ready to resume production in 1946 along with the rest of the major manufacturers. The company did, however, build a very few prototype models and offered a few production cars using the new Packard chassis and the pre-war bodies. The main reason that Henney did not build cars could have been the fact that Packard did not build a commercial chassis for its 1946 Clipper models. The few cars that Henney did build had to be mounted on the new Clipper auto chassis.

In Canada, Dominion Manufacturers Ltd. of Toronto, began to distribute Miller products along with its line of caskets and funeral sundries. Dominion had always played a prominent part in the coach field in Canada either through sales or with production of its own cars. When the horse-drawn vehicles were in vogue, it built these. With the switch to motorized cars, it began constructing these, too. After becoming an affiliate of the National Casket Co. of Boston, it handled both the National Kissel and later the National funeral coaches in Canada, which was its last involvement in the field until this year.

Because of the post-war material shortages and a great demand for cars, most of the makers did not offer a carved side model in the lines, but instead concentrated on rapidly producing a line of landau and limousine funeral coaches and ambulances, service cars and flower cars. Hess & Eisenhardt alone offered a carved model in the 1946 and 1947 model lines, to become the last of the major American manufacturers to offer this type car. Later the company offered some funeral coaches and ambulances with carved drapery panels set into the rear side windows, but this was to become the full extent of the carved type coach in the post-war U.S.

S & S changed ownership in the late 1930's and in 1942 the corporate name was changed to Hess & Eisenhardt, but the old traditional S & S name was retained. The S & S, Hess & Eisenhardt Model 46200 DeLuxe Arlington Combination limousine funeral car was priced at $5,480. Notice the high waist line and thin rear quarter pillars. The S & S date marks are still on the hood. The hubcaps are S & S as well.

Carved cars emanated from Canada for a few more years though, especially from the little shop of John C. Little in Ingersoll, Ontario. Mr. Little hand-built coaches in his garage to his customers specifications. He had formerly been employed by the Mitchel Hearse & Body co. and learned his trade well. Extending the frames of regular passenger cars and hand crafting bodies of his own design, the John Little funeral coach became quite popular throughout Canada. Other Canadian manufacturers involved in custom building hearses and emergency vehicles on an extended passenger car chassis in this era included Carmichael Motors Ltd. of Tillsonburg, Ontario, which built several coaches on Chevrolet and Pontiac chassis, and Ingersoll Auto Body Corp. of Ingersoll, Ontario, which operated more along the lines of John C. Little.

The S & S Model 46250 Victoria was a rear loading landau funeral coach in the best traditions of the original 1938 Victoria. In straight rear loading form, this car was priced at $5,720, while the combination ambulance/hearse version of this car sold for the same price. The S & S Victoria Parkway ambulance carried a price of $6,248 and was an ambulance version of this luxurious funeral car. The Parkway ambulance was designated as a model 46293. S & S production for 1946 began with the serial number 3,400,001 and went up from there.

Since the car manufacturers were virtually swamped with orders for private passenger cars, there was little emphasis placed on the building of special chassis for the professional car trade. Thus, even though substantial orders for new funeral cars and ambulances were on order, the lack of suitable chassis and occasional lack of raw materials kept the total production of professional cars down to 1,188 this year.

1946

Superior Pontiac funeral cars and ambulances were built on the Pontiac 8-cylinder chassis. Production began with serial number P8LA-1001 and went up from there. This Superior Pontiac ambulance is fitted with optional roof pod warning lights and a combination roof mounted siren warning light. This car was priced at $4,215.

The Superior Cadillac ambulance models were mounted on the same chassis as the funeral coach versions and were priced from $4,865 for the limousine to $4,940 for the landau. The siren warning light and roof pod warning lights on this car were optional extras.

After the war there was a great demand for new cars, and the makers were swamped with orders. Superior resumed production of its Cadillac and Pontiac funeral car and ambulance lines with very little styling change from 1942. The Pontiac funeral coach by Superior was available in two versions, the limousine hearse at $4,035 as seen here, and the landau hearse at $4,110. Both were mounted on the 8-cylinder Pontiac chassis.

Superior also offered their popular Pontiac chassis coaches in ambulance versions on both the limousine and landau styles. The limousine ambulance was priced at $4,215, while the landau version sold for $4,290. Note the ambulance decal on the rear quarter windows.

Contrary to many Packard experts, Henney did build a few professional cars in 1946. This picture was taken from a 1946 Henney advertisement, with the headline "Worth Waiting For." The world's largest manufacturer of funeral cars and ambulances built very few of these cars and refrained from building any further coaches until 1948, when they introduced a totally new model on the new Packard chassis with all new styling.

Completely equipped and ready to roll, the Eureka limousine ambulance seen here with optional roof lights and siren light was priced at $5,590, and in landau form for $5,740. This car complete with spotlights and rear quarter window decal, was quite an attractive coach. Eureka had a very good reputation in this field although they were still using wooden body framing while the rest of the industry had gone to all steel construction.

Eureka produced a fine line of funeral cars and ambulances this year, once again all on the Cadillac 75 chassis. This limousine funeral car had electric side loading and sold for $6,465, while the same car with a manual three-way table would sell for $5,450 and a straight end loader was priced at $4,953. Landau funeral cars were available in the same choice of loading versions starting at $5,103 and going up to $6,615.

Flxible returned to the professional car field with Buick chassis coaches built on the Buick Series 50 and 70 chassis. The Series 70 cars were called Premier models. This year the company built 161 Series 70 coaches and 80 on the Series 50. The production began with the serial number 15081 and proceeded to 15247 at the end of production. The Premier ambulance shown here sold for $5,177 and was designated a Model B76-6. The limousine funeral car on the Premier chassis was priced at $4,512 and designated as a Model B76-5. Landau funeral coach versions of the Premier, called a Model B76-5L, sold for $4,712. Also this year, Flxible offered a limousine combination designated the Model B76-5C for $4,737 and a service car, Model B76-8, for $4,512. Flxible continued to raise the hood line of the cars. This car is equipped with optional fog lights, spotlights, and roof mounted warning lights and siren.

1946

The artist that made these A.J. Miller illustrations took some unusual liberties in stretching the car to a point where this one looks a mile long and only four feet high. The 1946 Millers were claimed to have been worthy successors to their pre-war leaders. Miller offered a wide range of models this year, but they were now all on the Cadillac chassis only. The limousine styling was priced at $4,871. The limousine ambulance sold for $5,003 and the landau ambulance for $5,267. A Miller landau funeral coach was priced at $4,937 and a landau Duplex combination car at $5,135. These prices are for coaches with the GM Hydromatic transmission. Coaches with manual shift were approximately $200 less.

With the resumption of automobile production, Ford began to supply Siebert with chassis for their ambulances and funeral cars. The 1946 Siebert models were available as hearses, ambulances and combinations. The sedan type styling of these cars made them unique and interesting. Note the sliding glass partition between the compartments and the small drapes that are at the very top of the rear side windows only.

The Siebert professional coaches were available on either the Ford or the Mercury chassis, both of which were extended for these vehicles. Seen here is the Siebert Mercury funeral coach with sliding glass partition wall and small draperies on the rear side windows.

The Homes Funeral Home of Dresden, Ontario, took delivery of this pleasingly styled Packard sedan type ambulance built by John Little. This shot of the car was taken in front of John Little's Ingersoll, Ontario, body shop and ambulance works. John Little built all of these cars by hand in this little shop that combined his Shell service station. His creations became quite popular throughout Canada. Note the roof mounted siren and warning light and the all white finish.

John Little of Ingersoll, Ontario, built this sedan type Ford funeral coach for the Ford Funeral Home of Blenheim, Ontario. This car was both a side servicing and a rear servicing car. The styling of this hand built coach was very similar to that of the Siebert models, and looked somewhat like an elongated Ford sedan.

Another Little creation was this sedan ambulance built on a Dodge chassis that had been elongated. These hand built cars were side servicing models and some were built in funeral car form. Note the etched glass in the rear door windows and the roof mounted siren. The craftsmanship that John Little put into these cars made them last for a good many years of rough service.

1947

Production for 1947 was much smoother without the material restrictions and price fluctuations that plagued the last sales year. Because new cars were still very greatly in demand and were at a premium, the manufacturers did not add any new models, with the only identifiable style variations emanating from the minute trim changes that Cadillac and other chassis suppliers made to their products.

Henney did not produce any funeral service vehicles or ambulances this year, preferring to design and engineer a totally new model for introduction in 1948. Even Packard was, at the time, scrambling to gather dies and tools for a new car. Indeed, in the post-war auto boom there was only one car manufacturer, Studebaker, that displayed the ability to get a totally new car into the showroom in 1947. Manufacturers using other chassis would have to be happy with older designs and styled chassis.

In order to more adequately handle sales and service, Miller built an extensive nation-wide distributor organization. This operation was patterned after the standard passenger car dealerships and offered local, or relatively local, service to all owners and prospective customers.

With chassis and materials more available this year, the companies engaged in building professional cars started to turn out these vehicles just about as fast as was possible, and a total of 3,746 left the shops.

Two ambulance models were offered by Superior on the Pontiac chassis this year. The limousine version shown here was priced at $4,668 and a landaulet version at $4,747. The built in roof warning lights were optional at extra cost as was the combination siren warning light. Note the Cadillac type rear fenders and taillights and the frosted glass in the rear quarter windows. This car is equipped with Pontiac dealer installed fog lights and windshield pillar spotlight. The Superior nameplate is seen on the front fender.

Superior Body Co. of Lima, Ohio, continued to offer the Pontiac chassis coaches this year without any significant styling changes. So high was the demand for new cars that long waiting lists were established and cars were subject to price at time of delivery. The Pontiac funeral cars and ambulances built by Superior were offered in either the limousine or landaulet bodystyle. The limousine funeral cars on the Pontiac chassis sold for $4,289, with the landaulet version going for $4,367. A service car version was also offered at $4,289. All Superior Pontiac coaches bore a serial number from P-8 MB-1001 up. Notice the Cadillac rear fenders and taillights that were used on this year's Superior Pontiacs.

Superior continued to offer Cadillac chassis coaches in the same styling as last year. This landaulet funeral coach was the top offering and was priced at $5,535. Both the Cadillac chassis limousine hearse and the service car sold for $5,450 this year. This car had a two-tone paint finish. The styling was superb for the price range, and these cars won Superior many life long friends in the trade.

Superior also offered Cadillac ambulances in either landaulet or limousine styling. The coach shown here was built for the U.S. Government, but other than the insignia on the rear side doors, was just like all of the civilian models. The limousine ambulance on the Cadillac chassis went for $5,785, while the landaulet version carried a price tag of $5,865. This model is completely standard without any optional warning lights or other accoutrements.

The Isenhoff Auto Rebuilding Co. of Grand Rapids, Mich., offered this interesting sedan ambulance called a Dualance on the Chrysler chassis this year. The Dualance is shown here fully open to reveal the cot that rode beside the driver, who rode in a bucket seat. The attendants rode on the rear seat.

1947

The last of the fully carved side funeral coaches was very likely this S & S Macedonian carved funeral coach, offered this year. The panel of this car was of cast bronze, which was painted with a dull finish. The carved drapery styles had changed over the years and are now more modern and flowing. Carved cars were being phased out because the pontoon fenders of the new cars make it hard to apply the carved panels and retain a graceful flowing line.

Offered in both three-way electric and manually operated versions, this S & S DeLuxe limousine funeral coach was designated a Model 47230. In the electric three-way form, this distinctive car sold for $6,921, while the manual version was priced at $6,283. Other limousine coaches in the S & S line were called Arlingtons. They, too, were offered in a wide variety of loading styles and prices.

The S & S Florentine flower car was pleasingly styled and very versatile. This car was operated by the W.W. Chambers Morturay in Washington, D.C., and bore the large chrome Chambers nameplate on the doors. Note the high appearance of the car and the wide rear side doors. The flower tray was finished in stainless steel and surrounded by a stainless steel railing.

The Flxible Company of Loudonville, Ohio, continued to be the only manufacturer of Buick professional cars. This year it turned out 357 such vehicles. Mounted on the Buick Series 70 chassis, these cars carried body serial numbers between 15248 and 15599. This limousine funeral car was designated a Model B21-747 and sold for $4,992 in straight end loading form, $5569 in manually side loading form, and $5932 in electrically operated side loading versions. The limousine ambulance was designated a Model B22-747 and wore a price tag of $5,796. A combination ambulance funeral car, Model B23-747, sold for $5,301. There were no landau models offered this year. The hood lines were still raised for proportional appearance. This limousine coach was an end loading model.

Adorned with the special S & S hood marks and hubcaps, this was the S & S Victoria for 1947. The car's styling had changed very little from that introduced in 1942 and was available in either standard or DeLuxe versions. DeLuxe Victoria three-way electric sold for $7,171, while the three-way with a manually operated table was priced at $6,533. Both three-way cars were designated as Models 47280 and the electrically operated models had an E placed behind this serial number. Here is the Victoria straight end loading funeral coach Model 47250 that sold for $5,945. A combination version of this coach sold for $6,010 and carried the designation 4720C. The Parkway Victoria ambulance continued to be offered complete with leaded glass in the side windows and complete emergency equipment for $6,485. It carried the Model designation 47293.

Carried over from the 1942 models, the styling of the Flxible service car was still a very attractive coach. Designated as a Model B24-747, this car sold for $4,992 complete. Notice the optional bumper mounted fog lights and the spotlight. The chrome wreath and decorative strips were part of the service car package.

1947

Meteor continued to build a coach lamp into the front door pillars of their cars and to use a uniquely shaped front door window. The rear loading funeral car in limousine style, as shown here, was designated Model 470 and sold for $5,080, while the landau version, Model 470-L rear loading funeral coach was priced at $5,272. A service car was offered with a blanked rear body side decorated with large wreaths and cycus leaves. This car, designated Model 472, sold for $4,904.

Eureka offered only one style of ambulance this year, and it was the limousine type. This car in standard form sold for $5,863, but the price was subject to equipment. For instance, this car with the built in roof lights and special medicine cabinets would cost somewhat more than the standard model.

Carmichael Motors of Tillsonburg, Ontario, was a Chevrolet and Pontiac dealer which began to build a few extended frame ambulances and funeral cars this year. This long wheelbase Chevrolet chassis Carmichael ambulance featured extra wide rear side doors, special rear quarter windows, and roof mounted siren and corner warning lights.

Meteor styling was unlike that seen on any other make of funeral coach availalbe. Even the ambulances carried the door window and coach lamp. Seen here is the Meteor limousine ambulance complete with frosted glass in the quarter windows. The Meteor ambulance was offered in two body styles, limousine and landau. The limousine version, Model 471, sold for $5,464, while the Model 471-L landau was priced at $5,654.

The Eureka landau funeral car was also offered in three variations. The straight end loading car sold for $5,350, the manually operated side servicing coach for $5,867, and the power operated side servicing three way coach for $6,930. Eureka landau models could not be confused with any other make due to the unique styling of the upper rear quarters, with the ornate coach lamp and landau bows. The rear side windows were slanted forward also, to add distinction to the styling. Note the Eureka nameplate on the hood.

Eureka built all of its coaches on the Cadillac Series 75 chassis. This year Eureka offered the limousine funeral coach in three models; straight end loading for $5,141, the three-way side-servicing manually operated casket table model at $5,685, and the three-way side-servicing coach with a power operated table for $6,721. The car illustrated here is complete with Cadillac type fog lights with red lenses. The Eureka models had a distinctive Eureka styling touch on cars equipped with optional fender skirts.

This interesting Meteor limousine ambulance was supplied to the Asbury Park (N.J.) Fire Department and was equipped with a vast array of special equipment. Note the interesting fire extinguisher on the front fender, the roof pod warning lights, and spotlights on the windshield pillars.

1947

The most attractive offering in Miller's line for 1947 was the landau, which was available in five versions. The standard end loading funeral car was priced at $5,146, and is seen here ready for delivery to the George E. Logan Funeral Home in Canada. Other Miller offerings in this body style included a three-way manually operated coach for $5,887, a three-way loading coach with power assisted casket table for $6,444, a combination ambulance/hearse called a Duplex for $5,382 and an ambulance for $5,541. Note the pebble grain finish on the roof, the landau bows, and the ornamental torch on the rear quarters.

A.J. Miller again offered a line of stylish coaches on the Cadillac chassis only. This car, a straight end-loading funeral car, sold for $4,886. Three-way models were available in either manually operated versions for $5,626 or in a power operated version for $6,184. The limousine styling was also offered in the combination coach called a Duplex for $5,118 and as a straight ambulance for $5,277. Miller styling was pleasingly straight-to-the-point, without any unneeded ornamentation or frills.

John Little of Ingersoll, Ontario, also engineered, designed and constructed this attractive Chevrolet carved side funeral coach this year. The frame was extended several feet and the carved wooden panels were placed on a hand-formed metal body. The carved panel was of hand carved wood as was the body frame. There were windows between the carved draperies.

The Campbell Funeral Home of Chatham, Ontario, was the recipient of this attractive sedan type ambulance built by John Little of Ingersoll, Ontario, on a Canadian Monarch chassis. This coach was built on an extended frame and was hand finished throughout. Note the interesting pattern on the rear quarter window and the Federal combination siren, warning light on the roof.

The Weller Brothers of Memphis began to build coaches on the Chevrolet, Dodge, Pontiac, Chrysler and Buick chassis in this year. This small car was called the Standard Four Purpose Coach, and could be used as a small emergency ambulance, a child's hearse, a wagon-type flower car or a service car. This Weller Chevrolet was finished as an emergency ambulance, complete with warning lights on the grille and a combination warning light siren made by Federal on the roof. Notice the spotlights and the rear compartment air intake for the ventilation system on the rear body side. The elongated rear quarter window with the ambulance cross is also interesting.

The Monarch chassis that this John Little ambulance was built on was a distinctly Canadian product derived from the Mercury and Ford. This side view of the Campbell Funeral Home job illustrates the length of the extension used for this coach. Sedan type ambulances were loaded from the side doors as a general rule but the ones built by John Little could also be loaded through the rear. This car was finished in white with gold lettering.

1948

Things were again humming at the Henney plant in Freeport, Ill. In this year the company re-entered full-scale production of cars and had its biggest sales year to date. The year was to see Henney produce 1,941 units on the totally new Packard series 22 chassis. Using the slogan, "World's largest manufacturer of funeral coaches and ambulances," Henney announced the 1948 models that had a lot going for them. The 160 horsepower Packard engine put these cars ahead in traffic. The new cars had the best maneuverability of any coach in the industry and were mounted on Packard's 156-inch wheelbase commercial chassis. With these new cars Henney was able to get the jump on the rest of the manufacturers by introducing all new models a year before anyone else could. Price-wise Henney cars were not inexpensive. Henney prices were just below those of the S & S line (S & S were the leaders in high quality and high priced coaches) and above those of the rest of the industry. One of the most beautiful Henney models for 1948 was a white landau three-way coach built for the Wilson Sammon Company. This car, shown at the National Funeral Directors Convention, featured a novel neon sign above the windshield that read "funeral," whitewall tires, heavy light-colored draperies and spotlights.

Superior Coach built its first flower cars this year and mounted them on Cadillac chassis. They also ventured into using Chrysler Corp. chassis for the first time this year. Constructing funeral coaches, service cars and ambulances on Dodge, DeSoto and Chrysler chassis, Superior created another new line and price category for itself.

The sedan type ambulance had almost disappeared from the scene in the United States, but in Canada it was taken one step further. Sedan style hearses were being built in the garage of John C. Little of Ingersoll, Ontario. The most notable of these was a conversion carried out on a 1948 Lincoln sedan. The car was lengthened and a hearse table and interior put into it. The coach was side servicing and from the outside it was hard to tell that it was anything more than a lengthened Lincoln sedan.

The other manufacturers continued to use the standard Cadillac commercial chassis with very few changes in styling for this model year.

With chassis and materials now available, the professional car builders finally started catching up on the back orders. As a result, a record total of 4,727 funeral cars and ambulances were produced in the U.S. this year — a record that would not be broken until 1965.

In late 1947 Henney introduced the new post war models through elaborate displays at the Funeral director's conventions. The car in the foreground was a special unit built by Henney for the Wilson Sammon Co. It was a Model 14800-L finished in white. This landau had spotlights on the windshield pillars, and an interesting neon "FUNERAL" sign over the windshield. This landaulet, complete with Elecdraulic operated casket table and Leveldraulic to keep the coach level during loading or unloading, sold for $7,022 in standard form.

Henney serial numbers ranged from G-600001 this year. They produced a record of 1,941 cars on the Packard 2213 chassis. This landau, with manual side servicing, sold for $6,736. The same car could be equipped with Leveldraulic and Elecdraulic.

This interesting shot shows the front view of the Henney funeral car and illustrates how the Leveldraulic kept the coach level during loading or unloading operations. Experts will notice the extra row of grille bars that were added to Henney Packards.

The rear loading Henney Packard funeral car was offered in either landau or limousine body styles. This limousine coach was priced at $5,644 while the landau sold for $5,878. Notice the high body lines and the formal draperies on this car.

1948

This car is one of a fleet of Henney Packard ambulances operated by the Smith Ambulance Service. It features optional roof pod warning lights, spotlights, and a Federal roof siren warning light.

Henney also offered a combination coach on the new Packard chassis. These cars were rear loading only and could be changed from ambulance to funeral car in an instant. The combination in limousine form was priced at $5,908 while the combination in landau styling went for $6,146. The ambulance insignia in the rear quarter window was metal and snapped in quite rapidly. These cars were quite popular because many localities did not furnish a regular ambulance service and this job fell to the local funeral home, which many times could not afford two separate vehicles. The combination coach fulfilled this need in a smart-looking unit of dual usage.

Henney also offered a complete line of ambulances on the new Packard chassis in both limousine and landau styling. The new Henney Packard ambulance in limousine form sold for $6,119, while a landau version was priced at $6,357.

The most beautiful offering in the Henney Packard line for 1948 was this Flower car shown here ready to receive flowers. The low rear deck was stainless steel on both the inside and the flower deck outside. As the handling of flowers was an important part of the funeral service, these vehicles were a definite asset to any funeral home's cortege. Henney also offered a service car on the same chassis as all other Henney vehicles, but with no rear windows and a large stylized wreath and chrome strips on the upper rear body sides. The service car sold for $5,120, while the flower car shown here carried a price tag of $5,774.

The Flxible Premier funeral car with limousine styling was priced at $4,998 and the same car with landau styling sold for $5,198. A Premier service car was also offered for $4,998. This view of the standard Flxible Premier funeral car shows the raised hood line. The limousine version could also be had in ambulance form for $5,804.

Flxible continued to offer a strong line of ambulances, combinations, service cars and funeral coaches on the Buick Series 70 chassis. This year the Loudonville, Ohio, firm built a total of 646 Buick professional cars of all types. Body serial numbers went from 15600 to 16245 and the coaches were offered in both limousine and landau styling. This Flxible landau was priced at $4,998 plus $200 for the landau styling. Dynaflow transmission was also a $200 dollar extra. When equipped with a manual three-way casket table operation the car sold for $5,575, and the power operated model went for $5,932. Note the unique rear window treatment and the raised hood line.

1948

The Hess & Eisenhardt S & S Classic Arlington Service car was the preferred coach by those looking for classic formality. Carved panels replaced glass in the rear compartment windows of this model. The car as a whole shared the body of the Arlington limousine funeral coach. Priced at $5,595 this car was the sole remaining heir to the carved coach's demise, and this style would be around for quite a few years on service cars in America.

The classic S & S Victoria was as beautiful as ever in its new 1948 garb. Though styling had been altered very little, the car continued to have the grace and elegance that befitted this top-of-the-line model. The Victoria was offered in two series in this year. Called the DeLuxe Victoria and the Victoria, the line consisted of three models in each series. Shown here is the DeLuxe Victoria side loading coach that when manually operated was priced at $6,533. A power operated version sold for $7,171. A straight end loading DeLuxe Victoria was priced at $5,945, while the Victoria Parkway ambulance, a derivative of the funeral car but with ambulance equipment and window styling, was priced at $6,485. Standard Victoria models differed from the DeLuxe series in interior appointment and minor styling details and these cars were somewhat lower in price.

All of the manufacturers using the Cadillac chassis were now supplied with Series 76 chassis cowl units. This Hess & Eisenhardt S & S Deluxe Florentine Flower car was designed to be used as both a flower car and a service car. It featured a high waist line with a low rear deck. Note the rear deck tonneau and the S & S ornamentation on the hood. This car carried a price of $5,940 and was a distinctive, exclusive prestige-building addition to many funeral fleets.

The S & S Hess & Eisenhardt DeLuxe Arlington limousine funeral coach was offered in three versions: the straight end loading model shown here at $5,695, the manually operated three-way version for $6,283, and the power operated three-way version at $6,921. All three cars shared the same styling and exterior appearance with varying interior details and appointments.

Eureka offered four versions of the funeral car in this year. The landau seen here was priced at $5,341 while a limousine coach carried a price tag of $5,141. When equipped with a manually operated three-way casket table the limousine sold for $5,685. The price of a landau with the side servicing feature in either form was $200 above that for the limousine. Note the graceful curve of the rear window. The Eureka nameplate on the hood was unusual.

This S & S Speical Kensington Ambulance was owned and operated by the Milwaukee Fire Department and designated as Ambulance Number 1. In standard form, the S & S Special Kensington Ambulance sold for $6,512. This version is equipped with optional roof mounted Federal warning light siren and special interior features. Note the rather high waist line.

Eureka also offered its Cadillac ambulances in both landau and limousine version. The limousine version shown here cost $5,863 with the roof pod lights and the combination siren warning light on the roof being optional at extra cost. Note the AMBULANCE nameplate in the windshield and the red lenses over the Cadillac parking lights.

1948

Venturing into a new area this year, Superior attempted to attract buyers from the smaller custom makers with a line of coaches on Chrysler Corp. chassis. These cars were offered in limousine, landau, service car and ambulance styles on a choice of Chrysler, DeSoto, or Dodge chassis. This attractive landaulet is shown on the DeSoto chassis. All Chrysler frames were from the long wheel base, nine-passenger cars and did not need to be lengthened. Note the angular styling of the area where the doors and roof come together. This was a rear loading coach only but still a very attractive package. Superior built 602 cars this year.

The Superior limousine funeral car on the Cadillac chassis sold for $5,450 and could be ordered from mid-year with the new Superior Lev-O-Trol and an automatically power operated casket table. Superior's Lev-O-Trol was operated from the driver's seat with a single push-pull lever that when activated operated the automatic coach leveling and the casket table. The Superior Cadillac limousine is shown here with this new option.

Superior continued to offer an extensive line of Cadillac coaches through this year. This Superior Cadillac landaulet was offered in either ambulance or funeral car versions. This attractive coach in hearse form was priced at $5,535, while in ambulance form the landau sold for $5,865. Superior also offered a service car on the Cadillac chassis, and this car with blanked off rear side windows and service car exterior trim was priced at $5,450.

The Superior Chrysler 8 ambulance is shown here in limousine styling. Other models available on Chrysler Corp. chassis included combinations, service cars, and flower cars. All models were offered on either the Chrysler 6 or 8-cylinder chassis. The ambulance and combination as well as the funeral car was offered in a choice of limousine or landaulet styling. This was the only year Superior ever offered Chrysler Corp. products as standard chassis within their line.

This stylish Superior Dodge service car carried styling very similar to that found on the large Cadillac chassis Superior products. Note the flowing lines of this chassis.

This Chevrolet landau funeral car was the product of the Weller Brothers of Memphis. It featured a fabric covered roof. The car was a rear servicing unit only and was also offered in limousine and ambulance versions. Weller also offered coaches on the Dodge and Pontiac chassis.

The National Body Co. of Knightstown, Ind., was a small firm specializing in the extension frame type professional car. This National Chevrolet service car is a typical example of this firm's work. Note the long wheelbase and the decorative panels set into the window areas. Other models in the National range consisted of funeral coaches in both landau and limousine styling and ambulances.

1948

The Meteor landau was offered in five versions: Model 480-L was a rear loading funeral car (shown here) and priced at $5,598; Model 480-LC was a combination and carried a price tag of $5,711; Model 480-LH was the manually operated three way funeral coach and sold for $6,516; while a power operated three way coach was also offered and designated Model 480-LHE, priced at $7,057. The Meteor landau was also available as an ambulance for $5,871.

The Meteor Motor Co. of Piqua, Ohio, produced a fine line of funeral cars and ambulances on the Cadillac Series 76 chassis. This combination hearse/ambulance was a rear loading car only, was priced at $5,550, and designated a Model 480. The ambulance was designated Model 481, sold for $5,711 and did not differ substantially in styling. This combination has red lenses over the Cadillac fog lights and retains the Meteor coach lamp on the front doors.

The A.J. Miller Co. of Bellefontaine, Ohio, continued to offer a complete line of Cadillac coaches. This coach was offered in no less than five different models, as was the landau. The standard Miller limousine funeral car was priced at $5,219, while the landau carried a price tag of $5,486. These cars were also offered in Duplex combination versions, manually operated and power operated three-way versions, and as ambulances.

The Meteor limousine was also offered in three funeral car versions that were designated Model 480. It was a rear loading funeral coach, priced at $5,432. The Model 480-H was equipped with a manually operated three way casket table and was priced at $6,359. The top of the limousine funeral car offerings was the Model 480-HE that was equipped with an electrically assisted three way casket table. This sold for $6,889.

Meteor's Model 482 was this attractive Cadillac service car that sold for $5,174. Note the continuance of the odd-shaped front door windows, and the ornamental wreath and cycus leaf on the rear body sides. The carriage lamps on these cars were mounted on the rear sides.

The A.J. Miller flower car continued to be offered with styling as individual as the car itself. The Miller flower car had a fabric covered roof with dummy landau bows on the rear quarters. The low rear deck was finished in stainless steel with a railing around it. At the rear of the car was a tonneau for the flower deck.

1948

A triumph in the coach builders art was this attractive Lincoln sedan type side loading funeral coach built in the shop of John Little in Ingersoll, Ontario. Looking very much like an ordinary Lincoln sedan that had been stretched, this car was all built by hand. The interior was fitted with the highest grade mohair.

Little's Lincoln was a side loading coach only and the casket table would swing through either extra wide rear side door. Note the width of the rear side door and the fact that opening the front door was not necessary with this side loader. John Little was one of the last to build these coaches completely by hand.

Owned by the Nicholls Funeral Home of Wallaceburg, Ontario, this was another Canadian creation. This Chevrolet landau funeral coach, with side servicing, was built by the Ingersoll Auto Body Corp. of Ingersoll, Ontario. This extended frame unit was built from a panel delivery unit. Note the unusual landau area.

The Chevrolet chassis was quite popular in Canada, as is evidenced by this John Little Chevrolet limousine funeral coach. The Chevrolet frame was extended and the body hand formed by Mr. Little. It took approximately three months to complete a car for a customer.

John Little built many one-of-a-kind funeral coaches and ambulances in his time. This carved Chevrolet is but one example of his fine work. Mr. Little was born in Scotland and for many years worked for the Mitchell Hearse Co. He entered the business on his own in 1940 and soon his creations were eagerly sought after throughout Canada. Typically, a customer would describe what he had in mind and then Mr. Little would go at it.

This interesting Citroen combination pallbearer's car and hearse was owned and operated by the Furgeon Chapel in Paris. This chapel is operated by Masion Roblot, one of Paris' leading funeral homes. The rack on the roof was to carry floral tributes, and there was a seat behind the driver's seat for passengers. The area directly behind the passenger seats was reserved for the casket and this compartment was not partitioned off as in American funeral cars. The four side windows of this casket compartment featured drapes on them.

1949

The major manufacturers using the Cadillac commercial chassis finally announced their long-awaited post-war models with the new Cadillac style. Although the Cadillac Series 75 chassis was still wearing post-war type styling, the new commercial chassis had the more modern type front and rear styling that had been introduced on Cadillac passenger cars in 1948. This new chassis also featured a totally revised powerplant. The new Cadillac engine represented an entirely new concept in automotive engineering. It was smaller and lighter in weight while providing greater fuel economy, better performance and smoother, quieter operation. It was designed to take full advantage of the new higher octane fuels that were just now becoming available. The new engine was a 90 degree V-8 that displaced 331 cubic inches and developed 160 horsepower at 3800 RPM.

The major manufacturers using the new Cadillac commercial chassis took this opportunity to announce their first all new coaches and designs. Maintained were some of the styling trademarks of some of the manufacturers. Meteor cars all featured a coach light mounted on the front door window frame lending the window a strange shape. A.J. Miller landau models were available with an oval porthole window on the rear roof sides, and the Eureka Company landau models featured a large ornate carriage lamp in the landau area. These were, in many cases, styling trademarks that had been common with the respective makers in previous models but were carried on with the new cars for distinction and easy identification. In many cases these styling trademarks were to be carried right into the mid and late 1950's.

A number of smaller makers began to join the business in this year. Most were doing conversion type work extending passenger car chassis or converting station wagons. One firm to join the business in this era was Guy Barnett & Co. of Memphis. Building extended Chevrolet ambulances and funeral cars, Barnett was one of the first of a long line of Memphis based companies involved in this type of work.

After one tremendous production year, the business fell into a slump this year, and only 2,853 funeral cars and ambulances were produced in the U.S.

Henney continued to offer their beautiful flower car, now mounted on the Packard 2313-9 chassis. Selling for a moderate $5,120, this car was the optimum of style and utility.

When not in use, the flower deck could be covered by a special all-weather deck cover to protect the stainless steel of the flower deck. The all-weather deck cover was stored in the tonneau at the rear of the flower deck when not in use. Note the graceful lines of this beautiful coach.

The Henney Packard Model 14890 three-way side servicing funeral car with limousine styling was priced at $6,502 when equipped with an electrically operated casket table. The same car, when equipped with electrically operated casket table and Leveldraulic, was priced at $6,788. Note the smooth styling and soft rounded lines of this coach.

The least expensive offering in the 1949 Henney Packard range was the service car at $5,774. The Henney Motor Co. of Freeport, Ill., claimed to be the world's largest manufacturer of hearses and ambulances. This service car had a special Henney stylized wreath and chrome bars on the rear body sides. The Henney nameplate appears on the sides of the hood. Henney built a total of 380 cars in this year.

1949

Like all Henney Packards, this ambulance was powered by the famous Packard 106 horsepower cylinder engine. Selling for $6,119, this attractively styled ambulance was fully equipped and featured special rear quarter window decals and pull-down center window shades.

When seen from the rear, the Henney Packard ambulance in limousine styling was as pleasing as from the front. Note the large rear door opening. The ambulance was also offered in Landau styling and was priced at $6,357.

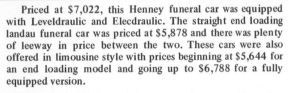

The Henney Packard, Model 14991, dual purpose ambulance and funeral car was the Henney combination car. It was offered in either landau or limousine styling. The limousine shown here carried a price of $5,908, while the landau style sold for $6,145. These cars featured concealed casket rollers and hardware and instantly removable ambulance signs for the rear quarter windows.

Priced at $7,022, this Henney funeral car was equipped with Leveldraulic and Elecdraulic. The straight end loading landau funeral car was priced at $5,878 and there was plenty of leeway in price between the two. These cars were also offered in limousine style with prices beginning at $5,644 for an end loading model and going up to $6,788 for a fully equipped version.

The Shop of Siebert of Toledo continued to offer a line of vehicles built on Ford chassis. This long wheelbase ambulance was built on the light Ford F-100 panel truck chassis, with ambulance type rear quarter windows, a large rear hinged center door, dual roof mounted fresh air ventilators and roof marker lights. The fender mounted warning light was optional.

Siebert also offered combination service cars and funeral cars on the Ford F-100 truck chassis with the most popular being this service car version. This vehicle could also be used as a flower car. Note the service car decorative panel on the rear quarter and the coach lamps. The Siebert nameplate appears on the lower half of the cowl just above the fender.

One of the most interesting vehicles that Siebert offered to the funeral industry was this extremely long wheelbase Ford six-door sedan. They suggested its use as a pallbearer's car or family limousine. It was offered with either the Ford 100-horsepower V-8 or the 95 horsepower Ford 6-cylinder engine. Siebert claimed that the long wheelbase made the ride of this car as comfortable as a limousine costing thousands of dollars more.

1949

The Meteor ambulance was also offered in both landau and limousine versions this year, with this version being designated a Model 491, selling for $6,316. The landau version of this car was priced at $6,477 and carried the designation 491-L. The warning lights on the roof and the Federal siren light were optional at extra cost.

Looking very high, the 1949 Meteor coaches were all new as far as the chassis were concerned but carried the same bodies and styling that had been seen on previous models. This Meteor rear loading funeral car sold for $5,942. When equipped with electrically operated side loading, the same vehicle sold for $7,185. Combination models were also offered starting at a price substantially above that of the regular funeral car.

Available in four versions, the Meteor landau carried styling similar to that seen on previous models. The end loading model shown here was designated a Model 490-L and sold for $6,099, while the same car fitted with side loading went for $6,775 and was given a designation 490-LH. When fitted with the power assisted side loading casket table, the car was given the designation 490-LHE and was priced at $7,342. There was also an ambulance version of this car available for $6,477 and this was designated a model 491-L. Note the formal draperies on this model as well as the unique Meteor door-mounted coach lamps.

At $5,756, the Meteor service car was the least expensive coach offered in 1949. It was called a Model 492 and carried all of the decorative ornamentation on the rear body sides that all Meteor service cars had for the past few years.

The National Body Manufacturing Co. of Knightstown, Ind., continued to offer a vast array of models on almost any chassis. This is the National Pontiac limousine ambulance complete with optional warning light pods and a roof mounted combination siren warning light. Note the rear marker lights on the roof and the ambulance pattern in the rear quarter window.

Meteor continued to offer coaches on the Chevrolet chassis. These vehicles were available in a wide variety of styles that included a landau (shown), limousine, combination and ambulance.

National also offered the Chevrolet service car this year, but with a much cleaner and more attractive looking car than last year. The service car decorations on the rear body sides were in the form of a central wreath. National was one of the smaller makers that were aiming at the lower price ranges of this field with some rather high quality vehicles. National had taken over the old Knightstown funeral car plant and was beginning to make its presence felt.

This 1949 A.J. Miller landau was delivered to the Foster & McGarvey Funeral Home of Edmonton, Alberta. The oval window on the rear landau panel was optional and was to be a popular Miller signature in the years to come. The rear loading landau this year was priced at $5,700 with an ambulance version going for $5,965. A combination Duplex was priced at $5,860, while a three-way manually loading coach was priced at $6,330 and a power operated three-way coach at $6,850.

Miller also offered both funeral cars and ambulances in limousine styling. The standard limousine ambulance as shown here carried a price tag of $5,700, while the funeral coach version of this car was priced at $5,440. A Duplex combination coach in this body styling was priced at $5,595 and the funeral car was offered in both a manually operated side servicing model for $6,070 or a power operated model for $6,590. The attractive lines of this car were complemented by the optional roof pods for warning lights and the frosted glass in the rear quarter windows.

The Hull & Son Funeral Home, Little Chapel of the Flowers, purchased three Superior coaches on the Cadillac chassis. All were finished in gray and black. This year Superior of Lima, Ohio, offered on the Cadillac chassis landau, six versions with prices beginning at $5,835 for the straight end loading landaulet. The landaulet with power side servicing was priced at $6,595 and the Lev-O-Trol was another $375.

This attractive Superior Cadillac limousine funeral car was supplied to the Regina Funeral Home in Regina, Saskatchewan. The standard rear loading funeral coach with limousine styling went for $5,750 and side servicing in either power or manual operation was available. Lev-O-Trol was also offered on side servicing cars for an additional $375.

Part of the Hull & Sons fleet of new Superior Cadillac coaches was this attractive service car. Notice the double stylized wreaths on the rear body sides. The Cadillac service car carried a price tag of $5,750 this year.

Also part of the Hull & Sons fleet of 1949 Superior Cadillac coaches was this attractive ambulance with special rear window motifs and unusual pod mounted roof warning lights with crosses in the lenses. This year the Superior Cadillac ambulance was offered in either landaulet styling priced at $6,165, or with limousine styling as seen here for $6,085. The series designation for the ambulance was 1486 with the landau designated a Model L-1496. Superior built 625 professional cars this year.

1949

The Hess & Eisenhardt S & S Knickerbocker was the company's limousine offering for the year. This car was offered in four versions plus one called the Londonderry ambulance. The straight end loading funeral coach carried a hefty price tag of $5,990 while the same car equipped as a combination went for $6,060. When equipped with a manually operated three-way casket table, the Knicherbocker was priced at $6,595, and the three way power operated casket table boosted the price to $7,271. The ambulance version of this limousine coach, called the Londonderry carried a factory drive-away price of $6,460.

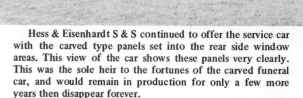

Hess & Eisenhardt S & S continued to offer the service car with the carved type panels set into the rear side window areas. This view of the car shows these panels very clearly. This was the sole heir to the fortunes of the carved funeral car, and would remain in production for only a few more years then disappear forever.

One of the most attractive of the S & S offerings this year was the series 400 Florentine flower car combination. The rear inside of the body was lined with satin polished aluminum as was the casket table and flower deck. The rear deck was unusually low for easy loading and the rear doors were wide enough for easy entry to the rear compartment. The flower tray on this car was entirely new and exclusive with S & S this year. It could be power inclined, level, or fully removed, giving a completely open rear space for a full capacity load. This car carried a price of $5,940.

Hess & Eisenhardt also offered a line of standard models at a lower price range. These carried all of the same convenience features and craftsmanship of the Series 400 or the Superlines. Here we see the Standard limousine funeral coach just as it was offered. Note the lack of extra chrome embellishments.

Eureka of Rock Falls, Ill., offered this all new DeLuxe flower car. It had a stainless steel deck with a hydraulic lift and chromaloid interior. A standard version of the car was offered with a canvas covered flower deck and plastic interior trim.

S & S Hess & Eisenhardt continued the Victoria with styling changes to fit the all new Cadillac chassis. The Victoria was offered in both the Series 400 and the luxurious Superline models. The series 400 Victoria Parkway ambulance (shown) sold for $6,730 and carried a model designation of 49493. The Victoria funeral car with the same styling and equipped with electrically operated three-way casket table was priced at $7,541. There were many other Victoria models in between.

The least expensive offering from Eureka was the Cadillac service car. Of interest is the styling of the rear body sides and the ornamentation that was unique to this firm for their service cars. The coach was a five door vehicle with a partition between the front and rear compartments. The service car and the flower car were announced after the beginning of the 1949 model year.

Restyling their entire line of coaches, Eureka again offered its Cadillac chassis ambulances in both limousine and landau body styles. This limousine ambulance sold for $6,732 while the comparable Eureka landau ambulance was priced at $7,032. The roof pod warning lights and the roof mounted siren warning light were optional at extra cost. Note the interesting rear quarter window pattern on this car.

1949

Eureka continued to offer a complete line of coaches on the Cadillac Series 86 chassis in both landau and limousine body styles. The landau shown here was completely new for 1949, with more modern and graceful lines. In standard rear loading form, this car sold for $6,125 and when equipped with a manually operated three-way loading casket table, the same car was priced at $6,889. The most expensive landau funeral car offered by Eureka this year was the version equipped with the three-way loading that was power assisted. This car sold for $7,487. Eureka continued to place an ornamental coach lamp on the landau models. This was becoming a Eureka trademark similar to the Meteor door mounted coach lamp and the new A.J. Miller oval window on their landaus.

The Eureka limousine funeral car styling followed that of the landau in overall lines. This artist's conception shows the side loading Eureka limousine funeral car in somewhat stretched rendering. All of the chassis supplied to the various coachmakers by Cadillac were the Series 86 which had a wheelbase of 163 inches. They were powered by the new Cadillac V-8 that produced 160 horsepower and displaced 331 cubic inches. This Eureka limousine funeral car sold for $5,825 in rear loading form, $6,503 as a combination, $6,599 as a manually operated three-way and $7,187 as a power operated three way coach.

New for 1949 and sold exclusively through authorized Chevrolet dealers, this was the Barnette-Chevrolet long wheelbase ambulance. Built by Guy Barnette & Co. of Memphis, these cars were made of all steel while some of the large makers were still using wooden body framing. Note the Barnette nameplate on the front fender.

Another new maker offered a truck type ambulance in this year. This is the Oltman-O'Neill van ambulance on the Dodge chassis. Note the large frosted glass window with the red cross, the fender mounted Funeral siren/warning light and the step-van type front doors. This car was also offered as a service car or a van-type flower car.

Building a total of 403 professional cars this year on the Buick chassis, Flxible offered a newly styled car to match the new Buick design. The Buick Premier three way landau was attracting favorable comment wherever it was shown. Flxible's three way coach featured a very low loading height which was a definite improvement in casket handling. The funeral car was offered in either limousine styling or with the increasingly popular landau type of body. The three way landau, shown here, carried a price tag of $7,713, which included the electrically operated casket table and a Flxible leveler for the car during loading and unloading. The limousine model was offered for $5,864 in the rear loading style, $6,235 as a combination, and $7,040 as a manually operated three-way. When equipped with electric casket table operation and leveler, the limousine went for $7,413. All of these styles were also offered in a landau version for an additional $200.

The least expensive model offered by Flxible this year was the Premier service car at $5,427. This car continued to carry a chrome wreath and chrome bars on the upper rear body sides and made a very attractive vehicle. Note that the hood line is not raised. The car carries the famous Buick portholes on the front fenders.

The Flxible Premier ambulance for 1949 was mounted on the new Buick Roadmaster chassis. This model in its standard form was priced at $6,716, but the two-tone paint finish, roof mounted warning lamps, and whitewall tires were optional at extra cost. Note that with the new styling, the necessity of raising the hood line was done away with.

1950

Cadillac, which had introduced a new chassis in 1949, restyled the bodies and, of necessity, the commercial chassis front and rear ensembles. While the manufacturers had completely retooled for the 1949 models, they found themselves again retooling for the 1950 model year. For the second year in a row, the trade was offered a totally redesigned line of professional funeral vehicles from which to choose.

Henney, on the other hand, offered a warmed-over version of the original 1948 style. While offering no new models, Henney did offer exterior trim distinctions that quickly identify 1950 models. Fender skirts appeared on some Henney cars for the first time this year.

The Meteor Motor Co. began to produce a line of lower priced coaches on extended Chevrolet and Pontiac models. These cars were conversions of panel deliveries or station wagons or extended frame passenger car chassis. The extended cars were offered in a complete line of hearses, ambulances and service cars. One of the most interesting offerings in this Meteor range was the Pontiac flower car built on an extended frame.

The Acme Motor Co. joined the ranks of funeral service vehicle makers this year with offerings on the Pontiac chassis. Acme built a line of ambulances, funeral coaches and service cars by reworking the Pontiac light duty sedan delivery models and in some cases extending the chassis. With these vehicles Acme, based in Sterling, Ill., entered the low price bracket funeral coach field.

For the second year in a row, the business was in a slump, and only 2,971 funeral cars and ambulances were produced in the U.S.

The most attractive of the Superior Coach Cadillac chassis offerings was the Coupe De Fleur flower car. This coach was available in two versions, the first with a manually operated flower tray, and the second with a power operated flower tray. Note the stainless steel rear deck and a stainless steel tonneau cover. The coupe roof line incorporated a new wrap around rear window that was becoming the vogue in the passenger car world around this time.

Showing the Superior Coupe De Fleur from another and perhaps more pleasing angle, this illustrates the new wrap-around rear window styling and the stainless steel rear deck. In the manually operated form this car carried a price tag of $5,970, while the hydraulically operated deck version cost $6,050.

Superior continued to offer a complete line of coaches in both limousine and landau styling through this year. Shown here is the Superior three way automatic loading limousine that carried a price of $6,910, while this same car with standard rear loading was priced at $6,050. Landaulet models with comparable equipment were priced at $6,260 and $7,115. Combinations were offered in both body styles beginning at $6,145 and soaring up to $6,630. Superior styling in this year was clean and crisp with modern formal draperies and no unnecessary chrome trim. Superior constructed a total of 717 cars in 1950.

Superior offered a complete array of ambulance models in both limousine and landaulet body styles again this year. This attractive limousine ambulance was owned and operated by the Oldsmobile Division of General Motors at their plant at Lansing, Mich. The roof pod lights were optional at extra cost as was the two-tone paint finish. In standard form the Superior limousine ambulance carried a price tag of $6,500, while the Landaulet cost $6,710.

Meteor again offered a line of 12 Cadillac chassis funeral cars and ambulances in both limousine and landau body styles. The landau was offered in four separate versions and the same number of price ranges. The standard end loading landau designated Model 500-L sold for $6,417, with a rear loading combination going for $6,740 complete with a removable steel hearse table, a heater built into the partition for the rear compartment, cot hooks, one fold away attendant seat, and rear bumper step. It was designated Model 500-LC. The landau funeral car was also available as a three-way car that sold for $7,086 in a manually operated form and $7,648 in electrically operated three-way form. Meteor styling continued the door mounted coach lamp.

Bedecked with every conceivable type of warning light system and a fire extinguisher built into the front fender, this Meteor ambulance was an impressive sight. Even the coach lamp has a red lens, as do the Cadillac parking lights. The Meteor limousine ambulance carried a standard price tag of $6,634 and was designated a Model 501. This car is equipped with many optional warning lights that include a revolving mars light on the roof, twin pod mounted warning lamps on the roof, fender mounted red warning lights and spotlights. Cadillac offered fog lights combined with their parking lights this year, and the red lenses on this indicate that they were probably made to flash as well as operate as parking lights.

With a healthy price tag of $6,079, only the larger and more affluent funeral homes could afford a Cadillac service car such as this Meteor version. The rear body sides were blanked off and a coach lamp was mounted in the center, from which massive chrome wings spread.

This interesting Meteor emergency ambulance featured an extra high headroom rear compartment as is indicated by the trough in the roof side and the extra extension on the windshield. It is not known if this car was ever built but it was an interesting proposition and one that preceeded today's modern high headroom ambulances. This car is fitted with the rear quarter window vents that were offered on all Meteor combinations and ambulances and that interesting fire extinguisher mounted in a well in the front fender.

The Meteor limousine was offered in the same number of versions as the landau but at substantially lower prices. The straight end loading limousine carried a price tag of $6,261 and was designated a Model 500. In combination form, the same car carried a price of $6,581 and in addition offered a no-draft vent window in the rear quarter windows. As a manually operated three-way coach, this car sold for $6,931, and as a power operated three-way it sold for $7,492. The coach illustrated was a three-way manually operated verison.

Offering a flower car again for 1950, Meteor built this attractive car without the stainless steel rear deck for flowers. The flower baskets and wreaths sat in the rear compartment and there was a tonneau to cover this large hole during inclement weather. Cars with this type of rear flower deck have come to be known as Chicago type due to the fact that this is where most of the cars of this style were sold. Meteor also went to the wrap-around type of rear window with a coupe roof that resembled that of the Cadillac Coupe De Ville. Standard flower cars built by Meteor carried a $6,130 price tag, but cars with this type of rear deck would have been somewhat less expensive.

1950

Designated a Model P-500 this Meteor Pontiac funeral car made up the lower end of the Meteor price scale and was offered in either 8 or 6-cylinder models. It was also offered as a landau funeral coach or a service car. The version shown here was built on the 8-cylinder chassis and was a straight end loading car.

The most economical coach line in the Meteor range for 1950 was this Chevrolet funeral car with limousine styling. Designated a Model C-500, this car was also offered as a landau, a combination, or an ambulance. These Chevrolet funeral cars carried all of the craftsmanship and excellence of the larger and more expensive Cadillac coaches, but at a lower or a more attractive price.

The S & S Florentine flower car with styling introduced originally in 1949 was carried over into this year with only mild restyling on the Cadillac front end and the roof line. S & S continued to utilize the interesting date marks on the hood and new chrome trim on the back part of the roof.

Meteor also offered a low priced flower car on the Pontiac chassis this year. The styling of this car was very similar to that found on the large and expensive Cadillac flower car with the exception that these Pontiac flower cars were of sedan type configuration and had a fixed pillar between the side windows. Once again there is the wrap-around rear window and the Chicago type flower tray.

Continuing with their distinguished styling, the 1950 Hess & Eisenhardt S & S models were available in either Series 400 or the luxurious Superline. The S & S Superline Knickerbocker limousine funeral car sold as a rear loading car for $6,352, as a combination complete with Duo-Floor reversible floor pieces with casket rollers on one side for $6,427, as a manually operated three-way side servicing coach for $7,157, and as an electrically operated three-way car for $7,733. Again offered this year was the famous S & S Victoria in no less than nine differing versions. Prices for the Victoria began at $6,723 and went to $8,004.

Continuing as the last vestige of the carved car was the S & S service car with stamped metal carved type inserts in the rear window areas. Note the interesting owner's initial plaque on the front door and the traditional S & S date marks on the hood.

Eureka offered a line of 10 Cadillac chassis coaches for this year in both limousine and landau styles. Here is the Eureka landau three-way that sold for $7,787 when equipped with an electrically assisted casket table and $7,199 when equipped with a manually operated table. A standard rear loading Eureka landau would fetch $6,769 this year, while comparable limousine models sold for $7,487, three way electric; $6,899, three-way manual; $6,469, rear loading, and $6,535 for a combination. Ambulance prices in this year ranged from $7,032 to $7,332. Note the continuance of the landau coach lamp on the Eureka car and the smooth sweep of rear end styling.

Shown in front of a Trans Canada Airlines' passenger plane and awaiting some air freighted human remains is the 1950 A.J. Miller landau funeral coach of the A.S. Bardal Funeral Home of Winnipeg, Manitoba. This landau is equipped with the oval rear quarter windows. In a straight end loading form it sold for $6,175. Called a Duplex, the Miller combination sold for $6,835. The most expensive Miller landau was the electrically operated three-way version at $7,380. Limousine models began at $5,915 and zoomed up to $7,120.

Available with or without the unique oval window in the landau area, the Miller landau was a dignified and well respected funeral car. This example was photographed when traded several years later for a new car. Miller also offered this car with or without a crinkle finish on the roof.

Appearing for the last time this year was the Packard and Henney styling that made its debut in 1948. Henney serial numbers ranged from 2313-5-2070 up, and they produced a total of 244 cars on the Packard 2313-5 chassis. The Henney Packard landau funeral car was availalbe in seven versions. This coach is equipped with optional fender skirts and had a crucifix in the rear door window. Small coach lamps were mounted on the door pillar and an extra row or grille bars were added to the Packard commercial chassis.

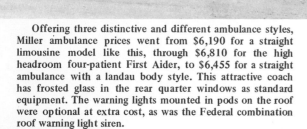

Offering three distinctive and different ambulance styles, Miller ambulance prices went from $6,190 for a straight limousine model like this, through $6,810 for the high headroom four-patient First Aider, to $6,455 for a straight ambulance with a landau body style. This attractive coach has frosted glass in the rear quarter windows as standard equipment. The warning lights mounted in pods on the roof were optional at extra cost, as was the Federal combination roof warning light siren.

The Henney Packard ambulances for 1950 were offered in either landau or limousine styling with prices beginning at $6,119 for the conventional limousine ambulance and going to $6,357 for a landau version. The spotlights, pod mounted warning lights, and central roof mounted warning light siren were all optional at extra cost, as was the two-tone paint finish shown here. The flower car and the service car were still represented in the Henney line.

From the side, the Henney Packard was a very attractive coach. Henney Packards carried price tags that went from $5,878 for a rear loading landau through $6,502 for a side servicing landau with an electrically operated table in a three-way casket table and Leveldraulic. This coach was delivered to the Billups & Sons Funeral Home.

The low cost and attractive Siebert utility service and flower car was built on the Ford F-100 light truck chassis. This car, fitted with roof vents and service car body side ornaments that included a coach lamp, was a very low priced unit that was a popular alternative to the large and expensive Cadillac chassis vehicles offered by the major manufacturers.

The Flxible limousine funeral car was offered in straight funeral coach, combination and ambulance styles at prices ranging up from $5,864 for the Premier and up from $5,394 for the Sterling. Here is the Premier limousine funeral coach in straight rear loading form. This car was also offered as a combination and an ambulance. Series numbers on the Premier series ranged from 5220972-7 up and on the Sterling from 56600955 up.

$4,892

Siebert continued to offer a strong line of coaches based on the Ford chassis. This attractive funeral car on a stretched Ford passenger car chassis was called the Aristocrat DeLuxe hearse and carried a price sticker of $4,892.90, with all taxes included, F.O.B. Toledo, and their porducts were widely distributed throughout North America.

Also offered within the new Acme line of Pontiac chassis funeral cars and ambulances was this attractive limousine funeral coach. Note the strikingly professional conversion job done by this new and relatively small operation. The firm also offered a limousine ambulance within the model line.

Flxible continued to offer a line of coaches on the Buick Series 70 and 50 chassis. The coaches mounted on the Series 70 Roadmaster chassis were called Premiers, while the ones on the Series 50 chassis were called Sterlings. Flxible styling was refined to match the new Buick chassis and production of Buick professional cars reached 366 units this year. The Buick Premier Landau rear-loading funeral coach, shown here, was designated a Model B21-750 and sold for $6,064. A Sterling landau, as a rear loading vehicle, carried a price tag of $5,594 and was designated a Model B21-550. Both series coaches offered three-way cars in both manually operated versions and power operated forms.

With interior dimensions equalling those found on Cadillac chassis vehicles, the Acme Coach Co. entered the funeral car and ambulance field with a line of Pontiac chassis vehicles. The Acme Pontiac landaus were built from the Pontiac sedan delivery units that were being marketed at the time. The Acme company simply extended the frame and built the landau body. Note the coach lamps on the landau panel.

The French have a tendency to make things flamboyant and extremely different, and this also applies to their funeral cars. This Renault hearse was a "domed" model operated by the Roblot Funeral Home of Paris. Note the dome on the rear compartment roof with the ornamental torches at each corner, and the initial plaque at the center of the roof sides. The draped curtain on the large rear side window was part of this DeLuxe hearse.

1950

The popularity of the Pontiac chassis could not be underestimated in this era. Small makers were relying on this chassis to supply the lower medium price range of the field. This Pontiac chassis offering was from the Guy Barnette Co. of Memphis. Mounted on the Pontiac eight chassis, this unit in limousine form sold for $4,462 complete and ready to go to work. The less expensive Pontiac chassis commanded enough prestige to make it a good selling chassis for funeral car and ambulance work.

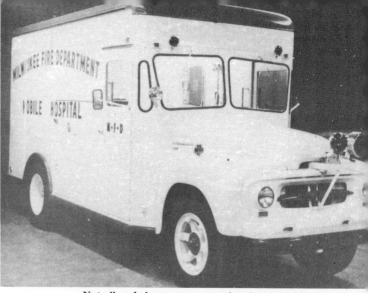

Not all ambulances were owned and operated by funeral homes or were convertible for use as funeral cars. This International truck served as a mobile hospital for the Milwaukee Fire Department and was completely equipped to handle almost any emergency case. The body builder is unknown, but the vehicle is an interesting diversion from the automotive type ambulances seen in this era.

The Greenwood Funeral Home in Stratford, Ontario, was the recipient of this attractive John Little Chrysler landau funeral car with a three-way casket table. This was but one of several coaches built by Little this year.

Complete with the Quebec type of roof ornamentation, this 1950 Meteor (a Canadian derivative of the Ford) flower car was built by John Little of Ingersoll, Ontario. These cars were in great demand in the Canadian province of Quebec, where they were a customary part of the funeral procession. Note the neat coupe styling and the low rear deck. The cross on the roof of this model was traditional on coaches used in Quebec.

Mounted on a Canadian Pontiac chassis (that was actually that of a Chevrolet with the exception of the trim) this John Little funeral car is unusual in that it has carved-type fixed draperies on the edges of the rear side windows. Mr. Little would execute almost any funeral director's idea into a vehicle, and a customer would only have to describe what he wanted. Three months later the finished product would roll out of Mr. Little's shop.

Built by Thomas Startin Ltd. of Birmingham, England, this Daimler chassis funeral car was typical of the styling that predominated in England at the time. Note the typical British razor edge type styling, the large rear window exposing the casket to total view, the flower rack on the roof, and the large headlamps. The car was quite high.

1951

The Miller flower car was considered part of the landau series and carried a price tag of $6,790. All Miller coaches were available with Cadillac Hydramatic drive for an additional $174. This stylish Miller flower car was the prized possession of the J.B. Marlatt Funeral Home of Hamilton, Ontario. Note the stainless steel deck and the wrap-around rear window.

A.J. Miller again offered a complete line of funeral cars and ambulances on the Cadillac chassis. This limousine hearse is a straight end loader which went for $6,270, while a Duplex combination model carried a price of $6,445 and a three-way manual coach was $6,970. The three-way electrically operated casket table version was the most expensive limousine funeral car offering at $7,550. Two ambulance models were also offered in the line. The straight ambulance with limousine styling went for $6,560, while the high headroom, emergency equipped First Aider carried a price tag of $7,220. Note the straight forward design of this car and the rather pleasing lines.

With the United States involved in the Korean conflict the manufacturers were swamped with a flood of orders. Funeral directors who had been "caught short" and had gone through World War II with old, and many times, inadequate equipment, began to purchase heavily against the possibility of another period of shortages.

Superior Coach expanded their Lima facilties to include 500,000 square feet of floor space and thus expanded their funeral coach and ambulance manufacturing operations.

Henney restyled its model line in this year for what was to be the last time. The new Packard styling, done in what was to prove a timelessly efficient layout, offered the Henney people some great space advantages. The Henney literature and advertisements boasted of the room in the driver's compartment and overall quality of the body design. Both chassis and body were completely new. Every feature of the body was checked with known measurements to insure a practical result in usability. The new interior styles and trim materials were available in a vast rainbow of reds, greens, oranges, and blues. The exterior styling was superlative, long and low and with Henney all-steel construction. The rear windows wrapped around the rear sides and were called horizon-wide vision. This new and unique style coach was designed by Richard Habib, a renowned stylist that Henney had hired away from General Motors.

Meteor made a slight change in its styling trademark by building the door-mounted carriage light into a more conventionally shaped window-door post. On ambulances this light was many times red to match the rest of the emergency lights.

The National Coach Company of Knightstown, Ind., (no relation to National companies seen previously) also entered the low-priced extended-frame type coach market. Offering coaches in full lines on any General Motors chassis, predominantly Pontiac, National in this year introduced its first flower car. Added to the long wheelbase Imperial series, this new Pontiac flower car was graceful in design and functional in its beauty. The car was offered on either the Pontiac or Chevrolet chassis and featured an electrically operated flower deck similar to that of the more expensive makes offered by other manufacturers.

Production in general was way up this year, with 4,177 funeral cars and ambulances leaving the various plants in the U.S. Of these, 2,140 were built on Cadillac chassis.

The new owners of a Miller landau proudly display the car in front of the McDougall & Brown Funeral Home of Toronto. It featured the crinkle finished roof and the small oval landau windows that had become a Miller trademark. The landau was offered in six models ranging in price from $6,545 for the straight end loading hearse through $6,725 for the landau Duplex combination, to $7,245 for a manually operated three-way coach. The top of the line was the electrically assisted casket table three-way version at $7,825. Ambulances were also offered in this body style with the straight version going for $6,840 and the First Aider for $7,500.

1951

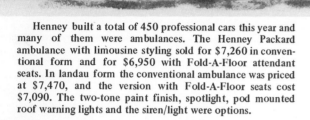

Henney built a total of 450 professional cars this year and many of them were ambulances. The Henney Packard ambulance with limousine styling sold for $7,260 in conventional form and for $6,950 with Fold-A-Floor attendant seats. In landau form the conventional ambulance was priced at $7,470, and the version with Fold-A-Floor seats cost $7,090. The two-tone paint finish, spotlight, pod mounted roof warning lights and the siren/light were options.

It is highly doubtful that Henney built very many of these attractive flower cars this year. The car didn't really look as long or as low as this horrible artist's rendering. This Henney model also has the wrap-around rear window and a new coupe type roof line. Henney claimed that there was nothing newer on earth than this Arbib styled flower car and that Henney was the one for 1951. The flower car cost $6,460 this year and continued to feature a stainless steel rear compartment and a stainless steel hydraulically operated flower deck.

Totally redesigned for what was to prove the last time, the new Henney Packard professsional cars were built on the Packard 2413 chassis. Packard introduced a completely new line of cars this year and the styling of these cars was carried over to the Henney coaches. When Henney learned of this totally new styling, they commissioned renowned stylist Richard Arbib to design new and innovative bodies for all of the 1951 models. This Henney Packard landau with Arbib styling was available in nine versions that included a combination coach and an ambulance. In its standard rear loading form this car carried a price tag of $6,970, while the more popular three-way coach was offered in either electrically assisted form at $7,900, or as a manually operated model at $7,510. The top offering in the landau funeral car line was the three-way coach with Elecdraulically operated casket table and Leveldraulic for the leveling of the vehicle. This car carried a price tag of $8,335. Exterior appearance of all of the landau models was similar, with the rakishly shaped rear side window, rear door mounted coach lamps and landau irons.

Probably the most revolutionary car ever introduced by Henney since the flower car was the 1951 models. The Richard Abib styling was an advance for the styling of funeral cars that has been followed to this day. This limousine funeral car was one of the most talked about styles in the profession and featured such radical innovations as a rear quarter window that wrapped all the way around the rear body sides and a rakishly shaped rear door window. Henney models this year were built on the Packard professional car chassis with serial numbers beginning at 2413-2001 and going up. This three-way limousine funeral car with Elecdraulically operated casket table and Leveldraulic cost $8,125, while a model with a manually operated casket table was priced at $7,715. A Henney Packard limousine coach as a straight end loading vehicle cost $6,760, a combination was $6,600, and a service car went for $6,985.

With sales slipping for both Henney professional cars and Packard passenger cars, due mainly to Cadillac, Henney introduced a new smaller, lighter model called the Henney Junior. Priced at only $3,333 F.O.B. Freeport, Ill., this car was the beginning of a trend that wouldn't really be taken up by other makers until the early 1960's. The short wheelbase funeral car was a totally new market at a somewhat more attractive price. The Henney Junior was a very popular car with the Civilian Defense organizations, small fire departments, or small, low budget funeral homes. The car was offered in ambulance, funeral car and service car styles. The unusual appearance of the car is due to the fact that there are no rear side doors. Note the continuity of the Arbib styling line on even this little low priced model.

$3333
F.O.B. FREEPORT
State and Federal Taxes in Addition

Not the most popular body style, but surely one of the most attractive, was this Superior Coupe De Fleur flower car, available in this year with the hydraulically operated flower deck only. This car cost $6,470 and again featured wrap-around style rear windows, coupe styling, and a stainless steel flower tray and rear compartment interior trim.

Another war was on and the funeral car and ambulance makers again began to build special vehicles to meet U.S. Government specifications. This ambulance, built by Superior for the U.S. Air Force, followed the pattern of blanking off the side windows and of being a three door coach only. This civilian type ambulance is equipped with both a roof light and a windshield pillar mounted spotlight. It was painted in Air Force blue.

Superior Coach Co. of Lima, Ohio, offered the popular landaulet funeral car in no less than five versions this year. Superior production soared to 886 in 1951, with the landaulet begin the most popular body style on funeral cars. Costing $6,675 in standard rear loading form, this car could be had with either a crinkle type or a high glass roof. The car was also offered as a manually operated side loader at $7,225, or as a power operated three-way car for $7,530. Limousine versions of the popular Superior funeral coach carried prices starting at $6,470 and going up to $7,325.

Offered at the same price as the limousine funeral car ($6,470), the Superior Cadillac service car carried a chrome stylized wreath and bars on the rear body sides. Note the SERVICE CAR windshield sign and the windshield pillar mounted spotlight. This car appears to have a roof with a crinkle finish.

Superior ambulances were among the most popular in the field and the 1951 models were no exception. Offered in both landaulet and limousine styling, these cars were available with optional roof pod warning lamps as shown and a myriad of other items. This limousine ambulance cost $6,920, while a landaulet version of this car was priced at $7,130. Note the frosted glass in the rear compartment windows.

Customers interested in a Chrysler chassis coach built by the Economy Coach Co. of Memphis would have to wait at least four weeks for delivery of one of these interesting vehicles. Economy Model 2400 landau rear loading funeral car is on the Chrysler Saratoga chassis. This car was powered by a 180 horsepower FirePower V-8 and sold for $5,675. The same car was offered on the Chrysler Windsor chassis that was powered by the 116 horse six-cylinder and sold for $5,275. Other models in the line included a service car at $5,675 on the Saratoga and $5,275 on the Windsor, an ambulance that went for $5,725 on the Saratoga chassis and $5,375 on the Windsor, and a limousine style combination for $5,875 on the Saratoga and $5,475 on the Windsor. The combination was also offered on the landau body style. These cars rode on a wheelbase of 139½ inches and were 228 inches from bumper to bumper.

1951

Hess & Eisenhardt introduced the 1951 S & S models as 75th Anniversary cars. With new Superline models in both landau Victoria and limousine styles, Hess & Eisenhardt presented this S & S Superline Knickerbocker in three funeral car versions. Prices for the Knickerbocker began with the rear loading version at $6,546 and went through $7,410 for the three-way manually operated casket table version, up to $7,996 for the model with electrically assisted casket table and three-way loading. Ambulance models were also offered in Knickerbocker styling. Two of these models were called the Kensington and were offered in straight ambulance form at $7,121, and as a high headroom Professional Kensington model at $7,524.

Called the best since 1876, the new Hess & Eisenhardt & S Victoria was offered in five funeral car versions and on ambulance model. Prices for the 75th Anniversary S & Superline Victoria began at $6,917 for the rear loading ca and went to $7,681 for the three-way manually operate casket table model and to $8,267 for the electrically assiste three-way coach. The Parkway Victoria ambulance was sti offered at $7,391 and a flower car was available in the S & line in this year, too. Called the Superline Florentine, this ca cost $6,651 and was one of the lowest priced flower car offered in this year.

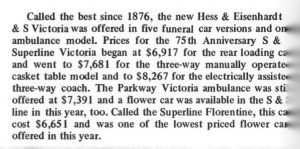

Flxible continued their concentration on the Buick Series 50 and 70 chassis, and produced a total of 375 cars this year. This Series 50 Sterling landau shared styling with the Series 70 Premier coaches but was mounted on a Buick Super chassis, while the Series 70 Premier models rode on a Roadmaster chassis. Styling differences were seen only in the number of Buick fender portholes. Premier models carried four such mouseholes, while the Sterling models had only three. The two-tone paint finish and the small red warning lights below the headlamps were optional. This Sterling landau carried a price of $5,848.

Also offered by Flxible was an array of ambulances on both the Buick 50 and 70 chassis. Here we see a Flxible ambulance on the Series 70 Premier chassis. This car is equipped with pod type warning lights on the roof and a roof mounted siren/warning light, which were optional. The Standard Premier ambulance sold for $7,003 and a Dynaflow transmission was optional at $169 on the Sterling Series 50 models. A Sterling ambulance of similar styling sold for $6,191. Other models in the Sterling line carried prices ranging from $5,648 for the rear loading funeral car, through $7,595 for a limousine with three-way side servicing and power operated casket table. A service car was also offered in the Sterling line for $5,216.

Eureka was one of the few major makers continuing to utilize a hand built wooden body frame for their coaches. This landau continues the styling that had become recognized as exclusive with Eureka. The smooth curved rear part of the roof, the sweeping landau irons, and the ornate coach lamp had all become a Eureka trademark. The landau was offered as a funeral car with end loading for $7,222, as a combination for $7,295, and as a three-way manually operated model for $7,645. A three-way electrically operated version topped the landau offerings and cost $8,358. Limousine models were offered also for about $330 less than comparable landau models.

The Buick Roadmaster chassis seen here wore a Flxible Premier combination car body and was finished in two-tone brown. The Premier limousine funeral car with rear loading cost $6,126, while the Premier combination seen here went for $6,492; a three-way manually operated casket table model carried a price tag of $7,313 with leveler. Also offered was a service car at $5,663. All of the Flxible Premier models were offered with landau styling for $200 over the limousine prices. Sterling combinations went for $5,874 and were on the Series 50 chassis.

1951

Continuing the popular styling introduced with the 1950 models, Meteor again offered a vast range of models and body styles. The Meteor landau was offered in five versions with prices starting at $6,675 for the Model 510-L rear loading car and going to $7,959 for the electrically assisted three-way coach, Model 510-LHE. The Meteor combination car in landau styling came complete with a removable steel casket table, a heater for the rear compartment built into the partition, cot hooks, one fold away attendant seat, no-draft vent windows in the rear side windows, and a rear bumper step. The combination was designed as a Model 510-LC and sold for $6,999. The three-way landau with manually operated casket table cost $7,377.

Meteor also offered a vast array of models on the limousine chassis. This combination coach carried a price tag of $6,841, while a straight end-loading hearse went for $6,520. The three-way manual coach sold for $7,221 and the power operated three-way for $7,803. Meteor continued to use the door mounted coach lamp. This car had red lenses over the parking lights.

Meteor designated the flower car as a Model 513 and it sold for $6,384. This rear view of the Meteor Cadillac flower car for the Wellman Funeral home of Louisiana shows the pleasing lines of these cars and the increasingly popular wrap-around rear windows. This is but one of many flower cars that Meteor delivered to the Wellman company over the years.

Meteor continued to offer the lower priced Pontiac coaches and this line even included a flower car on the Pontiac chassis. The flower deck is of the Chicago type and the tonneau is not stored in a stainless steel housing at the rear of the flower deck, but more like a convertible top. The attractive coupe styling is retained on this less expensive car, and this year it was made more interesting because it is of the hardtop type.

The Acme Company of Sterling, Ill., produced Pontiac chassis professional cars in ambulance, flower car, service car and funeral car versions and claimed that these coaches had dimensions equal to that of Cadillac coaches made by competing makers. The overall length of these Acme Pontiacs was 239 inches. This is the Acme Pontiac service car with graceful body lines and a simple distinctive service car decoration and coach lamp on the rear body sides. Acme Pontiac coaches were powered by the Straight eight engine that delivered 100 horsepower from 239-cubic-inch displacement.

1951

The National Body Manufacturing Co. of Knightstown, Ind. specialized on the economical chassis built by Chevrolet and Pontiac and produced some very attractive coaches this year. This long wheelbase Chevrolet end loading funeral car was a good example of their styling and work. Note the draperies and smooth body contours.

National offered the Chevrolet models in a choice of long or standard wheelbase lengths and in funeral car, ambulance, and service car styles. This is National's short wheelbase service car. These cars were usually built on the chassis of the popular panel delivery vehicles that both Pontiac and Chevrolet were building at the time. Note the unusually shaped rear side window and the SERVICE CAR sign in the windshield.

The National short wheelbase ambulance was an interesting looking vehicle with special exhaust ventilators for stale interior air on the rear body sides, an ambulance decal in the side window, and pod type roof warning lights. These cars were three-door vehicles, while the long wheelbase models were of five door configuration.

National offered their short wheelbase coach on the Pontiac chassis also, and this is an example of this vehicle as a funeral car. Once again notice the strangely shaped rear side window.

Continuing to build a line of vehicles on the Ford chassis, Siebert offered the new Aristocrat line that featured this end loading funeral coach. The Ford passenger car frame was extended and the Siebert limousine body was mounted. Notice the formal draperies and the chrome decorative strips on the rear quarters.

With a wheelbase of 156 inches, this Barnette Pontiac Eight ambulance was but one of the offerings from this small coach works in Memphis. Note the long wheelbase, the pod type warning lights on the roof and the frosted glass with the ambulance pattern in the rear quarter window. Barnette also offered these coaches on the Chevrolet chassis.

The Pontiac chassis was also popular for professional car work in Canada. The wheelbase of the Canadian Pontiac was lengthened from 115 inches to 137 inches for this special combination car built by St. Catharine's Auto Bodies of St. Catharine, Ontario. These cars were extended sedan delivery type vehicles.

The standard ambulance model from the St. Catharine's firm was a short wheelbase model seen here as an ambulance. The Canadian Pontiac utilized a Chevrolet chassis and engine and wore Chevrolet body panels with Pontiac trim. This ambulance wears roof corner marker lights, a spotlight, sun visor, a roof mounted combination siren warning light, and vent window in the rear quarter window.

1952

Photographed in a becoming setting, the 1952 Miller landau funeral car was a pleasingly styled vehicle. The landau's oval window had become as much a Miller trademark as the Meteor door mounted coach lamp, or the ornate coach lamps of the Eureka landaus. This A.J. Miller end loading landau funeral car carried a price tag of $6,930, while the end loading limousine funeral car cost $6,635. Other models offered this year included the landau combination Duplex at $7,120 and the limousine combination at $6,820, the manually operated casket table three-way landau at $7,680 and the limousine at $7,385, the electric three-way landau at $8,300 and the limousine at $8,005. A flower car was offered at $6,790 and ambulances began at $6,945 and went up to $7,650. Once again the landau could be ordered with the oval window deleted or without the crinkle grain roof finish shown here.

The Eureka limousine funeral car had a rakishly shaped rear quarter window. In its standard end loading form it sold for $7,376 and in three-way manual form for $7,810. The three way electric version sold for $8,536 and the ambulance for $7,945. This car was a three-way loader. Note the formal draperies and the sloped rear end styling.

The stunning Eureka landau funeral car shown here wearing Ontario plates, was finished in two-tone gray and was an end loader. The Eureka end loading funeral car cost $7,686 while the three-way manual landau was priced at $8,120 and the three-way electrically assisted model went for $8,846. A combination was offered in the landau styling for $7,760 and an ambulance for $8,255. Eureka's ornate coach lamp, like the Miller oval window, had become a landau trademark for this Rock Falls, Ill., firm.

This was another year of virtually no change in the styles of the chassis supplied to the manufacturers and very little change in the bodies offered to the trade.

Another new company joined the funeral coach makers' roster in this year. Economy Coach Co. of Memphis began production of funeral service vehicles and ambulances based on extended chassis. The first models were shown in three lines, Chrysler chassied cars in two complete series and Pontiac and Chevrolet models in seven body styles. The full-sized Chrysler line consisted of two series, one on the Saratoga chassis with the Firepower V8 engine rated at 180 horsepower and the other on the Windsor chassis powered by the six-cylinder Spitfire 119 horsepower motor, both with ambulances, service cars, combinations and hearses. Both the Economy-Pontiac and Chevrolet lines featured coaches on chassis extended from 18 inches on the junior models to 30 inches on the Master series.

This demand for smaller, less expensive coaches did not go unnoticed by Henney which entered its own smaller model in this year. Called the Henney-Packard Junior, this new vehicle was built on a shorter 127-inch wheelbase chassis and equipped with the 288-cubic-inch Clipper engine. Henney stressed the fact that this new car was not an extended pleasure car or panel delivery type car but a funeral service vehicle or ambulance designed as such. Along with the shorter wheelbase, lighter weight, and higher utility, the Junior came with a new lower price of $3,333, F.O.B. Freeport. The car had a very utilitarian look with very little chromium trim and only two doors. It looked as if it was all rear fender and door, and could not by any stretch of the imagination be called a styling masterpiece. Sales of this new little Henney were high and it helped the overall sales picture at the company. A total of 380 Juniors were sold this year with a large portion of these going to the government.

Another piece of news that came from Henney this year was the dynamic Packard Pan-American show car. Designed by Richard Habib and built by Henney, the Pan-American created such a sensation that Packard put a production version of it in the showrooms part way through the next year.

The Flxible Company of Loudonville, Ohio, which had begun funeral coach and ambulance production in 1925 and had, after the war, built exclusively on Buick chassis, ceased production. Flxible also was a prime supplier of metropolitan type busses and due to a large volume of orders for these vehicles, it was decided to abandon the production of funeral coaches and ambulances until they could do so without divided interests. Thus, with the end of the 1952 model run, Flxible was to quit the field for six years.

Total professional car production dropped substantially this year, to a low 2,622 vehicles. Of these, 1,771 were on Cadillac chassis.

1952

In their last year of producing funeral cars and ambulances for a period of seven years, Flxible also offered the Premier models on the Buick Series 70 Roadmaster chassis. This Premier combination in limousine styling sold for $6,627, while the rear loading funeral car was priced at $6,253, the ambulance at $7,152, and landau versions of these models were offered for $200 more. These cars were powered by the Buick straight eight engine coupled with Dynaflow transmission. Note the optional spotlight and the formal draperies.

Producing 132 professional cars this year, Flxible would resign from the field for a few years to look after numerous bus orders that they had to fill. This Sterling landaulet sold for $5,969. Other models in the Sterling line included the rear loading limousine funeral car at $5,769, the ambulance at $6,325, the service car at $5,325, and the combination at $6,000. The Sterling models were mounted on the Buick Series 50 Super chassis and were fitted with standard three-speed synchromesh transmission.

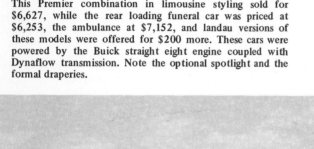

The Flxible Premier service car was a stylish vehicle that sold for $5,776. This car features a stylish chrome wreath and chrome bars on the rear body sides. It was a five-door style. Note the nameplate on the door and the four Buick Roadmaster mouseholes on the front fender. Flxible withdrew from this field after building 132 cars this year to look after large orders for their busses, and was not to re-enter the field until 1959.

Wearing their last total restyling that was carried over from last year, the Henney Packard models were a distinguished looking family. This flower car loaded with floral tributes was one of the most attractive cars in the range and sold for $6,650. The long, low rear deck was sheathed in stainless steel, with the tonneau at the rear of this deck. These cars were up from 2513-2001. Henney built a total of 320 cars this year.

Henney Packard landaus ranged in price from $6,950 for a conventional rear loading hearse through $8,510 for the landau with three way Electdraulically operated casket table and Leveldraulic for the leveling of the car during loading and unloading. In this year Henney offered no less than 21 different models in landau, limousine and flower car styling. These coaches retained the Richard Arbib styling that had made them so revolutionary when first announced last year. This view of the Henney landau with three-way electric servicing shows the highly detailed interior appointments and the fine mohair used for the upholstery.

The small Henney Junior three-door utility funeral car line was continued through this year. This is the Henney Junior fitted out as a service car, but the car could be used as a hearse, or an ambulance. Note the continuity of styling from the large models to the Junior. The price for the Junior was continued at $3,333 F.O.B. Freeport, Ill.

1952

The Meteor contribution to the Korean war effort was to supply the U.S. Government with ambulances built on the GMC light panel truck chassis. These van ambulances featured roof ventilators, small rear side windows, and fender mounted warning lights. The completed Meteor GMC in the foreground was sent to the Medical Service of the U.S. Army and had its registration numbers painted on the front door as well as Red Cross insignias on the roof. The GMC vehicles in the background are awaiting their turn to enter the factory for conversion to army service.

With some new styling touches, the new Meteor funeral coaches appeared at somewhat higher prices than the preceeding models. This new landau sold in its basic rear loading form for $7,105, and as a combination for $7,821, while the three-way coach with an electrically operated casket table was priced at $8,415. Limousine model prices began at $6,946 and went up to $8,256 in funeral car form. Ambulances were offered in either landau or limousine form with the landau selling for $7,563 and the limousine ambulance costing $7,402. The new styling touches on the new Meteor coaches were restrained to items like two-toning treatments and roof styling touches. Note the roof lines of this landau. Meteor continued to place the coach lamps on the front door frames.

Another Pontiac based funeral car line was offered by the National Body Co. of Knightstown, Ind. The professional cars were offered in the long wheelbase Imperial series or with a short wheelbase. This Imperial Pontiac landau is an excellent example of National's work on the Pontiac chassis and is in many ways more attractive than the Acme landau. National also built these bodies on Chevrolet chassis.

Meteor now offered its beautiful flower car on the Cadillac chassis only. All Meteor coaches were built on the Cadillac commercial chassis Series 86. The flower car this year carried a price tag of $6,808. A Meteor service car was also offered for $6,755 and was designated as Model 522.

The distinctive Hess & Eisenhardt S & S Victoria continued in production without any major styling changes. This vehicle was owned by the William Speers Funeral Home of Toronto, and is seen in front of that mortuary. Prices for the S & S Victoria ranged from $7,578 for the rear loading model up to $8,989 for the Victoria with three-way electrically operated casket table. The distinctive S & S Victoria Parkway ambulance continued in production and cost $8,062. Once again Hess & Eisenhardt offered the beautiful Florentine flower car and it carried a price tag of $7,216.

While the Victoria was fast becoming the Hess & Eisenhardt money winner, the firm continued to offer distinctive limousine styling with the S & S Knickerbocker line. The standard S & S Knickerbocker limousine funeral car with end loading went for $7,199 and was designated as a Model 52400. The Knickerbocker combination ambulance/funeral coach sold for $7,281 and the three-way manually operated Knickerbocker, designated the Model 52430, cost $8,086. The most expensive Knickerbocker was the three-way electrically operated funeral car at $8,711. The formal draperies shown in this car were optional at extra cost.

1952

Continuing to offer stylish coaches at low prices, the Acme Co. of Sterling, Ill., built on the Pontiac eight-cylinder chassis only. This distinctive Acme Pontiac landau funeral car represented the top of the firm's funeral car offerings and featured large partially painted landau bows on the upper rear quarters. The conversion of the Pontiac from panel delivery van to long wheelbase funeral car was very well carried out.

The Acme Pontiac ambulance was the low price class prize winner and was completely equipped with frosted glass rear quarter windows, tunnel pod roof lights, and rear corner roof marker lights. The Acme Pontiac funeral cars were very popular and a good many of them were seen operating in small towns and rural communities.

Economy Coach of Memphis continued to offer their line of Chrysler chassis funeral cars through this year. Again they were offered on either the Saratoga 8-cylinder chassis or the Windsor 6 chassis. This is the Economy Chrysler Saratoga combination car complete with red warning lights mounted below the parking lights.

The least expensive Economy offering was this service car on a Pontiac chassis. The Pontiac sedan delivery body had been lengthened slightly and the service car ornamentation that in this case consisted of chrome plated wreaths and chrome bars and a nameplate were added. For a car at the low end of the pricing scale this was a pleasingly styled vehicle.

This interesting 25-seat mourners' coach was a popular type of vehicle used in France in this year. Used by the Furgeon Chapel of Roblot, Paris, this vehicle was painted black with sparkling chrome highlights and a large roof rack for flowers. Note the fancy hubcaps.

The Superior Coupe De Fleur flower car again appeared in its stainless steel rear deck garb and with very few styling changes. The Coupe De Fleur went for $6,680 this year and was offered with the hydraulic flower deck only.

The National Funeral Directors' Convention is held annually in October and this is the traditional launching pad for all of the new models. The makers set up elaborate displays and place backdrops for their new cars, such as was prepared for the 1952 Superior models. In the foreground is a Superior ambulance and next to that a Superior combination coach. The center of the display is a white Superior landau, while a Coupe De Fleur flower car rounds off the exhibit. Superior built 579 cars in 1952.

The most impressive ambulance built by Superior was this model called the Rescue Ambulance. The Rescue Ambulance had a significantly higher roof line and could accommodate four patients on cots that could be suspended from the roof or placed on the floor. The car also carried two attendant's seats in the rear compartment plus medicine cabinets and oxygen equipment. This particular vehicle is equipped with pod type roof lights, spotlights, fender mounted warning lights and a combination Federal siren warning light on the roof. The lights and sirens were optional extras, and the basic Rescue Ambulance sold for $7,868. This car belonged to the Kerr Brothers Ambulance Service and was finished in white with gold lettering. This car was also equipped with radio dispatching equipment and the antenna for this is visible on the roof.

Less dramatic and more popular was the Superior limousine type ambulance that cost $7,167 and featured the same styling as seen on the limousine funeral cars. This car was also offered with landaulet styling for $7,400. This car has red lenses over the parking lights, spotlights, pod type roof lights and a Federal siren warning light on the roof. It was finished in two-tone green.

There really wasn't too much restyling that could be done with the traditional limousine funeral coach. The 1952 Superior limousine funeral cars were offered in five hearse versions starting with the rear loading funeral car priced at $6,676 and going up through the side servicing manually operated version at $7,288 and the electrically operated three-way at $7,629. Combinations were priced at $7,009 and $7,081. Once again Superior offered a service car and this was priced at $6,676.

The most dramatically restyled coach in the Superior line-up for 1952 was the landau. The three-way coach seen here featured all of the new styling items, including new landau bows and landau shields with a halo type of paint job that went around the side windows, with the roof finished in black crinkle type paint. As a straight end-loading funeral car, the landaulet sold for $6,095, while the landaulet with manually side servicing and three-way went for $7,518. The height of the landaulet line was the magnificent three-way electrically assisted casket table model priced at $7,858. Landaulet combinations were available in two versions priced at $7,247 and $7,314.

With very little change in chassis from the suppliers, the major manufacturers continued their model lines unchanged from previous years. Although there were no styling changes in the Henney line, the price for the Junior economy models was raised to $4,350 without any styling or equipment changes. Henney was feeling a sharp drop in sales and entered into the construction of Packard limousines to offset the deficit from the funeral coach and ambulance areas. The new 7-passenger limousines and sedans put Henney in a position shared with Derham as Packard's only custom body passenger car maker since World War II.

Two firms celebrated their 100th anniversary in this year. Miller, formed in 1853, and Siebert, also formed in that year, celebrated with centennial models and specialized advertising. Miller issued special velvet-covered brochures on its 1953 models stressing 100 years of American craftsmanship, while Siebert marketed a long wheelbase version of the Ford Aristocrat called the Centennial.

With the makers now concentrating on limousine and landau versions of hearses, only a few companies endeavored to make their landau offerings look different from the rest of the makers. Meteor, with its unique door mounted carriage lamps, added a new touch to the landau models with a chrome strip that ran from the waist line to the end of the landau iron. This added an exclusive touch to the appearance of the car. A few years earlier Superior had built a plant in Kosciusko, Miss., as a branch plant for producing more school busses and funeral cars. In 1953 this plant was turning out the first completely designed Superior Pontiac funeral coaches and ambulances on a regular basis. This plant was to become the prime supplier of Superior Pontiac models in the future and Superior was to become the largest producer of Pontiac coaches in the industry.

With more and more passenger car manufacturers featuring wrap-around rear windows on their models, the coach manufacturers were not about to be beaten by this new innovation. This year Hess & Eisenhardt (S & S) announced models with rear windows that wrapped around the roof sides and projected a new image of up-to-dateness. This was only the first step, as other makers were to carry this new style much further in the future.

Funeral coach and ambulance production went up slightly, with 3,034 such vehicles having been built in the U.S. of which 1,949 were on Cadillac chassis.

Henney's limousine styling continued with the wrap-around window that it introduced in 1951. This feature, popular on passenger cars of the age, was being copied by other makers in this year on their coaches. The Henney Packard end loading funeral coach shown here sold for $7,150. It could be ordered in three-way form at $8,075 without Leveldraulic but with Elecdraulic operated casket table, and for $8,475 with Electdraulic and Leveldraulic. As a combination car, the Henney limousine coach sold for $7,326. A service car was also built and sold for $7,150.

The Henney Packard flower car was one of this firm's best looking offerings for 1953. It sold for $8,000 even. Here the owner of the Carbondale Funeral Home of Carbondale, Ill., takes the keys to his beautiful new car from Robert Rossiter, who was then the assistant sales manager of Henney Motor Co. and is today the general manager of Miller Meteor in Piqua, Ohio. Note the expanse of stainless steel that made up the flower deck and the tonneau at the extreme rear of the flower deck.

Henney continued to offer funeral cars on the Packard 2613 Super 8 chassis built especially for funeral car and ambulance work. No less than seven versions of this attractive landau funeral caoch were offered, with prices beginning at $7,250 for the end loading car. The Henney Landau with Elecdraulically operated casket table and three-way servicing went for $8,250, while the landau with Electdraulically operated three-way casket table and Leveldraulic cost $8,650 and represented the top of the Henney Packard line for 1953. This car is seen here complete with rear side door mounted coach lamps, a unique halo painted section around the rear side windows, and a curvaceous landau bow. Henney Packard funeral cars were among the finest available.

1953

The little Henney Junior was also offered as an ambulance as well as a straight end loading funeral car, a combination or a service car. Seen here in ambulance form, the Junior seemed to have been an attractive little coach. This car was finished in cream with a red top and was fitted with optional pod type roof lights as well as a Federal siren warning light. Henney built a total of 380 Juniors in this year, making this line the most popular of the year. Total Henney production for 1953 was 696 units. This figure includes limousines built for Packard.

Distinctly different from the Henney Packard full sized cars using the 2613 chassis, the Henney Junior coaches were built on the Packard Clipper 2633 chassis. This is a 1953 Junior in combination form. Note the optional small window just behind the driver's door. The metal ambulance initials snapped into the rear quarter window easily. In combination form, this car sold for $4,350.

This rear view of the Henney Packard limousine ambulance clearly shows the wrap-around rear quarter windows that were an important part of these Richard Arbib styled cars. Note how the rear door sweeps into the roof for easier loading.

The Henney Packard ambulance offerings were made in both limousine, as seen here, and in landau styles. Henney ambulance prices began at $7,700 for the limousine style and went to $7,800 for the landau. The panel between the rear wrap around window and the rear side door window is used for the large chrome cross and nameplate. The tunnel roof warning lights and the siren light were optional at extra cost. The ambulance decal in the rear quarter window and the rear door windows were standard on all Henney ambulances.

Packard handed over to the craftsmen at Henney the responsibility for building their renowned limousines on the Packard 2626 chassis. These large, seven passenger sedans and limousines were now completely hand crafted by the Henney Motor Co. of Freeport, Ill. This year Henney built a total of 150 of these beautiful cars.

The Woodland Beach Volunteer Fire Department was one of the recipients of a Henney Packard ambulance this year, that saw only 166 Henney Packard professional cars built on the 2613 chassis. This car sold for $7,700 in standard form.

The Superior Pontiac landau was a very attractive car from any angle. This car features the same styling that Superior utilized on the large Cadillac models. Note the interesting roof treatment and the chrome plated landau bows. The chrome waist strip had the famous Pontiac Vee near the rear end. This car is finished in the optional two-tone paint.

After a few years absence from the market, the Superior Pontiac funeral coach and ambulance returned. This was also the year that Superior Coach Corporation of Lima, Ohio, celebrated its 30th anniversary. The new Pontiac coaches were built on extended frames and were some of the best conversions of the Pontiac to be offered. The Pontiac limousine funeral car shown here was also offered as a combination.

With a crinkle type roof finish, Superior offered this attractive Pontiac chassis service car. The car was complete with the Superior stylized wreath and chrome bars on the upper rear body sides.

Shown here without fender skirts, but with a set of "gee whiz" GM wire wheel covers, the Superior Pontiac ambulance was a very impressive sight. This car is complete with tunnel type roof warning lights and a Federal siren warning light. Note that the Superior Pontiac limousine body style features wrap around rear windows. The ambulance decals in the rear quarter windows were also new for 1953.

The Superior limousine funeral car was also offered in a range of six versions. The straight end-loading funeral car with limousine styling went for $6,724, while the three-way coach with manually operated casket table sold for $7,336, and the electrically operated three-way cost $7,677. Combination coach models were offered in two versions for $7,057 and $7,129. A service car was offered for $6,724 with the main styling differences being that the service car had blanked off rear body sides with ornamental chrome wreath and bars instead of windows.

Supeior's 30th anniversary Cadillac coaches continued with the popular styling announced last year. This distinctive landaulet was offered in six versions with prices beginning at $6,953 for the end loading landaulet, and going up to $7,566 for the landaulet with three-way manually operated casket table. The three-way coach with an electrically operated casket table cost $7,906, and the landaulet combination went for $7,295 in one version and $7,362 in another.

1953

The top ambulance in the Superior range was this Superior Rescue Ambulance on the Cadillac chassis. This car featured a noticeable step in the roof height for the rear compartment. The Rescue Ambulance sold for $7,916. This particular unit was delivered to the Milwaukee County Institutions, and features optional tunnel roof lights and a Federal roof siren light. Of interest is the large air intake mounted on the roof to supply fresh air to the rear compartment.

Superior's Coupe De Fleur flower car is seen here in an interesting view that shows the stainless steel rear flower deck. The railings partition off the deck into four sections that were hydraulically operated to lower in step fashion. This made for better floral arrangement and the baskets could be set in an upright position. The tonneau at the rear of the deck housed the tarpaulin cover for the deck. Note the wrap-around rear window and the wide side doors. The area under the flower deck could be used for casket carrying when the deck was in this position. This area was lined with stainless steel. The Coupe De Fleur sold for $6,724. This car was one of 760 professional cars of all types built this year by Superior.

Hess & Eisenhardt gave the 1953 S & S coaches a mild restyling that included the popular wrap-around rear window. This feature was successfully blended into the plush S & S Victoria, as seen here. The 1953 Victoria sold for $7,776 as a straight end loading hearse and for $7,857 as a combination car. The Victoria went for $8,562 as a three-way coach with a manual casket table, and for $9,232 as a three-way coach with an electrically assisted casket table. The Victoria was also offered in ambulance form and called the Parkway Victoria. It sold for $8,260.

Hess & Eisenhardt, continuing with the marque name S & S, again offered the Florentine flower car. All Cadillac chassis cars by all makers were mounted on the Cadillac Series 86 commercial chassis. This attractive flower car sold for $7,413 at the factory in Cincinnati.

The Hess & Eisenhardt S & S ambulance offerings in limousine form were called the Kensington. The standard Kensington ambulance as shown here sold for $7,983. There was a high headroom model capable of handling four injured patients called the Professional Kensington that sold for $8,397. This ambulance has the S & S date marks on the hood. The built in tunnel warning lights were optional at extra cost, as was the roof mounted siren light. Note that this car features the new S & S wrap-around rear windows.

Hess & Eisenhardt continued to offer the Knickerbocker limousine line of four funeral cars and combinations. New to the limousine styling of the Knickerbocker for 1953 were the wrap-around rear windows and thin pillar between the wrap-around rear windows and the rear quarter windows. The Knickerbocker sold for $7,397 as a straight end loader and for $7,479 as a combination. Three-way coaches were also offered with the manually operated casket table model going for $8,284 and the Electdarulically operated version costing $8,954. All S & S coaches came with Cadillac hydramatic transmission as standard equipment.

1953

The Meteor Motor Co. of Piqua, Ohio, this year offered a landau that was mildly restyled for 1953. The new landau continued to feature the traditional Meteor door mounted coach lamps and now had additional chrome trim on the landau panel to add distinction to the style. The landau, available in five forms, had prices beginning at $7,219 for the rear loading hearse and going to $7,628 for the combination. The three-way coach with a manually operated casket table went for $8,529. A landau ambulance was offered for $7,561.

Meteor offered its flower car in two distinctly different versions. One called the Eastern type had the regular stainless steel rear deck for flowers and sold for $6,922, while the other, seen here, was called the Western Type. This car was typically called a Chicago type and had no hydraulically operated stainless steel flower deck, but a large flower trough instead. This car cost $6,524. Note the small landau bows on the rear portion of the coupe style roof and the tonneau cover which is in place. This picture was taken outside the Meteor Motor Company offices in Piqua, Ohio.

Meteor also built coaches on the Chevrolet chassis, and this short wheelbase ambulance is a fine example of this type of Meteor coach. This car features a vent window in the rear quarter windows, fender mounted warning light, tunnel roof warning lights and a Federal light siren on the roof. Below the headlamps are additional warning lights. Note the ambulance decal in the rear quarter windows.

In Canada, the Pontiac chassis funeral car was also popular. This example was built by Canada's famous coach builder, John C. Little of Ingersoll, Ontario. It was built on a Canadian Pontiac chassis and hence the Chevrolet appearance. This car was of landau styling with a stylish halo effect around the side windows and a cast torch in the landau panel area. These cast metal parts were done by a small foundry in New Hamburg, Ontario, and these parts were later chrome plated. The styling of this car is very pleasing and, like all Little creations, was built by hand.

Another interesting Meteor extension of the Pontiac chassis was this ambulance with but three doors and a long wheelbase. Notice the ambulance decals in the rear quarter windows, the tunnel warning lights on the roof, and the optional Federal siren warning light combination. It was finished in white.

The Meteor limousine funeral coach in rear loading form sold for $7,060, while the limousine combination went for $7,467. This car was also available in three-way versions with the manually operated casket table model selling for $7,776 and the power operated casket table version going for $8,370. Ambulance versions sold for $7,400.

1953

Following an increasingly popular practice among progressive funeral homes, the Barnett Funeral Home of Wytheville, Va., purchased two matching Miller coaches in 1953. They used the Miller flower car as their hearse and the limousine ambulance was used in the firm's emergency aid service. The flower car, a very attractive Miller unit, sold for $6,825, while the ambulance was offered in three versions. The limousine ambulance like the one shown here sold for $6,945, while the landau ambulance carried a price tag of $7,245. The high headroom Miller First Aider ambulance was priced at $7,650. This Miller ambulance was competing with the Superior Rescue ambulance.

Miller funeral cars and Duplex combinations were well known for their durability and dependability. This Miller limousine funeral car is seen in front of the Paul Funeral Home in Brockville, Ontario. This car was the second Miller limousine purchased by the Paul Funeral Home and signifies the popularity and repeat customers that Miller enjoyed in these years. The standard Miller limousine rear loading funeral car sold for $6,635, the Duplex combination for $6,820, the manually operated three-way coach for $7,385 and the electrically assisted casket table three-way coach for $8,005. This car also has red lenses over the parking lights.

Miller of Bellefontaine, Ohio, continued to offer a line of 10 limousine and landau professional cars all mounted on the Cadillac Series 86 commercial chassis. This attractive oval window landau was offered in rear loading versions at $6,930, a combination Duplex at $7,120, an ambulance for $7,245, a manually operated casket table three-way coach for $7,680, and an electrically assisted casket table three-way coach at $8,300. All models were offered with or without the crinkle finish roof and the oval window in the landau area. A distinctive Miller addition on all landau models this year were the ornamental Cadillac louvers on the rear fenders.

Eureka offered their coaches in both limousine and landau body styles. This landau ambulance was owned by the Thomas Funeral Home of Terre Haute, Ind. It features all of the styling characteristics of the Eureka landau funeral coach plus large chrome crosses on the rear body sides beside the landau bow and the ornate Eureka coachlamps. This car also features funnel type roof warning lights and a roof mounted siren warning light. This ambulance in standard form and with landau styling cost $8,255, while a rear loading funeral car with landau styling went for $7,686, and the three-way landau with manually operated casket table went for $8,120. The power operated casket table three-way version cost $8,846.

Eureka offered this attractive limousine style funeral car in five versions for 1953. The rear loading funeral car in limousine styling cost $7,376, and the limousine combination went for $7,450. Three-way models were offered in either manually operated casket table models for $7,809, or electrically operated forms for $8,537. This car was also offered in ambulance form for $7,945.

Eureka still built a flower car of conventional styling. This car, seen recently in the southern U.S., was in immaculate condition and finished in dark blue. The rear deck was of stainless steel and hydraulically operated. The rear windows were of the wrap-around style so popular at this time and the cab was of coupe configuration similar to that seen on all other makes of flower cars offered. The Eureka flower car sold for $7,287.

1953

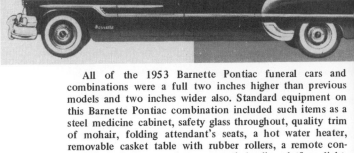

One of the most interesting Barnette offerings was this attractive and low priced Pontiac flower car with a hydra-electric flower tray. This car, patterned after more expensive flower cars from other makers, had a coupe type roof with wrap-around rear windows, a stainless steel flower deck and tonneau, and a stainless steel rear compartment interior. Here we see this attractive car with optional two-tone paint finish. It is loaded with floral tributes to illustrate the attractive appearance that this car had when loaded.

All of the 1953 Barnette Pontiac funeral cars and combinations were a full two inches higher than previous models and two inches wider also. Standard equipment on this Barnette Pontiac combination included such items as a steel medicine cabinet, safety glass throughout, quality trim of mohair, folding attendant's seats, a hot water heater, removable casket table with rubber rollers, a remote controlled cot fastener, and an automatic loading platform light. Note the optional formal draperies on this car.

The Economy Coach Co. of Memphis continued to offer professional cars built on the Pontiac chassis. This picture was heavily air brushed by factory artists to make the car a 1953 model. The styling of the Economy coaches were simple and straightforward. This was the Economy combination ambulance and funeral coach for 1953.

The ambulance offered by Barnette featured the styling of their limousine funeral car and combination but with frosted glass in the rear quarter windows and a complete ambulance interior. This car is seen with fender mounted siren warning lights, a set of tunnel warning lights on the roof and a revolving "gumball" light on the roof.

The Economy Pontiac ambulance was available with roof tunnel warning lights and bumper mounted red warning lights. The two-tone paint finish, whitewall tires and the Federal roof siren warning light were optional at extra cost.

National of Knightstown, Ind., continued to offer coaches on either the Pontiac or Chevrolet chassis and in landau or limousine forms. The long wheelbase versions were now called DeLuxe models and were available as funeral cars, combinations, and ambulances. The National DeLuxe landau on the Pontiac chassis, seen here, is a fine example of the work this firm turned out. This car also utilizes the rear fenders of the Pontiac station wagons.

The Shop of Siebert in Toledo, Ohio, celebrated its 100th anniversary this year and did so with a full contingent of models on the Ford and Mercury chassis. The long wheelbase version of this attractive little funeral car was called a Centennial. It was a five-door car with unusual rear window styling.

1954

Reorganizing their motor coach business this year, the [Wa]yne Works Inc. of Richmond, Ind., purchased the [Me]teor Motor Car Co. Wayne had operated the Meteor [co]mpany under a lease arrangement since early in 1953, [an]d in January of 1954, purchased the factory and its [int]ernal property for $230,000. Wayne, one of the [cou]ntry's largest manufacturers of school busses and [de]livery trucks, purchased this Piqua based firm and [co]ntinued operations and marketing of Meteor coaches. [Wa]yne had been involved in the construction of school [bu]s bodies for 117 years, and employed between 800 and [1,0]00 workers in its Richmond plant.

With the introduction of a newly styled chassis by [Ca]dillac, all of the manufacturers launched new models [fo]r the first time in three years. The wrap-around rear [wi]ndow, now common to most passenger cars, was [in]troduced in funeral coaches and ambulances by S & S [in] 1953 and in an attempt to go them one better, Miller [in]troduced a totally redesigned line of limousine models [in] 1954 with wrap around rear windows. Massive expan[s]es of glass were used on the sides of these new [li]mousines. The rear side window wrapped around the [ba]ck side of the car where it was met by a thinly tapered [pi]llar.

This year Eureka began the transition from the wood [fr]amed body to one of all steel. This was done by the [ot]her large manufacturers many years earlier but was [on]ly begun in this year by Eureka. This project, coupled [wi]th a great model change, proved to be more expensive [an]d time-consuming than any other project the company [ha]d ever undertaken. The main reason for the change on [E]ureka's part was to lighten the total weight of the car [an]d to cut expenses, wood being somewhat more [ex]pensive than steel. Then too, steel was much more [re]adily replaced than wood in case of an accident.

Although 1953 was a very good year at Henney, 1954 was not. Sales were off by 40% despite a longer and more tidy appearing style that came with the '54 models through a restrained use of chrome trim. While generally similar externally to previous models, the 1954 Henney's embodied many minor improvements over the last models. The new models were equipped with Packard's large 212-horsepower 359-cubic-inch engine and offered a whole new range of interior color trim selections. At the end of the model year, it was decided to cease production of all funeral coach and ambulance operations at Freeport. So, although the funeral directors liked the Packard coaches and Henney was the only manufacturer building on the Packard chassis, they were to be no more, and an 86-year history of quality coachwork ended.

Superior built a few prototype coaches that sparked their design studio and those who viewed the cars to induce the company to begin production of a newly conceived hardtop style funeral coach, flower car and ambulance line. Some were built and marketed around mid-year with the big advertising and promotional push, for these innovative vehicles were coming with the new model year.

A total of 2,880 professional vehicles were produced this year, with 1,756 of these being on Cadillac chassis.

With modest restyling of the roof line around the side windows, Henney announced the new Henney Packard landau models in no less than five versions. The Henney model seen here was mounted on the Packard 5413 chassis and in its rear loading form cost $7,283. In three-way Elecdraulic form it went for $8,198 and with a manually operated three-way table the car went for $7,883. This marked the first time since the early 1940's that Henney offered a manually operated three-way coach and did not offer Leveldraulic. The '54 Henney Packard flower cars sold for $8,000 and a service car sold for $7,150.

Built on the Packard 5433 Clipper chassis, the Henney Junior also enjoyed its last year of production with 120 units being constructed. This economy combination coach sold for $4,350 in 1954 and was offered as only one model called the Model 2733. Total Henney production for 1954 amounted to 325 professional coaches and 100 Packard limousines.

This 1954 Henney Packard DeLuxe ambulance was delivered to the City of Toronto for use in Department of Public Health, Division of Morgue and Ambulance Operations. The car was of limousine styling, and in ambulance form sold for $7,700. Other models in the limousine body style included a rear loading hearse for $7,150, a manually operated three-way car for $7,750, and a three-way funeral car with Elecdraulically operated casket table for $8,075. Henney built 205 cars on the Packard 5413 chassis in this year and at the end of the model year retired from the business altogether. With this retirement the Packard chassis funeral car and ambulance that had enjoyed considerable popularity over the years would end forever.

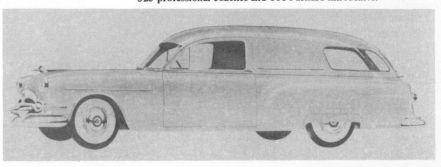

Superior offered the industry's most interesting and appealing medium priced landau funeral cars. This Superior Cadillac landaulet carried a price beginning at $7,498 for a rear loading car. The unique Superior landau shield area and roof styling made this car one of the best sellers in the field. In this year Superior built a total of 590 Cadillac chassis professional cars.

The Superior landaulet, like the limousine funeral car, was offered as a three-way loading car also. This Superior Cadillac landaulet was a three-way car. The car carried a price tag of $8,000 when it was manually operated, and $8,475 if it was power assisted. Superior coaches were also offered as combinations and the landaulet combination was offered in three versions. The first, for $7,846, featured reversible rubber casket rollers on the floor. The second had a removable tubular casket rack with rubber casket rollers and went for $7,846 also. The third version of the Superior combination with landaulet styling sold for $7,913. This car had special sections in the floor that were reversible to reveal casket rollers on one side and a flat linoleum covering on the other.

A totally revolutionary experimental car was unveiled by Superior Coach Corp. of Lima, Ohio, in the mid-part of this year. Called the Beau Monde Floral Coach, the new Superior car was a copy of a popular style of passenger car body. The Beau Monde was the first hardtop funeral car ever devised, and the demand and interest in the car was so overwhelming that the company built a limited number of these cars in 1954.

Superior's pleasingly styled limousine funeral car was one of the firm's most popular cars. This particular car belonged to the V.P. Sweeney Funeral Home of Yarmouth, Nova Scotia. The limousine was offered as a straight end loading vehicle for $7,264 or as an end loader with the DeLuxe Moderne package for $7,372. The side servicing models were priced at $7,891 for the standard form and $8,349 as a Moderne. Combinations were also offered with prices beginning at $7,602 and going up to $7,784.

This Superior limousine funeral car featured a glass roof section near the front of the top just above the driver's compartment. This special job was built for the Downard Funeral Home, and was just one of the numerous types that could be built at the customer's request. Naturally, these features raised the cost of the finished car, but also added additional novelty to a fairly common type of funeral car.

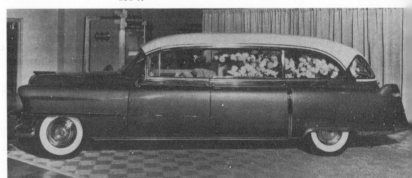

This year's Superior Coupe De Fleur flower car had a hydraulically operated flower tray as standard equipment and sold for $7,744. Note the wrap-around rear window.

1954

Some funeral directors thought that the service car was the most important vehicle in their fleet, and demanded dignified vehicles for service cars. They did not want ostentation, while at the same time they did want to reflect an impression of unquestioned excellence, of everything about their service being well done, even if some things did not attract much notice. If the first call was handled with an air of distinction, an excellent impression was thereby created for the entire funeral service. These were the things that the makers had to consider when they designed a service car. Superior's new Pontiac service car was designed with all of the distinction and elegance that befitted the profession. It featured a stylized chrome wreath and chrome bars on the upper rear body sides. This model also has a crinkle finish roof. This car was a full-length five-door style similar to the hearse on the Pontiac chassis.

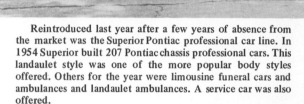

Reintroduced last year after a few years of absence from the market was the Superior Pontiac professional car line. In 1954 Superior built 207 Pontiac chassis professional cars. This landaulet style was one of the more popular body styles offered. Others for the year were limousine funeral cars and ambulances and landaulet ambulances. A service car was also offered.

Superior offered five ambulances in both landaulet and limousine styling. An added feature of the standard limousine ambulance was the availability of the Moderne styling and appointment option. The standard limousine ambulance cost $7,767, while the Moderne limousine ambulance went for $7,878. A landaulet ambulance was offered for $8,006 and the two Rescue high headroom models similar to the one depicted here went for $8,328 and $9,367. The latter one is shown here as it was prepared for the Detroit ambulance fleet.

Miller of Bellefontaine, Ohio, completely redesigned their funeral car and ambulance line for 1954, with the most dramatic effects being seen on the limousine offerings. The new Miller limousine models featured wrap-around type rear quarter windows that wrapped around so far that they left only a very thin rear roof pillar. This vast expanse of glass area was a major improvement over previous styles and afforded a totally new look. The standard Miller limousine funeral car with rear loading cost $7,399, while the combination Duplex went for $7,589, and the three-way car with manually operated casket table sold for $8,179. The three-way coach with an electrically assisted casket table sold for $8,819. Landau models ranged in price from $7,704 for the end loading job to $9,124 for the three-way electric.

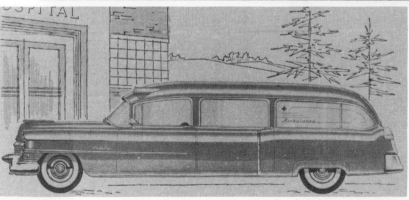

Miller offered a complete line of ambulances, as competition in this area was becoming tougher than ever. The straight limousine ambulance went for $7,719, while the landau version sold for $8,029.

Where the ambulance competition was really becoming tough was with the high headroom emergency vehicles. This Miller First Aider sold for $8,734 complete.

Meteor dramatically restyled their line of professional cars this year. The new bodies incorporated wrap around rear windows as seen here on this distinctive Meteor landau. Also restyled was the entire roof line and landau panel. The Meteor landau was offered in five versions including an ambulance. The Meteor Model 540-L was a rear loading landau that sold for $7,573, while the Model 540-LH was a three-way car with an electrically operated casket table that cost $9,233. A combination ambulance/funeral car was also offered in the landau body style, was designated a Model 540-L, and sold for $8,049. The Meteor landau ambulance cost $7,927.

With the restyling of the Meteor line of funeral cars and ambulances, the firm dropped the now famous door mounted coach lamps. The new coaches featured wrap-around type rear windows on the corners and extremely thin rear pillars. The limousine funeral car was offered in five versions with prices beginning at $7,484 for the end loading coach. The Model 540 rear loading combination coach went for $7,971, while the three-way side servicing car with a manually operated casket table cost $8,508, and the three-way electrically operated version went for $9,102. The limousine ambulance went for $7,767.

The interior fittings of the Meteor landau were neat and plush. The draperies, walls, and door panels were appointed in rich mohair and the floor was covered in linoleum with the hardware chrome plated. This picture illustrates the interior of the Meteor straight end-loading landau funeral car. Notice the metal casket rollers built into the floor, the small interior coach lamp on the side wall and the draperies placed on the blank landau panel.

Meteor offered the Pontiac line in two separate wheelbases. The full length line, similar to the limousine coach, and this short wheelbase three-door style. The Meteor Pontiac short wheelbase ambulance was complete with funnel type roof warning lights and special ambulance decals in the rear quarter windows. These cars were esxended very little. The short wheelbase car was offered in ambulance, service car and utility funeral car styles.

The Meteor flower car was dubbed the Envoy and was offered in either the eastern style, similar to the car shown, or the western style Chicago car with the flower trough and no hydraulically operated stainless steel deck. The eastern type car sold for $7,705 and the Chicago type cost $7,308. Both cars were designated Model 543. Note the coupe de ville type styling. A service car designated as a Model 542 was offered and went for $7,573.

Appearing for the last time was the Meteor line of extended frame Pontiac professional cars. The new Meteor Pontiac line for 1954 carried the styling that had made them so popular in earlier years and were offered in landau, limousine, and service car styles. This is the Meteor Pontiac limousine funeral car, complete with fringed draperies and optional whitewall tires.

1954

Also making its appearance for the last time was the Meteor Chevrolet line of economy professional cars. The ambulance seen here was one of the slightly lengthened versions and is complete with optional tunnel roof warning lights and a rear quarter window ventapane. Both the Pontiac and the Chevrolet Meteor coaches were built from standard panel vans.

While the Meteor short wheelbase ambulance of the Pontiac chassis had an ungainly appearance, the Meteor short wheelbase Pontiac service car was a very attractive coach. Notice the chrome wreath and bars on the upper rear body sides and the three door styling.

Hess & Eisenhardt's S & S Knickerbocker was the firm's standard limousine coach offering again for 1954. With prices beginning at $7,895 for the rear loading funeral car and going to $8,782 for the three-way Knickerbocker with manually operated casket table, the car was the S & S prestige limousine entry. The most expensive Knickerbocker was the three-way electrically assisted casket table model at $9,452. The car was also offered in combination form and this version carried the same price tag as that born by the rear loading funeral coach.

The Hess & Eisenhardt S & S Victoria continued without any styling changes for 1954, and retained the wrap-around type of rear windows that were the one new addition to the car's styling for 1953. The straight end loading S & S Victoria funeral car sold for $8,326, while the Victoria with combination car equipment (removable casket table with rubber rollers) went for $8,416. A Victoria in three-way form with a manually operated casket table cost $9,112; and with an electrically operated casket table, this type of car cost $9,728. The Victoria Parkway ambulance continued to be offered for $8,835.

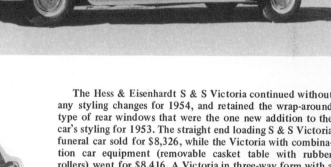

The Hess & Eisenhardt S & S Kensington was available in two forms, the standard limousine version with styling similar to that of the Knickerbocker, and the high headroom version called the S & S Professional Kensington. The standard Kensington ambulance sold for $8,508, while the Professional Kensington like the one shown cost $9,034. The raised roof carried a piece of blue plexiglass in its forward section and tunnel warning lights in the rear. Note the S & S hubcaps on this car. S & S was now building the most expensive and exclusive coaches in the field.

The Hess & Eisenhardt S & S Florentine flower car was the firm's only offering in this class and sold for $7,954. It is not known if this car is a standard representative of the Florentine model or not. It has the Chicago type flower deck and a convertible roof over the driver's cab. This car was the beginning of a new era in professional car styling and innovation.

1954

Eureka continued with the successful styling formula for the landau. The rounded curve of the roof and the rear door added to the air of dignity already established by the car's graceful landau bow and ornate coachlamps. This Eureka landau in its standard rear loading form went for $8,070, while the Eureka three-way manually operated coach cost $8,723, and the electrically assisted version of the three-way went for $9,451. Limousine versions of these cars went for $330 less than the landau models. Eureka coaches were the second highest priced Cadillac cars offered to the trade. Only Hess & Eisenhardt offered a more expensive line of coaches with their S & S models.

The Eureka Company of Rock Falls, Ill., was the only major manufacturer of funeral cars and ambulances that continued to fashion their body frames from wood. Other makers had long since abandoned this type of construction in favor of the all-steel type. This year Eureka offered the ambulance in only the limousine styling, which sold for $8,303. This Eureka limousine ambulance features optional two-tone paint and tunnel type roof warning lights. The ambulance decal in the rear quarter window is of a distinctive design.

The National Coach Company of Knightstown, Ind., also offered Pontiac chassis funeral cars and ambulances this year. While the cars were professionally finished, they were still sold at prices below that of a similar model from Superior or Meteor. National also offered these cars on the Chevrolet and Buick chassis.

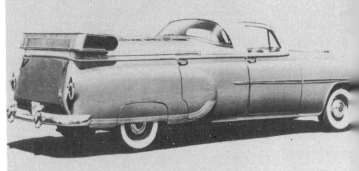

The Economy Coach Co. of Memphis was building professional cars on the Pontiac chassis, and this year offered this attractive Pontiac flower car. Note the graceful coupe styling and the high flower deck that was finished in stainless steel. This was a very low priced car that utilized Pontiac station wagon type taillights, and standard wagon rear side doors.

In France some funeral homes were using these little Renault bearers' coaches that could accommodate 12 passengers and a small casket or a pile of chairs or flowers. The rack on the roof was also for the flowers and the spare did not crowd these. The car was painted black and had very little chrome trim.

1955

The hardtop passenger car style had been introduced in 1949 on some makes and was by this time quite fashionable and popular. Superior shook the industry stylists with a new line of coaches that featured this type of hardtop styling. This was the most innovative design to be announced to the trade in many years. Called the Beau Monde, these cars were available as combination floral coach, hearse, combination hearse and ambulance or as a straight ambulance or hearse. The floral coach and combination floral coach featured a large flower tray in the compartment below which was a regular, lighted, casket compartment. When flowers were placed on the tray, they converted the coach to a very attractive, and distinctive looking closed-type flower car that could be used as a straight flower car or as the hearse and flower car combined. Far and away ahead of the competition from a styling point of view, the Beau Monde hardtop coaches were also expensive. The standard flower car combination carried a price tag of $8,612 and the sales figures reflected something less than spectacular for this model line. The remainder of the Superior model range featured mildly revised styling and new lavish interiors.

Another new model to make its debut in this year came from Meteor. Called the Crestwood, this new line of funeral coaches featured imitation woodgrain trim on the lower body sides. Similar to the treatment on some modern station wagons, this woodgrain was available on the Crestwood in two choices of wood types, walnut and mahogany. Delivery of the new Crestwood was restricted to only one such car in any community or locality, thereby assuring its owner, for the duration of the model year, all of the benefits derived through the use of an outstanding funeral coach. Meteor, now a division of the Wayne company, also built an interesting station wagon using the standard limousine style body with Crestwood type woodgrain appliques and three rows of seats. Looking very much like a Crestwood type limousine funeral coach without draperies, this station wagon was used by Wayne at the Richmond, Ind., plant and only one was ever built.

Economy Coach had been building coaches along parallel lines and the models had been similar. Together they offered extended frame coaches on almost any chassis. Meanwhile, another firm was formed in Memphis. Founded by Waldo J. Cotner and Robert Bevington, the new Comet Coach Co. built high quality, low cost vehicles on extended Oldsmobile chassis. Although the first Comet products were somewhat crude, as compared with other makes of the era, and their styling was quite similar to those of the other Memphis based companies, the Comet products were to outlast their hometown contemporaries with ease.

The Acme Company, founded in the early fifties, had not had an easy time in the crowded field, and had only one year that could be called good, which was 1953 when sales hit the all time Acme high of 87 units. Though Acme sold cars in 39 states from coast to coast, the tough competition of the field and lagging sales figures closed the company at the end of the 1955 model year.

Production of funeral cars and ambulances again sagged, with only 2,661 such vehicles being produced in the U.S. Of these, 2,040 were on Cadillac chassis.

The Memphis Coach Co. of Memphis had absorbed the Economy Coach firm of that city and was now producing luxury versions of the Economy line. This stylish Memphian Pontiac combination was an excellent example of the coachwork turned out by this small firm. Note the high headroom roof.

The Memphis Coach Co. would build a coach for a customer on almost any chassis desired. This Ford ambulance with a slightly lengthened wheelbase and three-door format was an example of this type of work. Note the vent windows in the rear quarter windows, the tunnel roof lights, gumball light, and the Federal fender mounted siren warning light. The car was most likely furnished to some local fire brigade. It was painted red and white.

Having been absorbed by the Memphis Coach Co., Economy was relegated to building only low priced coaches. Note the similarity between this car and the Memphian. Differences are in trim, mostly on both the exterior and interior. Red lights are mounted on the hood, signifying that it, too, was a combination coach. When compared with the Memphian, this car's lack of ostentation can be clearly seen.

1955

Superior offered the trade the most revolutionary new coach of the age with the Beau Monde Floral Coach. This car with four side doors and hardtop styling, was a vast departure from the traditional limousine. It was offered as a straight flower coach, a rear loading funeral car or as a combination ambulance/funeral car. This is the flower car version, which was a practical new type of car because not only were the flowers in full view but they were under the cover of a steel roof in case of inclement weather. Notice the wrap around type rear corner windows.

The Superior landaulet was offered in nine versions that were all similarly styled. The interiors of these cars varied with their purpose as did the prices. The straight end loading landaulet like the one seen here sold for $7,902, while the landaulet with a manually operated three way casket table and Superior Lev-L-Matic for leveling the car cost $9,328. A straight three-way coach with manually operated casket table went for $8,756, while the electrically assisted casket table version went for $9,196 and for $9,768 when this same car was equipped with Lev-L-Matic. The landaulet was offered in three distinctly different combination versions that ranged in price from $8,150 to $8,217. The landaulet ambulance went for $8,336 and completed the line up of offerings in this body style. The car seen here was finished in white with a black crinkle finish roof. Comparable limousine models sold for $235 less and Modernes for $144 less than the landaulets.

Occupying the traditional place of honor in the Superior line was the beautiful Coupe De Fleur flower car. This year the car sold for $8,242 and continued to feature a hydraulically operated flower deck made of stainless steel. Note the flower railings on the rear deck of this coach.

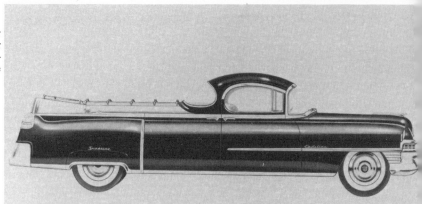

Ambulances were offered in four limousine styles, two of which had high headroom roofs and were designated as Rescue ambulances. This standard limousine ambulance cost $8,097 with a Moderne version going for $8,208. The Moderne limousine featured a deluxe interior decor and wrap around rear corner windows. The Rescue ambulance cost $8,706 and the Super Rescue went for $9,626. This particular limousine ambulance is fitted for service with a fire department and has bells on the front bumper, fender mounted warning lights, four roof mounted tunnel type warning lights, a "mars" light on the roof and a "gumball" type warning light also mounted on the roof. Note the fire extinguishers mounted in special wells on the rear fenders. This car was finished in red with red and white window decals.

Superior funeral coaches and ambulances were also offered on the Pontiac chassis. This landaulet funeral car was among the models available. Others included limousine and landaulet ambulances and a limousine funeral coach. Note the styling similarity between this Superior Pontiac and the Cadillac models.

1955

Superior built a total of 766 Cadillac chassis funeral cars in 1955. The new Beau Monde was the most innovative style funeral car to hit the field in many years and was offered in flower coach, rear loading funeral car, or this combination. Note that this car does not have the flower deck in the rear compartment that was part of the design on the flower car and rear loading funeral car. The Beau Monde in floral coach form sold for $8,612.

Also offered on the Pontaic chassis was the service car with chrome bars on the upper rear body sides and a fixing point for the owner's nameplate.

The now famous and popular A.J. Miller oval window landau continued to be offered. In standard form this car sold for $7,984. As a combination, or Duplex as Miller called them, this car went for $8,179, while the three-way coach with a manually operated casket table cost $8,764. The three-way car with a power operated casket table went for $9,404 and an ambulance version of the landau cost $8,309. This car has optional chrome trim on the lower rear fenders and a crinkle finish roof. This particular coach was the fourth Miller vehicle purchased by the J.B. Marlatt Funeral Home of Hamilton, Ontario. Miller cars were distributed by the Dominion Manufacturers in Canada at this time.

Miller offered a complete line of ambulance models that ranged in price from $7,919 for the limousine to $8,309 for the landaulet version. This Miller high headroom emergency ambulance was called the First Aider. This car sold for $8,524 and offered the highest roof height of any ambulance in its class. Note the built in tunnel warning lights on the roof, the spotlight, and the unique side of the roof and frosted glass in the rear quarter windows.

This Miller limousine funeral coach was owned by the Hulse & English Funeral Home, another Ontario mortuary that purchased Miller coaches religiously. The Miller styling had not changed for 1955 and continued to offer the wrap around rear quarter window and thin rear roof pillar on the limousine models. In the standard rear loading form the limousine sold for $7,599, while the Duplex cost $7,789 and the three-way coach with a manually operated casket table went for $8,379, while the three-way with a power assisted casket table was $9,019. Miller continued to offer a flower car within the line, and in this year it sold for $7,939.

1955

Seen in a quixotic atmosphere this was the new Meteor Crestwood landau. The Crestwood was a new and exclusive version of the Meteor landau and featured imitation wood grain paneling on the lower body sides. This was offered in two versions. One had a Mahogtrim finish while the other had a Walnutrim finish to the wood grain. The crinkle finish roof was offered in three colors as well as black. The delivery of Crestwoods was restricted to but one such car in any community or trade area, thereby assuring the owner of exclusivity during the model year. The chrome trim down the body sides was also found on the Crestwood coaches only, and was called a Starstreak. A crest with the owner's initials was placed on the rear side doors just above the dip in the Starstreak. Note the wrap around rear corner windows.

Meteor again re-designed the landau and made the wrap around corner windows optional on the standard landau. The rear loading landau sold for $7,692, while a landau combination went for $8,086. Three-way models were offered with prices beginning at $8,592 for the version with a manually operated casket table, and going to $9,175 for the version with a power assisted table. Note the new styling of the landau area and the combination airline type drapes on this car.

The standard Meteor ambulance with limousine styling sold for $7,963, while a landau version of this car went for $8,126. The Meteor limousine funeral coach was offered in rear loading styles at $7,563, a combination for $7,921 and a three-way side servicing manually operated table version for $8,463. The limousine funeral car with the three-way electrically assisted casket table went for $9,046. Meteor series designations for 1955 were 550 for limousine models and 550-L for landaus. The letter H designated a manually operated three way coach and the HE designation behind the model numbers designated a three-way electric. Ambulances were designated Models 551, the service car that sold for $7,652 was designated a Model 552, and the flower car a Model 553. This year Meteor listed only the Eastern type flower car and that was priced at $7,961.

Borrowing from the Crestwood's lower body paneling, Meteor built this interesting station wagon this year for use around the plant. The car featured three rows of seats and stainless steel flooring. The seats folded flat to convert the car into a regular utility van for shop errands or parts carrying. From the outside the car looks very similar to the limousine hearse. Only one of these coaches was built.

Just entering the funeral car and ambulance field was the Comet Coach Co. of Memphis. This company, formed by Waldo J. Cotner and Robert Bevington, was to become one of the large firms in the future. Building on the Oldsmobile chassis, they offered a line of attractive and economical vehicles. This is the Comet limousine combination.

Comet's ambulance was this attractively styled Oldsmobile coach with finely etched rear quarter windows, tunnel type roof warning lights, and a centrally mounted roof siren warning light. Comet built on the Olds chassis mainly but did build a few cars on the Pontiac chassis.

1955

Continuing to be the most luxurious and most expensive limousine funeral car in the field, the S & S Knickerbocker had prices beginning at $8,144 for the rear loading hearse and going to $8,234 for the Knickerbocker combination. The Kinckerbocker went for $9,031 as a manually operated three-way coach and for $9,756 in three-way Electdraulically assisted form. The S & S Knickerbocker was selling for a full $399 above its nearest limousine competitor, Eureka. Ambulance coaches built by Hess & Eisenhardt in this year ranged in price from $8,693 for the Kensington limousine (a car derived from the Knickerbocker body) to $9,324 for the deluxe emergency ambulance called the S & S Professional Kensington. S & S flower cars were offered in three versions that ranged in price from $8,081 to $8,318.

Hess & Eisenhardt continued to build S & S coaches in the ultra-premium price range, with nothing but the best in styling and appointments. The legendary S & S Victoria was offered this year in seven funeral car styles plus an ambulance. The Victoria with electric side servicing cost $10,022, making it the most expensive coach on the market. Other S & S Victoria models included the rear loading car at $8,511, the combination hearse ambulance for $8,601, the three-way coach with a manually operated casket table for $9,297, and the Parkway ambulance for $9,020. The distinctive S & S Victoria was continued without change for 1955.

The Eureka limousine coach continued with the styling that had made this car so popular in years past. The slanted "C" pillar and curved rear quarter windows added to the car's modern streamlined styling. This car, in a rear loading form, sold for $7,966 and as a three-way with a manually operated casket table cost $8,629. The three-way with an electrically assisted table went for $9,357, while the combination with a roller rack sold for $8,046, and the combination with a set of reversible rollers sold for $8,196. The combination with reversible floor pieces sold for $8,248 and the ambulance for $8,525. The car in this illustration is a rear loading funeral car.

Eureka was beginning to acquire the machinery needed to transform the body frames from wood to the more modern steel type. In the meantime, all of the coaches continued to be built with the wooden body frames by old master craftsmen. This dignified and attractive Eureka landau was offered in six versions. The straight end loading landau went for $8,296, while the landau combination coach went for $8,381 with a roller rack, for $8,531 with reversible floor rollers, and for $8,583 with reversible floor pieces.

Selling for $8,291, the beautiful Eureka flower car attracted few buyers this year. The car, with a dignified coupe driver's cab and wrap around rear windows, had a large hydraulically operated stainless steel flower deck and a stainless steel casket compartment below the flower deck. This was equipped with casket rollers and the car could be used as a hearse as well as a flower car or a service car.

1955

The Barnette Co., also of Memphis, also offered a line of professional cars on the Pontiac chassis. The styling differences between this car and the Memphian-Economy examples is clearly seen. This car featured wrap around rear side windows and a large "B" insignia on the rear fender. The keen competition among these Memphis based companies offered the consumer a good selection of high quality, low priced coaches and an alternative to the expensive models offered by the large makers.

The interior of the Eureka landau was luxuriously appointed with the finest mohair and velvets, as seen in this landau three-way coach. The three-way landau with a manually operated casket table went for $8,959, while the three-way with an electrically assisted casket table (shown) went for $9,687.

Siebert of Toledo, Ohio, announced a new line of Ford funeral cars and ambulances called the Centennial series. These cars were long wheelbase limousine or landau coaches. This landau shows the extremely economical lines that had made the Siebert so popular with small firms. Note the vent in the rear door window.

The National Coach Co. of Knightstown, Ind., continued to offer their coaches on either the Pontiac or the Chevrolet chassis. This was the beautiful National DeLuxe combination on a Chevrolet chassis. Note the smooth lines and the luxurious long wheelbase. The firm also offered landau and ambulance styles on both the Chevrolet and Pontiac chassis.

Two other models in the Siebert line included this Centennial long wheelbase ambulance and the small Aristocrat Service car at the top. Both were mounted on the Ford chassis that had been especially lengthened by Siebert. Note the vent window in the rear door of the ambulance and the wreath and coach lamp on the service car.

French flamboyant is the only way to describe this Peugeot chassis hearse used in France. The car carried the pallbearers in the cab and the casket in the highly draped, plumed and gilded rear compartment. This car was typical of the French funeral coach of the era.

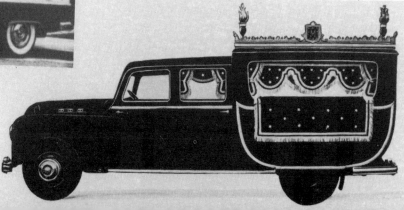

The Weller Brothers, also of Memphis, offered yet another alternative to the large coaches with a line of cars that could be mounted on any desired chassis. The extremely long Weller extension given to this car gave it superior rear compartment room. Note the unorthodox styling of this car and the Ford type hub caps. This car was a combination coach. Note the warning lights on the front bumper and the spotlight.

1956

The Wayne Works and its funeral coach and ambulance division, Meteor, had one of its biggest years since they formed a union. The Meteor Company received a large order for eleven special ambulances from the city of Chicago. The ambulances had been ordered from Meteor after receiving bids from other companies. While the Meteor bid was not the lowest, the city officials were impressed with the overall quality of the Meteor products. With most of the cars going to the Chicago Fire Department, they were all built to the special specifications of the city and the order was filled in two months for a grand total of over $100,000. The red and black ambulances were driven from the Meteor plant in Piqua to Chicago by members of the Chicago Fire Department, where they were put into immediate service. They were all equipped with the latest life saving paraphernalia including oxygen tanks and two-way radios, and could carry four ambulatory patients.

On March 19, 1956, the Wayne Works announced the purchase of the A.J. Miller Co. of Bellefontaine, Ohio. Miller, another leading producer of ambulances and funeral coaches, had been established in 1853 and had been one of the styling and innovation leaders for many years. The purpose Wayne gave for buying Miller was to bring this company into the Wayne family and, with Meteor, accelerate the progress in development of these vehicles. The two firms were to gain important advantages in the industry by way of mutual interchange of technical information and methods, according to Wayne. For the duration of this model year the two companies continued to do what they had done for years, compete fiercely with the other. This was the last year in which they would do so.

Production of funeral cars and ambulances sagged again this year, dropping to the lowest point since 1946. Only 2,281 professional cars were built in the U.S., with 1,799 of these being on Cadillac chassis.

The Meteor company unveiled its 1956 models that utilized the same body styling as the previous year's line. The same models were offered as in 1955 but the prices increased an average of $400. In addition to this attractive limousine with the Meteor Starstreak on the side and a two-tone paint finish, the company continued to offer the unique Crestwood landau with its unusual imitation wood grain side panels. This was the last appearance of this body styling and the Meteor name as an independent company. Meteor was owned by the Wayne Corp. of Richmond, Ind.

Part of a large consignment of Meteor ambulances purchased by Chicago is seen here. This car was painted a special Chicago gray with black roof and a special mars type roof light. These cars were especially selected because of their merits and equipment. They were all oxygen equipped and ready to meet any emergency.

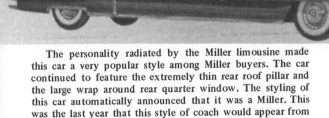

The personality radiated by the Miller limousine made this car a very popular style among Miller buyers. The car continued to feature the extremely thin rear roof pillar and the large wrap around rear quarter window. The styling of this car automatically announced that it was a Miller. This was the last year that this style of coach would appear from Miller. Appearing for the last time was the Miller oval window landau. This feature had become a trademark for the landau products built by the A.J. Miller Co. of Bellefontaine, Ohio, and was to dissappear next year along with Miller as an independent company. Late this year Miller was absorbed by the Wayne Corp. of Richmond, Ind., and in 1957 the Miller name was to be linked with another Wayne owned ambulance and funeral car maker, Meteor. Once again the Miller landau was offered with or without this oval window and with or without a crinkle finish roof. This particular vehicle was the second Miller coach purchased by the D.J. Robb Funeral Home of Sarnia, Ontario. Standing next to his new coach is Mr. Robb.

1956

Hess & Eisenhardt built this attractive S & S Florentine flower car on the mildly restyled Cadillac commercial chassis. This attractive coach featured a very low rear flower deck and a tonneau at the rear of the deck that was not as prominently high as some of the competition. Note the newly styled roof line of this car. Hess & Eisenhardt was producing the most expensive Cadillac chassis coaches in the field, in a wide variety of models and styles.

The traditional S & S Victoria continued through the year without any major styling changes. The car continued to feature a fabric covered landau roof with wrap around rear windows and curved rear side windows. The Victoria was offered as a rear loader, a three-way manually operated casket table version, or with an electrically operated table, as an ambulance, or a combination. The Victoria styling complemented the classic beauty of the original Victoria landau design.

Once again the S & S Knickerbocker represented the Hess & Eisenhardt limousine coach offering. This car, wearing styling basically unchanged form 1955, was offered in a choice of straight end loading funeral car, three way funeral car with either manually operated or electrical casket table, combination ambulance and hearse, or as an ambulance. The limousine ambulance version of this car was again called the Kensington. Air conditioning was offered as an option for the rear compartment and was very popular in the combination and ambulance versions of this car.

Eureka continued to offer its landau funeral car with unchanged styling for 1956. This was the last appearance of the Eureka coach with a wood framed body. Once again Eureka landaus had a curved rear side window, rear end and an ornate coach lamp on the landau panel. Eureka claimed that this car was the greatest quality line of coaches ever presented, and that these new cars had finer appointments and refinements than ever before.

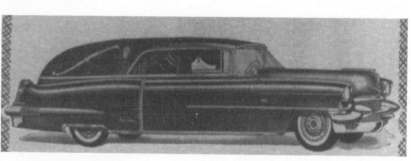

The Eureka limousine funeral coach and ambulance shared the same body with alterations to the rear compartment and roof. Note the sloped "C" pillar, the curved rear quarter window, and the wrap around rear corner window. The large manufacturers were now all using the Cadillac chassis for their coaches. These chassis were especially built by Cadillac and were supplied to the makers complete with the front and rear fenders and cowl.

Selling for $8,755, the Eureka limousine ambulance was mounted on the Cadillac Series 86 commercial chassis and blended the all new and exclusive lines of the Cadillac chassis with the graceful lines of the Eureka limousine coach. For over 69 years Eureka craftsmen had always developed brilliant, custom made, quality built coaches. Note the built-in tunnel type roof warning lights and the optional roof siren light.

1956

Built on the standard wheelbase Ford chassis, this Memphian ambulance featured a three-door type body, special rear quarter windows with vents, tunnel type roof lights, a Federal roof siren warning light, red lenses over the parking lights, two-tone paint job, rear quarter window ambulance decals, and roof marker lights at the rear. The Memphis Coach Co. built cars on almost any desired chassis and to the customer's specifications.

This Memphian limousine ambulance was mounted on an Oldsmobile chassis and featured a roof mounted gum ball type warning light and wrap around rear corner windows. This car was built on an extended Olds Super 88 chassis and was fitted out with all of the ambulance emergency equipment.

Continuing to manufacture coaches for the low price segment of the market, Memphis Coach offered a wide variety of styles and variations on almost any desired chassis, such as this Pontiac limousine combination coach. Note the wrap around corner windows, the ambulance type draperies, optional spotlight, whitewall tires and two-tone paint job. This car was also offered in a landau body style ans as a straight end loading funeral coach and ambulance.

Offering a complete line of coaches on either the Chevrolet or the Pontiac chassis, National Coach Co. of Knightstown, Ind., built this stylish limousine combinatin funeral car and ambulance and the straight ambulance on the Pontiac chassis. Note the clean flowing lines and the smooth conversion. The roof of the ambulance features tunnel lights and a combination siren warning light. The National nameplate appears on the front fenders.

John Little of Ingersoll, Ontario, built this attractive little short wheelbase funeral coach utilizing the basic body of a Meteor three door station wagon. Mr. Little added a landau type roof line and landau bows, small rear side windows and coach lamps. The roof had a crinkle type finish and there was a nameplate in the rear side windows that read J.N. Black. The Meteor was a Canadian version of the Ford sold by Mercury dealers.

The National landau funeral car was also offered on a variety of General Motors chassis. This Pontiac was an end-loading funeral car only. The company did not offer three-way coaches. These cars were aimed at the lower segment of the coach market, a segment that demanded stylish, dignified cars at a low price. National also built cars on Oldsmobile and Buick chassis, but the Pontiac and Chevrolet chassis were by far the most popular.

The Beau Monde hardtop floral coach and funeral car appeared for the last time this year. This model was a combination ambulance and funeral coach. The folding attendant's seats in the rear compartment are flipped up ready for use and the gumball light has been affixed to the roof. The styling of this car was extremely pleasing, but sales were disappointing.

Superior sales and production rose to 770 Cadillac units in 1956, with the model selection being similar to that of 1955. Prices of the new models rose approximately $421 for each model this year, with the rear loading Moderne limousine selling for $8,214 and the standard rear loading limousine going for $8,106. This is the attractive Superior Moderne limousine funeral car/ambulance combination, finished in two-tone paint with whitewall tires. Part of the additional cost of the Moderne series was reflected in the wrap around corner windows.

Continuing with a very attractive and popular style, Superior again offered this attractive landaulet in a variety of versions that included both rear and three-way loading funeral cars, combinations and an ambulance. The prices for the 1956 models were about $500 above the 1955 prices. Note the distinctive landau shield and the crinkle finished roof. This car was a rear loading funeral car only.

Superior's Rescue ambulance featured a high headroom type rear compartment roof with built-in tunnel type warning lights at the rear and the front. Note the swell in the roof behind the driver's compartment, the optional spotlight, and gumball light on the roof. Most of these cars were supplied to the emergency rescue departments of large fire brigades.

The beautiful and innovative Superior Beau Monde floral coach and funeral car continued through 1956. With the side windows rolled up it looked just like any ordinary limousine funeral car. The model selection continued to include a combination ambulance and funeral coach in addition to the standard floral coach and rear loading funeral car versions. This car has the optional crinkle top finish.

Wearing a fancy optional paint finish and chrome mouldings, the Superior standard limousine ambulance was the most popular vehicle of its type in the company's line. In addition to the chrome strip and unique paint scheme, this car features optional roof warning lights, spotlights, and whitewall tires. Note the ambulance pattern decals in the rear side windows, and the emergency switch panel on the dash.

1956

A distinctive looking car, this Superior Pontiac landaulet funeral coach features a two-tone paint finish, whitewall tires, and straight combination type draperies. Superior built on a total of 292 Pontiac cars this year.

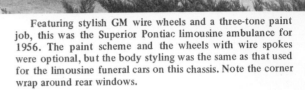

Featuring stylish GM wire wheels and a three-tone paint job, this was the Superior Pontiac limousine ambulance for 1956. The paint scheme and the wheels with wire spokes were optional, but the body styling was the same as that used for the limousine funeral cars on this chassis. Note the corner wrap around rear windows.

The Welles Corp., a large maker of school busses, engaged in some ambulance conversions. Located in Windsor, Ontario, they built this attractive little Ford ambulance from a three door station wagon. Modifications included a new roof with built-in tunnel type warning lights, ambulance decals in the rear side windows and a completely revised rear compartment.

Daimler and the large Austin Princess were the most popular chassis used by the major English hearse makers. These were long wheelbase limousine chassis. British hearses were offered in two versions. One (pictured here) was a straight funeral coach with a large casket compartment while the other, called a bearer's hearse, was a four door model with a rear seat for the pallbearers. This strikingly handsome Daimler hearse was built by the notable English hearse maker Woodall-Nicholson of Halifax. British hearse styling had continued to feature a large plate glass window on the sides that left the casket open to full view. Note the roof mounted flower rack.

Post-war Germany's auto industry could not feed the hearse builders with chassis sufficiently heavy or long enough to handle the coach bodies that they built. For this reason most German coachmakers began to import American chassis for this use. The most popular was the Cadillac commercial chassis that the North American hearse makers used. This example of German coach craft was built on just such an imported chassis and built by the Conrad Pollmann Norddeutsche Karosseriefabrik of Bremen. Note the American styling that prevailed on these Pollman coaches. The high roof line and the frosted glass in the rear side windows were typical of German cars of this era. Note the fender mounted parking light and the dual fender mirrors.

1957

This was a year of vast changes in coach styling and almost every major maker had something totally new to show. One of the highlights of the year was this unusual coach by Hess & Eisenhardt which featured a landau style when the rear compartment roof was raised. This car featured a collapsible rear compartment roof section that made this stylish flower car a semi-convertible landau.

For the first time in 19 years, the S & S Victoria was totally restyled within the original concept. The new Victoria featured squared off lines with a chrome landau shield behind the rear side door window. The landau bow was straighter than the old "S" type bow and the car now featured special trimmings that were more modern than any previous models. Note the special S & S side flash and the S & S nameplate on the rear fin. The car was called the master achievement of 81 years of S & S coach building.

Among the totally restyled Hess & Eisenhardt S & S vehicles was the Knickerbocker limousine funeral coach. This car had formal hardtop styling and featured wrap around side windows. S & S continued to offer the most expensive and exclusive coaches in the field. This car was offered as a straight end-loading funeral car, a three-way coach or a combination coach.

Cadillac revised its chassis and the manufacturer introduced new models based on the Cadillac commercial chassis. Superior again scored with totally redesigned models with styling that was a vast departure from anything ever seen before. All models from Superior mounted on the Cadillac chassis wore Criterion styling that reflected functional beauty and fine automotive styling in long, low funeral coaches and ambulances. Limousine models featured angular "D" pillars with wrap around rear windows and wide doors with vast expanse of glass.

Wayne began to market the coaches under the name Miller-Meteor, thereby combining the talents of both companies. The Crestwood models were again offered in an expanded line that now included a Crestwood flower car. The new flower car featured the same imitation woodgrain lower body work as the earlier Crestwood limousine and landaus.

Hess & Eisenhardt introduced a radical new innovation this year with the convertible flower car. Looking like a standard landau funeral coach when the top was up, the coach featured a convertible top that lowered to reveal a full floral display with ¾ of the coach still under a metal roof. This was truly one of the most beautiful coaches of the era and one of the most unique.

Memphian was now offering a line of funeral coaches, ambulances and service cars on Pontiac, Chrysler, Dodge, DeSoto, Chevrolet and almost any other chassis the customer asked for along with a line of limousines and seven passenger sedans. Comet survived building Oldsmobile chassied cars plus a few on other chassis with some success.

Production was up this year, and 2,917 funeral cars and ambulances were produced. Of these, 2,182 were built on Cadillac chassis.

The S & S Kensington was called the aristocrat of ambulances. It was certainly the most expensive limousine funeral car offered. This car, like the Knickerbocker from which it was derived, also received a major restyling. The wrap around rear side windows were gaining popularity this year. Note the tunnel type roof lights, spotlights, unique S & S side chrome treatment and the two-tone paint scheme.

1957

The big news from the industry's largest maker of funeral cars and ambulances was a new styling idea. Superior called these cars Criterion styled, and they featured crisp angular lines with chrome bright work. The new styling of this Superior limousine funeral car is most clearly seen in the rakish angle at which the rear roof pillar is sloped. The car also features wrap around type corner windows and an optional chrome quarter moulding. Production of Cadillac chassis coaches zoomed to an all time high of 949 vehicles at Superior.

An enhancing addition to any funeral home's fleet was this Superior Criterion flower car. This was the first really new styling idea applied to the design of the flower car since this vehicle's introduction in 1937. Note the rakishly slanted rear roof pillar and window and the low stainless steel rear flower deck. Superior production soared to 949 Cadillac chassis units this year, the popularity undeniably due to the striking new Criterion styling.

Superior ambulances also received the new Criterion styling as seen by this Rescue model. The roof was raised over the rear compartment and this raised portion was slanted in front to conform with the overall styling theme. Note the unique chrome treatment on the body sides and the two-tone paint finish. The built-in roof warning lights were a standard part of the Rescue package, but the gumball light and the Federal siren were optional at extra cost.

The Superior Criterion landau funeral coach featured all new styling of the landau area and new landau shields behind the rear door windows. This is the Superior Cadillac landau as a three way coach. This car does not have the optional chrome rear quarter panels seen earlier on some coaches.

Superior continued to offer service cars on both the Cadillac and Pontiac chassis. This is the Superior Pontiac service car for 1957. Note the long wheelbase, blanked off body sides with a stylized chrome wreath, and bar ornamentation on the upper body sides. This car was finished in black. The Pontiac service car was the least expensive car in the Superior model range. Superior built 356 Pontiac chassis cars this year, more than ever before.

Superior's fabulous new Criterion styling was not afforded to the firm's Pontiac models this year. Superior Pontiac styling remained consistent with the pattern set in the past few years, and that which was so popular on earlier Cadillac models. This car was finished in a light green with a black crinkle type roof finish. The funeral car on the Pontiac chassis was also offered in a limousine style. These cars utilized the Pontiac Star Chief chassis.

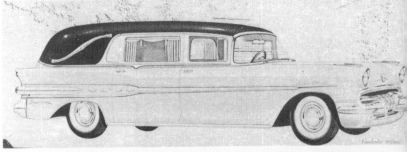

1957

The Wayne Co. took their two coach makers, A.J. Miller and Meteor, and formed a new company in this year called Miller-Meteor. Also new for this year was a line of totally restyled bodies for the new company. Continued was the unusual Crestwood, which still had a woodgrain applique on the lower body sides which was offered in two grain types, walnut and mahogany. New was the roof and body styling with wrap around rear corner windows, revised landau area with landau shields, paint schemes and chassis styling. This car was finished in a deep metallic bronze, with a mahogany woodgrain side panel and a light beige roof.

The new Miller-Meteor Co. was based in Piqua, Ohio, and combined the coach making skills of two of the country's oldest and most innovative firms. The 1957 models were all new from stem to stern and the styling was a smooth combination of traditional dignity and modern lines. This was the Miller-Meteor landau Traditional, and did not feature the wrap around corner windows of the Crestwood. Although these windows were not standard equipment on the landau, they were offered as an option. Note the new landau shields and the style of the chrome moulding on the side of the car.

Looking something like a watermelon with lights and wheels, the Miller-Meteor First Aider ambulance was deliberately designed to surpass all other ambulances in the field in the area of meeting the needs of rescue squads, fire brigades, and police departments. This high headroom, 4-litter ambulance was a veritable hospital on wheels. Among its standard items were an eight-cubic-foot cabinet for the storage of supplies, resuscitator, oxygen, hot water heater, bunk with reversible top for fracture board, and a huge divided storage compartment. Although this car is shown in Futura series trim, it was also offered in the Classic or Crestwood series trims. All Futura models featured the wrap around rear corner windows.

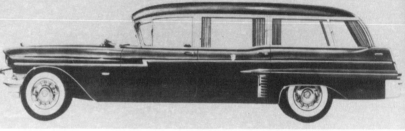

Miller-Meteor featured a limousine line of professional cars that were totally restyled in a manner that highlighted the new low wide Cadillac chassis. The length was more noticeable through the new larger glass area that wrapped around the rear body sides. Note the chrome moulding on the side of this coach, the initial crest on the rear door and the new Miller-Meteor crest on the rear fender. This car has combination "air line" style draperies.

Miller-Meteor offered this tastefully planned service car in the Classic series. This car was finished in a dark blue with distinctive chrome highlights, a chrome wreath and crest on the upper body sides, and a blanked off rear quarter. This car was a very utilitarian vehicle and although it was the least expensive Cadillac coach offered by Miller-Meteor the price it commanded was far above that of a Pontiac or Oldsmobile chassis service car.

Borrowing liberally from the Crestwood, the new Miller-Meteor flower car also featured imitation woodgrain panels on the lower body sides. The flower car was offered in two distinctive styles, the Crestwood as shown and the Classic without the side trim. This car features the Crestwood Mahogatrim and was painted a tropical green. Note the rakish slant of the rear portion of the roof and the high flower deck.

1957

The Memphis Coach Co. built any type of professional car desired on almost any chassis. This is a Memphian seven-passenger limousine built on a Chrysler Windsor chassis. This car matched their Chrysler funeral cars, and therefore the company was able to offer matched fleets of funeral cars and limousines. This car was also offered on other chassis, with three doors on each side, and as a nine-passenger vehicle.

Offering a complete line of professional vehicles on almost any desired chassis, the Memphis Coach Co. built this stylish long wheelbase Chrysler landau funeral car. The chassis was extended and the roof made of aluminum, with a high roof line, classic coach lamps, and landau bow on the upper rear body sides. These cars were rear loading vehicles only.

Memphian also offered a complete line of ambulances on any desired chassis. This high head room Memphian ambulance on a DeSoto chassis has built-in tunnel type roof lights, an optional Federal roof siren light, and rear roof mounted warning lights. Memphian ambulances of this vintage were mounted on such chassis as Pontiac, Dodge, Chrysler, Chevrolet and DeSoto. This car is equipped with the new dual headlamps that were offered as an option by Chrysler Corporation in this year, dual rear view mirrors, and a two-tone paint job.

Another offering from the Memphis based Memphian Coach Co. was this attractive long wheelbase Chevrolet service car. Note the extremely clean styling applied to the upper body sides with chrome bar highlights and a chrome wreath. The chassis was extended quite a bit and the car was painted dark gray and black. Memphian would construct a car on almost any desired chassis.

The Miller-Picking Corp. of Johnstown, Pa., offered this unique type of flower car called the Pickway. This car, based on the Ford Rancharo, had a removable flower tray over the pick-up bed and could be used for removals, vault service, flower car, or as a hearse. This car offered a low initial investment and could be used for virtually all purposes. These cars were only advertised for this one year and were not very popular.

For only $1,500 a convertible could be converted into a pleasingly styled flower car complete with a fiberglass coupe roof. This snap on unit was just one of many such flower car, passenger car conversions offered in this era. This unit is mounted on a DeSoto convertible.

1957

Since 1886 Eureka quality had been a symbol of craftsmanship, leadership, and quality. It was not surprising that the 1957 Eureka coaches were again considered among the best money could buy. Although the styling had not been as greatly refined as some of the competition, the limousine coach line featured more chrome around the windows and a smooth rounded rear roof line. Rear wrap around corner windows were optional at extra cost on all limousine models. This car was available as a straight end-loading funeral car, a three-way coach, a combination or as an ambulance.

The big news from Eureka was a totally steel construction now used on all models. Unlike other makers, Eureka did not completely revise their styling for 1957. The rear door window frame was now more rakishly sloped and the ornate coach lamp was retained. Even without a great deal of restyling, the Eureka coaches maintained a dignified appearance and their second most expensive coach status.

Eureka premiered its first all steel models this year. This was one of the first all steel flower cars built by this old concern. Note the classic coupe roof line, the crinkle type roof finish, landau shield, and large stainless steel rear flower deck.

The Comet Coach Company of Memphis continued to offer distinguished looking vehicles on the Oldsmobile chassis, as well as others, and this year offered this attractive limousine funeral coach, ambulance combination car. The standard Oldsmobile 98 chassis was lengthened and the smooth lined Comet body was dropped on. This car was finished in metallic gray with a white roof and white draperies.

Wearing completely different styling from the Comet Oldsmobile, this Comet DeSoto combination coach has wrap around rear corner windows and a somewhat higher body line. Once again the chassis was lengthened substantially. The rear door opens high into the roof line.

Mounted on a short wheelbase, this Dodge combination coach was also a product of the Comet craftsmen. This car was built on a slightly extended Dodge Coronet chassis and the roof has a high headroom. The back door opens into the roof. All Comet coaches were offered in a choice of limousine or landaulet styling.

Continuing to build custom professional cars for a few customers the shop of John Little in Ingersoll, Ontario, turned out this interesting limousine style funeral car on a Canadian Pontiac six-cylinder chassis. This car was a three-way loader and was built from a Pontiac station wagon that Mr. Little had lengthened. It uses a crinkle finish roof and formal draperies.

1958

Superior, celebrating its 35th anniversary in this year, added its unique Criterion styling to the Pontiac line. The Superior Pontiac models were offered in a complete line of Criterion styled coaches. Within the Pontiac line for this year were landau and limousine combinations, and partly styled Criterion service cars.

The largest single order in the history of the funeral car industry was made in this year. The Miller-Meteor division of the Divco-Wayne Corp. received an order for 40 funeral coaches from Chateau Auto Livery Inc. of Chicago. The units were delivered through Miller-Meteor's local distributor, Martin Vehicle Sales Inc. and were all painted in the famous Chicago grey with black tops. The delivery included three Futura limousine duplexes, eight flower cars, fifteen three-way coaches and fourteen Cadillac Series 75 limousines. The unique feature of this delivery was the tie-in with Chicago's "drive safely" campaign. Preceded by a Chicago police escort, the 40 cars formed a "cavalcade of cars" around Chicago's Lake Shore Drive. The event was widely publicized in the local newspapers, television news, and radio newscasts. Wire service photos were sent to many other newspapers throughout the United States and were published in the interest of accident prevention.

The Hess & Eisenhardt Co. (S & S) announced a whole range of models for this year. Consisting of the now famous Victoria at the top of the funeral car range, other models were the Park Row in four limousine versions, the Park Plaza in four limousine versions, the Park Hill in four versions in limousine style and the Park Place offered in both flower car and service car versions. The Professional ambulance line was offered in 16 standard model versions with others available to special order.

The Automotive Conversion Corp. of Troy, Mich., had been offering the trade custom conversions of station wagons for quite a number of years. They converted these standard station wagons into ambulances, funeral coaches or service cars without extending the chassis. This type of car appealed to the funeral director, ambulance operator, or village that could not afford some of the offerings of the commercial manufacturers. In 1958 they offered a conversion on the year's most talked about new car, the Edsel.

Again production was down, and only 2,452 professional cars were built in the U.S. Of these, 1,943 used Cadillac chassis.

The Superior Coach Corp. of Lima, Ohio, continued through this year with their successful Criterion styling and mounted this on the restyled 1958 Cadillac chassis. This year, all of the Superior limousine offerings were called Beau Monde, a name borrowed from earlier hardtop coach models. The Beau Monde three-way coach with an automatic (electric) three-way casket table as shown here sold for $11,277, while the manually operated three-way coach went for $10,739 and the straight end-loading Beau Monde cost $9,704. Combination versions of this car were offered with prices beginning at $9,956 and going up to $10,186.

Superior's Cadillac landaulet model L-604 was an attractive landau funeral car with straight end-loading capability. It sold for $9,863, while the same car designated as a model L-602 was offered with automatic three way loading only and went for $11,431. Superior did not offer a manually operated three-way coach in the landaulet series this year. The landaulet combination sold for $10,116 and featured reversible casket rollers embedded into the floor. A landaulet ambulance with roof height of 43 inches (standard height) sold for $10,230 and was the only landau ambulance offered by Superior.

The least expensive vehicle in the Superior Cadillac model range was the service car, Model 608. This attractive coach went for $9,588 but many funeral directors were beginning to feel that this was a bit too expensive for a vehicle that was used as a workhorse. Many operations began to use station wagons in place of the large and expensive service cars, and the sales of these cars began to decline. This year Superior production was down from the 1957 high, and only 802 vehicles were built on the Cadillac chassis.

The Superior Cadillac flower car, as seen here in a proper setting, sold for $10,267 this year. It featured an electrically operated stainless steel flower deck and Criterion styling. The flower car carried a model designation of 609. This car has the optional rear fender chrome mouldings.

1958

The Beau Monde ambulance was offered in three distinctive roof height versions this year. This was the Superior Cadillac Beau Monde 43-inch ambulance that sold for $10,076. A 48-inch version went for $10,306 and a high headroom 52-inch model cost $11,334. Note the unique roof corner warning lights that only Superior used, and the stylish Beau Monde design. The scoop on the rear fender was for a rear compartment air conditioner.

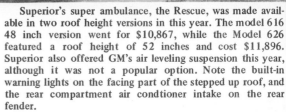

Superior's super ambulance, the Rescue, was made available in two roof height versions in this year. The model 616 48 inch version went for $10,867, while the Model 626 featured a roof height of 52 inches and cost $11,896. Superior also offered GM's air leveling suspension this year, although it was not a popular option. Note the built-in warning lights on the facing part of the stepped up roof, and the rear compartment air condtioner intake on the rear fender.

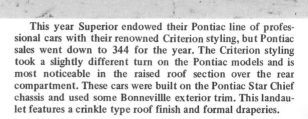

This year Superior endowed their Pontiac line of professional cars with their renowned Criterion styling, but Pontiac sales went down to 344 for the year. The Criterion styling took a slightly different turn on the Pontiac models and is most noticeable in the raised roof section over the rear compartment. These cars were built on the Pontiac Star Chief chassis and used some Bonnevillle exterior trim. This landaulet features a crinkle type roof finish and formal draperies.

Receiving a patient flown in by plane, this Superior Pontiac limousine ambulance featured all of the standard limousine styling features plus built-in tunnel type roof lights, rear roof corner lights, spotlights, and a roof mounted Federal siren.

One of the most attractive offerings in this year's Superior Pontiac line was the limousine funeral car, with its prominent step of the roof just aft of the driver's compartment, a central partition wall, and the back door opening into the roof.

Superior also offered a Rescue version in the popular Pontiac chassis line. This car had a higher rear roof line and built-in tunnel lights. It had ambulance decal patterns on the rear quarter windows, cot and other ambulance implements on the inside, and a fender mounted spotlight.

1958

281

Multi-tone colors were all the rage with the Miller-Meteor product line this year. Although black was one of the most popular and common colors applied to these vehicles, the company offered the total range of GM paint schemes and some colors of their own. Two and three-tone schemes were featured on the Miller-Meteor cars of this year. This Futura limousine coach sold for $9,717 in standard rear loading form, $10,754 as a three way car with a manually operated casket table and for $11,295 in the electrically assisted three-way version. The Duplex combination sold for $9,971 in this body style.

Miller-Meteor did not restyle their cars for 1958, but refined them to match the redesigned Cadillac chassis. This Landau Traditional was offered in four forms with prices beginning at $9,927 for the end-loading coach and going up to $10,962 for the three-way car with a manually operated table and $11,499 for the three-way with an electrically assisted casket table. This car was also offered with the modern wrap around style corner window and with this option, these cars were called Futura models.

Wearing slightly modified styling and a swept back rear window design, the 1958 Miller-Meteor flower car was offered in both Futura and Crestwood styles. The Futura flower car shown here sold for $10,218 and featured a high electrically operated flower deck made of stainless steel.

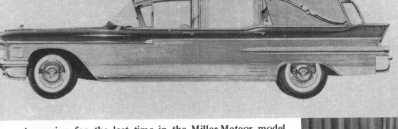

Appearing for the last time in the Miller-Meteor model line up was the woodgrain side panel Crestwood landau funeral car. This car featured imitation woodgrain side trim, a crinkle finish roof, and Futura type wrap around corner windows. The Crestwood was an exclusive car that was sold to only one funeral director in any one locality in any model year. It sold for $12,074 in three-way electrically assisted form. Also offered were rear loaders, three-way manual and combinations.

Miller-Meteor ambulances were available in both landau and limousine styles. The company offered these vehicles in three roof height models. This car was the ambulance with a 42½ inch roof, which sold for $10,090, while a similar car was offered in a landau body style for $10,295. The unusual roof lights were a Miller-Meteor exclusive.

Featuring a very high headroom type roof line, Miller-Meteor offered two special rescue ambulances this year. They were offered in roof heights of both 48 and 52 inches as seen here. The 52 inch model was adorned with a landau type style of the lower part of the roof and built-in tunnel lights. The side mounted warning light was also part of the 52 package. This car could carry four litter cases at a time and still have room for two attendants to work on these cases. The 52-inch model went for $12,084, while the 48 inch version cost $10,434.

1958

Hess & Eisenhardt offered a complete range of professional ambulances in a wide variety of roof heights. This Kensington ambulance was designated a Model 58443 and cost $11,003. Another S & S ambulance model was the Parkway, a derivative of the distinguished Victoria that sold for $11,357 and was designated a Model 58498. Note the interesting application of the two-tone paint scheme, the rear quarter window decals, and the roof lights.

Called the Hess & Eisenhardt Park Row, this car was a new style that replaced the older S & S Knickerbocker limousine funeral car. Although the car followed the styling trend set by the old Knickerbocker, it was given a new look through the tasteful use of chrome trim and the new Cadillac chassis styling. The Park Row was available in four versions with prices beginning at $10,721 for the end loading model designated a Style 581000. The Park Row combination with Duo Floor cost $11,037 and was designated a Model 581000D, and the three-way car with a manually operated table went for $11,627. The three-way car with an Electdraulically operated casket table sold for $12,227 and was designated a Model 581030E.

Appearing for the last time was the special Hess & Eisenhardt combination flower car, service car and hearse with a convertible rear roof portion. Called the Park Place, this car was designated a Model 581457 and sold for $10,996. There were very few of these cars produced even though they were very interesting vehicles.

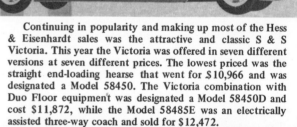

Continuing in popularity and making up most of the Hess & Eisenhardt sales was the attractive and classic S & S Victoria. This year the Victoria was offered in seven different versions at seven different prices. The lowest priced was the straight end-loading hearse that went for $10,966 and was designated a Model 58450. The Victoria combination with Duo Floor equipment was designated a Model 58450D and cost $11,872, while the Model 58485E was an electrically assisted three-way coach and sold for $12,472.

Eureka, continuing with their distinctive and traditional styling, offered 12 models in 1958. This landau version was offered in three funeral car versions, three combination models, and as an ambulance. The straight end-loading funeral car in landau form sold for $10,633, while the manually operated three-way car went for $11,464 and the three-way electrically assisted casket table version went for $12,254. Landau combinations went for $10,624 to $10,842 and the prices depended upon the type of casket rollers specified. A full line of limousine models was also available.

The Shop of Siebert in Toledo was still building custom coaches on Ford and Mercury chassis. This Mercury Aristocrat long wheelbase funeral car is a fine example of the type of specialized work carried out by this firm. Other products built by Siebert included ambulances and service-flower cars.

The Weller Brothers of Windsor, Ontario, were building a few conversion type professional cars from station wagons. This Weller Pontiac was photographed on Riverside Drive in Windsor, with Detroit in the background. Note the unusual roof lights and the rear window ambulance pattern. Called a Weller Ambulatte, this car is built on a Canadian Pontiac Patherfinder six-cylinder station wagon chassis.

1958

Continuing to offer their classic station wagon conversion, Automotive Conversion Corp. offered this interesting amble-wagon on the ill-fated Edsel Villager chassis. Amblewagon conversions were based on standard station wagons without any lengthening of the wheelbase. The rear floor was modified as were the rear quarter windows and in some cases the roof heights.

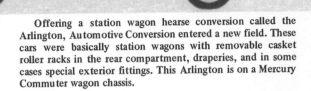

Offering a station wagon hearse conversion called the Arlington, Automotive Conversion entered a new field. These cars were basically station wagons with removable casket roller racks in the rear compartment, draperies, and in some cases special exterior fittings. This Arlington is on a Mercury Commuter wagon chassis.

Featuring big car preformance with a small car price and maintenance, these luxury custom built Comet Oldsmobiles were dramatically styled and very modern. The cars were offered in both landau and limousine body styles and in funeral car, combination, and ambulance versions. This Comet Olds combination features air line type draperies, and Olds 98 chrome trim, and chassis components.

Offering custom built coaches on almost any make of chassis, the Memphis Coach Co. built this distinguished looking Chrysler landau on a New Yorker chassis. The company called their products Memphians and these cars appealed to the upper middle price range of the field. Note the large and ornate coach lamps on this car and the unusually high roof line.

Memphian coaches were offered on either the standard passenger car wheelbase or the elaborate lengthened one shown here on a DeSoto chassis. This ambulance features the same characteristics as the landau funeral car, with a high roof line and modern styling. Chrysler chassis were very popular with the Memphis Coach Co., and a good number of these vehicles were built by them.

National of Knightstown, Ind., continued to build coaches to special order on almost any GM chassis. This National Pontiac combination featured the finest craftsmanship for the dollar invested of any low priced coach. Note the thin rear quarter pillars and the stylish roof lines. National also offered this car on a Chevrolet chassis.

Utilizing the body components of a Chevrolet sedan delivery, John Little of Ingersoll, Ontario, built this attractive little funeral car with end loading capabilities only. The rear door was hinged at the top and the upper side panels featured small quarter windows, imitation coach lights, and landau bows. The roof was finished in crinkle grain.

1959

The normal run for a funeral coach or ambulance body style is up to three years or more and when the manufacturers were faced with the Cadillac commercial chassis for 1959, not all of them were ready with new bodies to go with the radically styled, highly-finned new chassis.

It was Superior that again shook the stylists within the industry with the introduction of totally new bodies that reflected a new, more modern dignity to the radical Cadillac chassis. They introduced the most dramatic change ever seen in the seemingly timeless landau body style with the Crown Royale Landaulet. This new coach was a take-off on the Criterion styling introduced in 1957. But the style was greatly modernized for the new chassis. The Crown Royale Landaulet featured a rear roof pillar that slanted forward and continued up and over the coach's roof, lending new definition to the landau area. The landau bow, traditionally S-shaped, was now almost straight. This new type of styling was carried over to the remainder of the line using the Cadillac chassis. Superior coaches on the Pontiac chassis wore a cleaner more sophisticated version of last year's Criterion styling and were most attractive vehicles.

Miller-Meteor was not quite as ready as Superior with new bodies to compliment the Cadillac chassis and so used warmed-over versions of the bodies introduced in 1958. Eureka had not, to any noticeable extent, altered the overall style of its cars since the early fifties, yet they always reflected a quiet dignity and a pleasing line. They retained the large ornate carriage lamp on the landau models, as it had become a Eureka styling trademark.

The Flxible Company had ceased production coaches in 1952 to fill large orders for the company famed busses, but in 1959 it re-entered the field and brought in an innovation. Building on Buick chassis only, Flxible announced a full line of funeral coaches and ambulances in two long wheelbase series, Premier and Sterling, and a new short-wheelbase Flxette line of vehicles. Back in 1952, Henney had introduced a short wheelbase economy model with the Junior, but the idea never really caught on with the rest of the industry. But with the introduction of the Flxible Flxette, on the standard Buick wheelbase, the industry was sent humming. Flxible styling was extremely attractive, with a full range of landaus and limousines being offered in both wheelbases. Even the short Flxette was an attractive coach due to the distinctive styling given it by the Flxible designers.

Production went up substantially this year, moving up to the highest point that it had been in the past eight years. A total of 3,367 funeral cars and ambulances were built in the U.S., with 2,184 of these being on Cadillac chassis.

Within the new line of Flxible funeral cars the company offered a new and distinguished type of combination car. This vehicle could double as either a funeral coach or an ambulance. It featured a removable panel that could be applied over the rear side window and had a dummy landau bow on it. This quickly converted the car from a limousine ambulance or funeral car to one with the landau styling flair. The roof lights could be snapped on or off in just a few seconds.

After a seven year absence from the ranks of the funeral car and ambulance makers, the Flxible Company of Loudonville, Ohio, re-entered the field with a new range of Buick based coaches. Flxible built a total of 49 coaches this year, offering these high quality cars on two separate wheelbases. This Premier limousine funeral car was built on the lengthened Buick Electra chassis. It used formal draperies, full wrap around rear corner windows, and smooth graceful lines.

Reviving an idea originally introduced in the very early 1950's by Henney, the Flxible Co. introduced the new Flxette line of professional cars. These cars were built on the standard passenger car length Buick chassis with styling similar to that found on the full length models. These new short wheelbase coaches were priced significantly lower than the larger cars, and with this model Flxible started a completely new vogue in the field. These cars were offered in both landau and limousine styling and as both funeral cars and ambulances as well as combinations.

Perhaps the most daring departure in landau styling was the Crown Royale Landaulet introduced by Superior this year. The Crown Royale featured a rear door pillar that sloped forward and was shrouded in chrome that encircled the car's roof. This added a new definition to the landau area of the roof. The new look was heightened with a landau bow that had almost no curve at all. Available in four versions, the landaulet ranged in price from $10,781 for the rear loading coach, through $11,863 for the three-way with a manually operated table, to $12,427 for the automatically operated three-way model. Superior also offered this innovatively styled landau in a combination version for $11,047. Superior built a total of 955 Cadillac chassis cars in 1959.

Classified in the Royale series, this attractive Superior flower car was one of 23 such cars built this year. The flower car, complete with a hydraulically operated stainless steel flower tray, went for $10,785. Superior production of Cadillac chassis cars was dramatically up for 1959. They built 124 side service funeral cars, 221 end loading hearses, and 393 combination coaches.

Superior built a very few Super Rescue versions of the Pontiac ambulance. These cars wore the straight limousine style that was so popular on the ambulance models, but with a special fiberglass raised roof area over the patient's compartment. Special warning lights were built into this raised roof and the sides of this panel bore decorative gold applique.

Wearing a more subtle styling line, the 1959 Superior Pontiac models featured the crisp angular rear corner window similar to that on the Crown Royale Cadillac models. These cars did not have the Crown type of "over-the-roof" chrome trim, but utilize a decorative landau shield in its place. This car is a combination ambulance hearse. The rear fender mounted air intake is for the rear compartment air conditioning.

Superior built a total of 194 ambulances on the Cadillac chassis this year. This car is a Royale Rescue ambulance with a 48-inch headroom, and it sold for $11,552. Other Rescue ambulances featured a 54-inch roof height for $12,999 or a 42-inch roof height at $10,746.

The low head room Superior Pontiac ambulance was by far the most popular ambulance version of this popular vehicle. Note the interesting two-tone paint scheme, the roof mounted lights, and the complete ambulance equipment in the rear compartment. This car was air conditioned at least in the patient's area.

1959

With Cadillac's revolutionary new and highly finned chassis, this attractive Miller-Meteor flower car was built with a Chicago type flower deck. Miller-Meteor was now adding more Superior type styling to their flower car's roof line, with the rakishly swept rear window. The flower car sold for $10,799 with the stainless steel hydraulically operated flower deck, and for significantly less with the Chicago type flower trough.

Offered in five versions, the Miller-Meteor landau carried prices beginning at $10,482 for the straight end-loading funeral car and going to $10,694 for the Duplex combination with the customer's choice of either a removable floor casket table or a reversible floor with casket rollers built in. The landau ambulance went for $10,814 and featured a roof height of 42½ inches. Three-way models were offered in both manually operated versions for $11,516 or electrically assisted for $12,079.

Called the Futura Amblandau, this attractive Miller-Meteor ambulance had the Futura type landau panel sandwiched between the side door windows and a wrap around rear corner window. This panel was decorated with a large chrome plated cross and a chrome landau shield. These versions were priced somewhat higher than the ordinary limousine ambulance.

Miller-Meteor offered limousine ambulances in two roof heights, 42½ and 48 inches. This attractive 48 ambulance features built-in roof tunnel lights and an interesting two-tone paint scheme. The 48 ambulance went for $10,962, while the 42½ version cost $10,600.

All Miller-Meteor landau models were offered as Futura models in addition to the standard landau roof style. The Futura versions featured small wrap around corner windows with a small landau area sandwiched between these windows and the rear door window. The Futuras were offered in all variations and at a slight additional cost.

The Miller-Meteor limousine funeral car had side doors that extended well into the roof line for additional loading, entry and egress room. This car is an example of the limousine combination ambulance/hearse called a Duplex and sold for $10,474. The straight end loading funeral car in this body style cost $10,207, while the three-way coach with a manually operated casket table went for $11,297 and the three-way with an electrically operated table cost $11,865.

1959

Signifying a discerning appreciation for leadership and enjoying the eminent satisfaction of building the most distinguished and expensive funeral car in the industry, Hess & Eisenhardt continued to offer the attractive S & S Victoria in four funeral car versions and one ambulance model. The straight end-loading Victoria went for $11,654, while the three-way manually operated funeral car version cost $12,651 and the electrically operated three-way version went for $12,284. Combination versions of the Victoria were offered with Duo Floor disappearing casket table rollers and went for $12,042, while the straight ambulance version of the Victoria cost $12,125.

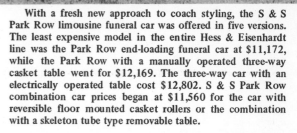

With a fresh new approach to coach styling, the S & S Park Row limousine funeral car was offered in five versions. The least expensive model in the entire Hess & Eisenhardt line was the Park Row end-loading funeral car at $11,172, while the Park Row with a manually operated three-way casket table went for $12,169. The three-way car with an electrically operated table cost $12,802. S & S Park Row combination car prices began at $11,560 for the car with reversible floor mounted casket rollers or the combination with a skeleton tube type removable table.

Hess & Eisenhardt continued to offer the distinctive Park Hill line of professional cars styled with a roof band just above the driver's compartment, with the rear portion of the roof covered with fabric. This distinctive coach was offered in five versions with the least expensive being the Park Hill end-loading funeral car at $11,427. The Park Hill with a manually operated three-way casket table cost $12,469 while the version with an electrically operated table went for $13,102. The Park Hill combination coach with Duo floor reversible casket rollers cost $11,860 as did a version with skeleton type removable casket table. Note the distinctive S & S chrome side flashes and the thin chrome strip above the Cadillac logo on the front fender.

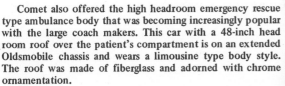

The Comet Coach Co. of Memphis continued to offer high quality funeral cars and ambulances on either the Oldsmobile or the Buick chassis. The Comet Oldsmobile was offered in both landau and limousine body styles, and as either a straight hearse or as an ambulance. This Comet Oldsmobile landau funeral car wears the modern landau panel styling sandwiched between vast expanses of glass.

With a vast glass area, the Comet limousine coach was a very modern looking unit when mounted on an extended Buick Electra chassis. The Comet Coach Co. had become one of the fastest growing small coach companies to join the field since the war.

Comet also offered the high headroom emergency rescue type ambulance body that was becoming increasingly popular with the large coach makers. This car with a 48-inch head room roof over the patient's compartment is on an extended Oldsmobile chassis and wears a limousine type body style. The roof was made of fiberglass and adorned with chrome ornamentation.

1959

The Eureka limousine funeral car or combination was offered with "Full-Vision" styling (rear wrap around corner windows) at no extra cost this year. This was all part of the new Eureka look of lasting quality. This Eureka limousine professional car has the optional Full-Vision styling. The limousine as a straight end-loading funeral car with an electrically assisted table sold for $12,636. The Eureka combination with limousine styling ranged in price from $11,006 to $11,224.

Eureka gave their cars a mild restyling for 1959, but the old Eureka trademarks were still there. The beautiful Eureka landau was offered in six versions. The end-loading landau cost $11,260 and the manually operated three-way casket table version went for $12,091. The fully powered three-way car with an electrically assisted casket table cost $12,091. The fully powered three-way car with an electrically assisted casket table cost $12,881, while the Eureka combination coach ranged in price from $11,251 to $11,468 with the price varying according to the type of casket table or rollers that were fitted. Eureka maintained their standing as the second most expensive line of professional cars in the field.

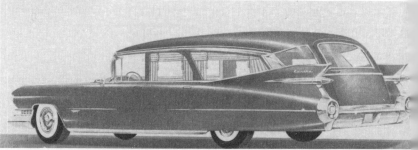

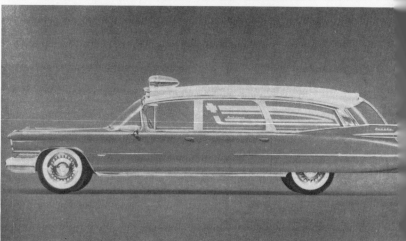

The Franklin Body & Equipment Corp. of Brooklyn, N.Y., offered this Franklin ambulette which turned out to be their most popular conversion. Franklin produced a strong line of mobile hospitals, bandmobiles, bookmobiles, armoured trucks and police vans as well as both truck type and passenger car style ambulances. This Chevrolet sedan delivery ambulance conversion was by far the most popular Franklin ambulance conversion and offered all of the advantages of a large ambulance service at a fraction of the cost.

Wearing a price tag of $11,552, the stately looking Eureka ambulance featured modern Eureka Full-View styling that consisted of wrap around corner windows at the rear quarters, built in roof tunnel lights, and an optional Federal roof mounted combination siren light. The contemporary Eureka body design maintained the ease of loading that Eureka had always held in such high esteem.

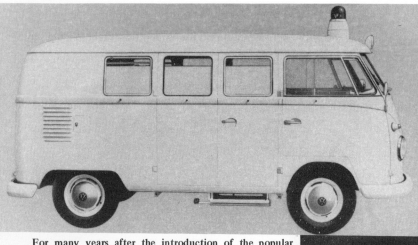

Built by Europe's elite coach builder, Pollmann Karosseriefabrik of Bremen, Germany, this Ford was a product of the hard times in the German automotive industry. Most German makers could not supply sufficient chassis for professional car work and for this reason the large American chassis were utilized for hearses. This Pollmann Ford funeral coach features a high roof line and frosted glass in the rear side windows.

For many years after the introduction of the popular Volkswagon transit van series, this large West German car maker offered a factory made ambulance based on this economical vehicle. These cars were complete with frosted glass in the lower half of the side windows, a choice of rear compartment interior plans, flashing blue roof warning light and an illuminated ambulance cross above the windshield. A good number of these reliable vehicles were imported into North America.

1960

The Comet Coach Co. had successfully operated since 1955 building extended frame coaches mounted on most any chassis, but primarily on the Oldsmobile unit. It had moved operations to Blytheville, Ark., and this year it changed its name. In mid-1959, the company had been approached by Ford Motor Co., which wanted to purchase the rights to the name Comet for a new car that it was planning to market. The company sold the rights to the Comet name for a considerable profit and changed its name to Cotner-Bevington to honor the firm's founders. The Cotner-Bevington name plate first appeared on the firm's Oldsmobile coaches this year and with the funds from the sale of the Comet name, the company was placed on an even better financial footing.

Superior, which had introduced new styling on the 1959 models, continued this line-up with very little change to the Cadillac chassis coaches. However, the new styling was carried over to the company's Pontiac line this year. At Miller-Meteor styling was refined for this model year with some new innovations. Siebert offered extended frame ambulances and funeral coaches on the Ford chassis with bubble type roofs and unique side styling.

Miller-Meteor launched a new line of ambulances with the top of the Guardian looking like a lighted easter egg. With headroom 52 inches high, the Guardian had a four-patient capacity. It featured an angular weather shield over the rear door that contained an overhead spotlight and allowed the rear window to be lowered for ventilation. Other Miller-Meteor ambulance offerings included an attractive "ambulandau" model available in either 42 or 48 inch headroom models. This model featured the dignified styling of the landau, with closed rear quarters, with a chrome cross replacing the landau bow in this area.

Production was up again this year, with 3,854 funeral coaches and ambulances being produced in the U.S. Of this number, 2,194 were on Cadillac chassis.

The car that started the new trend toward the short wheelbase professional cars was the Flxible Flxette, introduced in 1959 by this progressive Loudonville, Ohio, firm. The Flxette was offered this year in two distinct wheelbase sizes called the Flxette and the Flxette Shortline. In this year Flxible produced a total of 70 Flxette models and 40 Shortline Flxettes. This Flxette landau funeral car carried a price tag of $7,226 in straight funeral car form and $7,489 as a combination funeral car and ambulance Note the similarity in styling to the large Premier models.

The Flxible Company of Loudonville, Ohio was now back in full production of Buick chassis professional cars and offered a complete line up of models and body styles. This year saw 86 full-size Premier coaches built by this progressive manufacturer, with all coaches built on the Buick Electra chassis that had been lengthened. The Premier landau was offered in two versions with prices beginning at $9,014 for the end loading funeral coach. The landau combination funeral car and ambulance went for $9,216.

Also offered in the Premier series of long wheelbase models was this attractive limousine coach. This body style was offered in four versions with prices beginning at $8,785 for the end-loading funeral car. The limousine combination coach went for $8,987, while the limousine ambulance cost $9,408. The Flxible Company also offered a high headroom version of the limousine ambulance that carried a price tag of $10,275.

The Flxette short wheelbase funeral car line was built on the wheelbase of the Buick Electra passenger car and was priced significantly lower than the full-sized Premier models. This Flxette limousine coach was offered in hearse, ambulance and combination versions. The limousine funeral car went for $7,163, the combination for $7,252 and the ambulance version cost $7,449. Also offered in this model line was a distinctive service car version that carried a price tag identical to that of the limousine funeral car.

1960

Carrying the second most expensive banner again for this year, the Eureka line of professional cars continued with the popular styling originally initiated in the early 1950's. The Eureka limousine coach was offered in no less than six versions with prices beginning at $11,305 for the rear-way funeral coach and going to $12,136 for the three-way manually operated table version. The Eureka limousine when fitted with an electrically assisted three-way table cost $12,962 and combination versions of this car ranged in price from $11,296 to $11,471. The wrap around rear corner windows seen on this car were part of the Full-Vision styling option and were extra at no additional cost.

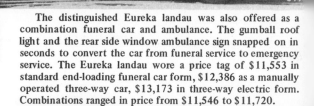

The distinguished Eureka landau was also offered as a combination funeral car and ambulance. The gumball roof light and the rear side window ambulance sign snapped on in seconds to convert the car from funeral service to emergency service. The Eureka landau wore a price tag of $11,553 in standard end-loading funeral car form, $12,386 as a manually operated three-way car, $13,173 in three-way electric form. Combinations ranged in price from $11,546 to $11,720.

Very few of these attractive limousine ambulances with a high roof line were built by Eureka this year. This type of unit was very popular with emergency squads and fire departments all over North America, and were veritable mobile hospitals.

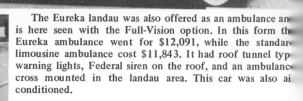

The Eureka landau was also offered as an ambulance and is here seen with the Full-Vision option. In this form the Eureka ambulance went for $12,091, while the standard limousine ambulance cost $11,843. It had roof tunnel type warning lights, Federal siren on the roof, and an ambulance cross mounted in the landau area. This car was also air conditioned.

The newly named Cotner Bevington Corp. offered a complete selection of professional cars on the Oldsmobile chassis. These cars were offered in both limousine and landau styling and as funeral cars, combinations, and ambulances. This attractive Cotner Bevington limousine funeral coach features smooth flowing styling with a large amount of glass area and formal draperies. Note the new Cotner Bevington star emblem on the lower part of the rear fender. With the new name, the company moved to new quarters in Blytheville, Ark.

The Comet Coach Company of Memphis sold the rights to the name Comet to the Ford Motor Company in 1959 for use on a new compact car that Ford wanted to market under the Comet name. The 1960 vehicles from the old Comet firm were renamed to honor the company's founders, Waldo Cotner and Robert Bevington. The new Cotner Bevington Corp. produced a line of medium priced funeral cars based on the Oldsmobile chassis. These cars were built in both the landau and limousine styles with consistent quality and built-in prestige. This is the 1960 Cotner Bevington landau funeral car that was competing directly with the Flxible Premier landau.

Superior continued to produce the innovative and beautiful Crown Royale landaulet, unchanged from last year with the notable exception of the restyled Cadillac chassis components. The 1960 Superior Crown Royale landaulet model DL-604 rear servicing funeral car carried a price of $10,781 while the same car with a manually operated three-way casket table went for $11,863 and the three-way electrically assisted car cost $12,427. The Crown Royale combination in landaulet styling commanded a price of $11,047, while the landaulet ambulance with a 42-inch headroom went for $11,166.

The Superior Crown Royale series was the very top of the Superior Cadillac line for 1960 and the Crown models featured the exclusive over-the-roof chrome trim bow and chrome sheathed rear quarter panels plus assorted other exterior and interior luxury trim. This attractive Crown Royale limousine funeral car was available in a wide variety of models and versions at prices beginning at $10,614 for the rear servicing funeral car, $11,701 for the three-way manually operated car, $12,267 for the three-way electrically assisted table version and with combination versions going for $10,879. The Crown Royale ambulance in limousine form sold for $11,006. Superior built a total of 977 Cadillac chassis professional cars this year, a full 22 units more than in 1959.

The Royale limousine funeral car and ambulances were called Beau Monde models and ranged in price from $10,194 for the end loading version to $11,281 for the three-way manual car. The Royale Beau Monde three-way electrically assisted casket table model carried a price tag of $11,847, while the Beau Monde combination car went for $10,460.

The beautiful and graceful Superior flower car was put into the Royale series under the new system of series designation for all Superior Cadillac models. The 1960 flower car commanded a $10,785 price and only 10 were built. Superior built 128 side service cars, 221 end loading funeral cars, 393 combinations and 194 ambulances on the Cadillac commercial chassis in 1960.

The standard Superior funeral car line on the Cadillac chassis was called the Royale series. These cars lacked the ostentation and luxury trim of the Crown Royale models and the styling of the cars differed greatly, too. The Royale landaulet, shown, sold for $10,367 in standard form and for $11,443 as a manually operated three-way car. The Royale three-way electrically assisted car sold for $11,847. The Royale combination car prices began at $10,627 with the Royale landaulet ambulance going for $10,746.

Superior built a grand total of 194 Cadillac chassis ambulances in 1960 and offered two of the emergency high headroom type Rescue ambulances this year. The Royale Rescue ambulance had a 48-inch roof height and went for $11,552 while the Royale Super Rescue with a 54-inch roof (shown here) cost $12,999. The extra roof height was obtained through the application of a fiberglass addition over the patient's compartment. This roof addition was complete with warning lights and wiring for additional roof lights. These cars were capable of carrying four litter patients at any one time.

1960

After 22 years, the S & S Victoria, of Hess & Eisenhardt of Cincinnati, Ohio, was still the original landau funeral car design. Not only that, but the Victoria was the most expensive landau funeral car offered to the trade and in some respects the most luxuriously appointed. The Victoria prices began with the end-loading model that was priced at $12,140. The three-way car with a manually operated casket table went for $13,178 and the three-way with an electrically assisted table sold for $13,836. The Victoria was also available as a combination car with prices beginning at $12,545, while the straight Victoria ambulance called the Parkway sold for $12,630.

National Coaches of Knightstown, Ind., offered a complete line of professional cars on almost any General Motors chassis. This Oldsmobile is an excellent representative of their Minute Man ambulance. This body was also mounted on Chevrolet and Pontiac chassis and offered a headroom of 42 inches.

For prestige leadership at a sensible price, National offered the DeLuxe series of professional coaches on the Oldsmobile, Chevrolet and Pontiac chassis. The DeLuxe series features landau type styling on both the ambulance and the funeral coach and the differences between the two types is clearly illustrated in the two cars above. Note the wrap around type corner windows at the rear and the straight air line style draperies. "Born to lead the Field" was the National slogan in this year.

Continuing to offer professional vehicles on extended Ford and Mercury chassis, the Shop of Siebert Associates introduced this long wheelbase landau type funeral coach this year. The roof was raised significantly through the addition of a fiberglass pod and the rear windows were blanked out with landau panels. The extension of the chassis can easily be seen in this view. The landau shield just behind the rear side door carried the owner's nameplate.

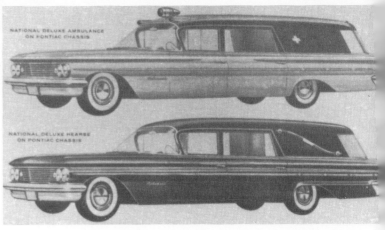

In addition to excellent ambulances on a passenger car chassis, National Coaches offered funeral cars and combinations similar to the one seen here on a Chevrolet chassis. They also built squad type ambulances on General Motors truck chassis, as seen in the lower portion of this photo.

Siebert also offered a long wheelbase Ford ambulance with a significantly raised roof line. This limousine style coach featured a plaque at the belt line between the rear windows with a large chrome ambulance cross, and the roof featured built-in warning lights. This car could carry three stretcher patients, one being suspended from the roof.

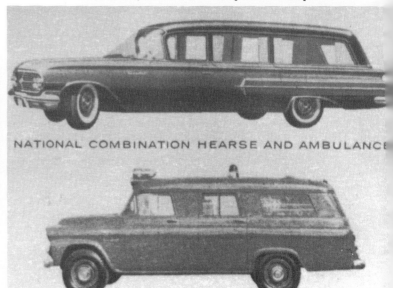

Miller-Meteor's attractive limousine funeral car line was the company's most popular style. The least expensive coach in this line was the end loading funeral car at $10,207. The three-way manual coach was priced at $11,297, while the three-way power assisted car cost $11,865. Again the company offered a combination called a Duplex for $10,474. Here a Miller-Meteor end loading funeral coach receives an air freighted casket from a Seaboard & Western Airlines plane.

Miller-Meteor continued the successful styling approach introduced in 1959 on their complete range of models. The landau was offered either without the side corner windows or with them as seen here. The inclusion of this option did not affect the final price of the car. The landau was offered in five versions with the least expensive being the end loading funeral car at $10,428. The landau with a three-way manually operated casket table sold for $11,516, while the three-way with a power assisted table cost $12,079. A Duplex combination with M-M landau styling was offered for $10,694 with the customer's choice of either a reversible floor or a removable tube type table. A Miller-Meteor landau ambulance was also offered and had a roof height of 42½ inches and cost $10,814.

The Miller-Meteor flower car continued to take a styling page from the Superior design book and featured a canted rear roof line with the rear window at the same angle. The two-toning was a distinctive Miller-Meteor touch. On this car the fin portion of the body was finished in gold with the rest of the car being painted white. The 1960 Miller-Meteor flower car sold for $10,799 complete with a stainless steel flower deck and a stainless steel compartment under the flower deck.

Looking somewhat like an illuminated Easter egg, the new Miller-Meteor Guardian ambulance featured a roof height of 48 inches and had a rear roof line that was sloped like that of the flower car. This huge sloped portion of the roof hung over the rear door to provide shelter for the attendants and the patients in inclement weather and also had a spotlight mounted in it. This car carried a price of $10,962, while an ambulance with a more conventional limousine style and a straight line type of roof styling and a 42½ inch headroom went for $10,600. Also offered was a landau ambulance with a 42½ inch headroom that sold for $10,814.

Memphis seemed to have been a breeding ground for small coach builders and this city had seen many come and go during the fifties. The Weller Brothers was also located in Memphis, and the funeral car and ambulance manufacturing capacity of this firm was an offshoot of a large school bus trade. This Weller Plymouth combination offered a high roof line and extra length, a light swinging rear door, folding attendant's seat, and rapidly removable emergency equipment.

Another of the small custom coach companies that built coaches on almost any specified chassis was the Pinner Coach Co. of Memphis. Pinner, a new firm, began by building cars on a specially lengthened Pontiac chassis. The Pinner Pontiac combination with limousine styling had a relatively high roof line and angular wrap around rear corner windows.

1961

When Flxible re-entered the field in 1959, it brought in a compact line of professional vehicles called the Flxette. The idea caught on with some of the other manufacturers. Superior introduced a line of compact or regular wheelbase models on the Pontiac chassis in this year. Called the Consort series, the new Superior model line consisted of limousine and landau funeral coaches and combinations, ambulances and a service car. Cotner-Bevington also introduced a line of regular wheelbase models with the Seville series. Although the company built primarily on Oldsmobile chassis, the compact models were constructed on the Chevrolet Biscayne, Pontiac Catalina and Star Chief as well as Oldsmobile. Flxible continued marketing the Flxette, now wearing styling that was a copy of that of the long wheelbase cars. The new compact cars were introduced to fill the demand for a more economical, lighter, less expensive coach. This was primarily noticeable in the area of the service car or casket wagon as they were called in the early years. Having been built on the long wheelbase chassis in past years, the cost of these utility vehicles had soared and many funeral directors were turning to less expensive station wagons that could be easily converted for many uses.

Superior restyled its complete model range with new more angular styling. Miller-Meteor gave its models a mi restyling, lowering the overall height of the complete lin Flxible gave the Buick chassied line a new look with "C pillars that were sloped and included the Buick cres Landau models featured new thinner landau irons and vee-shaped area behind the "C" pillar bearing the sam crest.

A total of 3,449 professional cars were built this yea with 2,331 being on Cadillac chassis.

In the timeless Eureka tradition, the 1961 Eureka landau preserved the formality of line that had made this car so popular in the past. The 1961 Eureka line was subtly restyled to match the new Cadillac chassis. The landau was available as a straight end loading funeral car for $11,689, a three-way manually operated casket table model at $12,521, a three-way electrically operated casket table hearse for $13,307 or as a combination that sold for $11,682 to $11,857.

The 1961 Eureka custom crafted limousine masterfully portrayed a years-ahead design and was a dramatic example of Eureka's renowned coachwork. The newly styled limousine was offered in six versions with prices beginning at $11,445 for the straight end-loading hearse. The three-way car with the manually operated table went for $12,276, while the three-way electrically assisted car sold for $13,063. Combination ambulance funeral car versions of the Eureka limousine were offered at prices beginning at $11,436 and going up to $11,612. Air conditioning and an air leveler were made available at extra cost on all Eureka models.

The culmination of years of Eureka advanced professional car engineering was seen in the 1961 Eureka limousine ambulance with the full vision rear side window option and the standard 42 inch roof height. Other ambulance models were available with either a 48 or 52 inch roof height and suitable styling. The straight limousine ambulance as pictured here went for $11,843, while a landau version carried a price tag of $12,086.

Automotive Conversion Corp. offered Amblewagons on any station wagon chassis, and they did not miss the opportunity to offer this unique and interesting ambulance conversion of the Chevrolet Corvair Greenbriar chassis. Called as modern a concept as tomorrow, the Amblewagon Corvair was not what one would call a screaming success, although the design of the vehicle did make it quite maneuverable in congested city traffic.

1961

Hess & Eisenhardt advertising for this year brought out the point that a funeral director's professional car should be a symbol of his professional reputation and that his car shouldn't reflect second best. The Hess & Eisenhardt S & S professional car was the most exclusive and expensive coach that money could buy. This Park Row limousine funeral car went for $12,131 as a straight end loading coach, $12,706 as an end loader with an extension table, $13,188 as a manually operated three-way and $13,860 as a three-way with an electrically assisted table. The Park Row when utilized as a combination car carried a price tag of $12,550 with Duo Floor and $12,550 with the skeleton casket tables. Note the unique S & S gas cap on the rear fender of the 1961 cars.

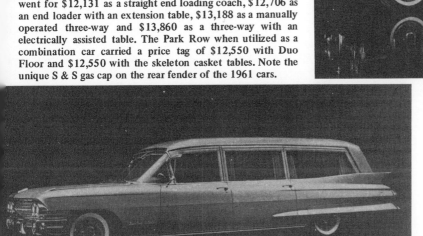

The distinctive and stylish Hess & Eisenhardt S & S Victoria continued to be offered in its traditional garb and without any major restyling to the body. The S & S Victoria body always seemed to be in complete harmony with any style of Cadillac chassis. This year the Victoria was offered in five funeral car and combination models and an ambulance. The lowest priced Victoria was the straight end loading car that went for $12,651, while an end loading car with an extending casket table went for $13,226. There were again two versions of the three-way car. The manually operated version sold for $13,710, while the electric model cost $14,381. The Victoria combination with Duo Floor and disappearing casket rollers went for $13,072, while the combination with a skeleton type of casket table sold for $13,072 also.

For 85 years S & S had designed and constructed some of the most efficient ambulances in the trade and the 1961 version was true to the tradition. Perfection had always been a byword at S & S, and was carried on this year with the new models. This S & S Parkway ambulance cost $12,639, while a more elaborate version with a high roof line and a considerable amount of life-saving apparatus was called the Custom Professional Parkway and sold for $18,787 in standard form. The S & S Parkway Victoria ambulance with landau styling cost $13,159.

The beautiful and luxurious S & S Custom Superline flower car was a car that saw very limited production in this year. Fewer than 10 of these stylish cars were built by Hess & Eisenhardt in 1961. This particular car went to the Immole Funeral Home. It carried a price tag of $12,872 and was designated a Model 61420. Note that the rear deck is covered with the tonneau and the style of the back part of the roof.

Continuing to offer extremely unusual looking long wheelbase professional cars on the Ford and Mercury chassis, the Shop of Siebert of Toledo introduced their 1961 models. These cars were a somewhat toned down version of the uncanny looking cars they built last year. Here we see the Siebert Ford Aristocrat ambulance with a high headroom rear compartment.

Knightstown, Ind., had been known for fine funeral cars and ambulances since 1900, and 1961 was to be no exception. The National Custom Coach Co. of that city offered a complete line of professional cars built on almost any chsssis and to the customer's specifications. This National Minute Man ambulance is seen mounted on a Chevrolet chassis and features a slightly lengthened wheelbase and all of the ambulance accoutrements. National also offered a landau funeral car called the Knighton.

1961

Following the recent lead of the Flxible Company and the earlier groundwork laid by Henney, Superior launched a line of passenger car wheelbase professional cars this year. Built on the Pontiac chassis the new short wheelbase cars were called Consort. In this inaugural year Superior built a total of 132 Consorts of which 13 were straight hearse models, 81 were combinations, and a further 38 were ambulances such as shown here.

The distinctive new Consort caught the imagination of the trade and within a few years there would be a wide variety of these short wheelbase cars on the market. Here we see an attractive Consort service car complete with the Superior type of stylized chrome wreath and chrome strip bars. All Consorts were constructed on the standard Pontiac chassis.

The new styling was called Criterion and an excellent example is seen here on the 1961 Superior Pontiac Super Headroom ambulance. This car featured a roof height of 4 inches and was capable of carrying several patients. Superior built a total of 264 long wheelbase Pontiac professional cars in this year and a total of 396 Pontiac chassis cars for the year.

Continuing in popularity, the long wheelbase Pontiac professional cars by Superior made their debut wearing all new styling. Notice the thick D pillars adorned with chrome trim and a star. This attractive limousine was one of 163 Superior Pontiac long wheelbase combination cars built this year. They also constructed a total of 20 straight Pontiac funeral cars and an additional 81 ambulances.

The Crown Royale was also available as a limousine and this style also featured the over-the-roof chrome band that was an integral part of the landaulet's design. The new Criterion styling is seen in the rear portion of the roof. The straight end-loading Crown Royale cost $10,614, while the manually operated three-way car sold for $11,701 and the three-way with electrical assistance went for $12,267. The Crown Royale limousine combination cost $10,879 and an ambulance version of this car was priced at $11,006 and had a roof height of 42 inches.

Superior introduced their 1961 Cadillac chassis models in a wide variety of new Criterion styles and in four series. The Royale and Crown Royale featured the wrap around rear corner windows, while the new Sovereign and Crown Sovereign series had this area closed in. Here we see the Crown Royale landaulet that was offered in five versions. The basic end loading car sold for $10,781, while the manually operated three-way version of this car carried a tag of $11,863 and the electric three-way version went for $12,427. The Crown Royale landaulet was also offered as a combination at $11,047 and as an ambulance with a 42-inch headroom for $11,616. The new roof styling and the chrome rocker panels were an integral part of the Crown Package.

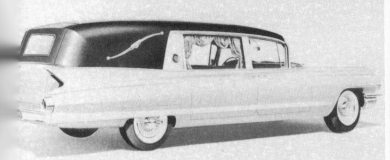

The styling difference from the Royale series to the Sovereign series is easily seen in the rear portion of the roof line and the chrome trim and ornamentation. Once again the Sovereign was offered in both standard and Crown versions. The Sovereign landaulet as an end-loading car sold for $10,549, while the Crown version went for $10,969. The three-way car with a manually operated table cost $11,631 in standard form and $12,051 as a Crown landaulet. The electric three-way coach sold for $12,035 in the Sovereign landaulet form and for $12,615 as a Crown Sovereign Brougham landaulet. Both series offered a complete selection of limousine styles, ambulance, and combination versions.

The standard Royale landaulet had styling that differed greatly from that seen on the Crown version. This car in rear loading form carried a price tag of $10,361, while a rear loading Royale Beau Monde limousine went for $10,194. The three-way car that was manually operated cost $11,443 as a landaulet and $11,281 as a limousine. The three-way coach with electrical assistance went for $12,007 as a landaulet and $11,847 as a limousine. Other models in this series included ambulances and combinations in both landaulet and limousine styles.

Considered part of the Royale series, the Superior flower car continued with its distinctive styling line and sold for $10,785. The Cadillac service car was also considered a part of the Royale series and carried a price tag of $10,071. Note the continuity of styling from last year and the clean flowing lines that this car exhibits. Superior built a total of 37 flower cars this year. Their total Cadillac chassis cars amounted to 1,055, the most cars that Superior had ever built in one year on the Cadillac chassis. Of these, 135 were side service funeral cars, 221 were end loading funeral cars, and 469 were combinations.

Superior built a total of 193 Cadillac chassis ambulances in 1961. This Royale Super Rescue ambulance went for $12,999 and featured a headroom of 54 inches. The Royale 48" ambulance went for $11,552 and the Royale 42" sold for $10,962. The door opens well into the roof for easier loading. The Rescue versions also featured a built-in bumper step.

Mounted on one of the most attractive Oldsmobile chassis ever built, the 1961 Cotner-Bevington professional cars were very attractive. All of the large Cotner-Bevington models were now built on an extended Oldsmobile Ninety Eight chassis and were called Cotingtons. 1961 saw a totally new styling approach by Cotner-Bevington and the cars were very tastefully appointed and styled. The Cotington landau in funeral car form sold for $8,088, while the landau combination went for $8,221. The Cotington ambulance sold for $8,354 in landau form.

Called the Seville, these short wheelbase Cotner-Bevington professional cars were offered on Chevrolet, Pontiac Catalina, Pontiac Star Chief, and the Oldsmobile chassis. Here we see the Cotner-Bevington Pontiac Star Chief Seville in combination form. This car sold for $5,613, while a straight funeral car version went for $5,506. A straight ambulance version of this Pontiac Seville sold for $5,719, while a service car version was offered at the same price as the straight funeral coach. Prices for cars on other chassis fluctuated according to their value, i.e., Oldsmobiles cost more than the Chevrolet chassis vehicles with the same bodies.

With completely new styling and a new innovation in rear doors, Flxible announced their 1961 professional car line. The new cars were available with a novel type of rear door called Flxi-door. This unique option featured the ability to open from either the right or the left side and had two rear door handles to facilitate this novel option. This type of door would adapt to any type of local conditions for loading or unloading. The attractive Premier landau shown here was offered in both straight funeral car form for $9,014 and as a combination for $9,216. The interesting landau plate incorporated the Buick shield insignia.

Featuring the stylish Buick shield insignia on the C pillar and a clean flowing side window line, the 1961 Flxible Buick limousine was a very attractive package. The Premier limousine in straight funeral car form went for $8,785, while the limousine combination cost $8,785. All Flxible Buick professional cars were mounted on an extended Buick Electra chassis.

Looking like a miniature version of the large Flxible Premier landau, the Flxible Flxette landau styling borrowed liberally from the large cars. This car also features the Buick shield emblem on the landau shield and a posh looking landau bow. In this year Flxible built a total of 84 Flxette coaches with both landau and limousine styling. These cars were available as funeral coaches, ambulances, combinations and service cars.

This experimental vehicle was built by a St. Catharines, Ontario firm for the Ministry of National Defense in Ottawa. It was a combination hearse and gun carriage. The casket was placed on an elevator that would lift it to the raised position as seen here for the cortege, while it was lowered for unloading. It was built for the Ministry with a view to replacing the horse drawn cassion for military funerals across Canada. It is not known whatever came of the car or the idea.

Flxible also offered a complete selection of ambulances on both the Premier long wheelbase chassis and the Flxette short wheelbase frame. The straight Flxible ambulance with a 42-inch headroom and limousine styling sold for $9,408, while the special high headroom Premier ambulance cost $10,275. Note the interesting paint scheme on this car and the roof load of warning lamps.

Continuing without any major restyling, the Pinner Pontiac combination car was offered by the Pinner Coach Co. of Memphis. The roof gumball light could be applied in minutes for ambulance duty or removed for funeral car service. Note the pronounced wrap around rear corner windows and the sharp roof crease at the rear.

1961

Miller-Meteor continued to market their most successful limousine with the vast expanse of glass area. Called the Futura, these limousines were very attractive and available in as many as three colors on the car. The Miller-Meteor limousine funeral car prices began at $10,532 for the end loading job, with the manually operated three-way car going for $11,654. The three-way coach with an electrically assisted table cost $12,239. The Duplex combination car with a reversible floor or with a skeleton type casket table that was removable sold for $10,806.

Illustrating the diversity of the landau body style is the standard Miller-Meteor landau funeral car. Note the rear roof styling and the three-tone paint finish. The Miller-Meteor landau in manually operated three-way form went for $11,880, while the three-way car with electrically assisted casket table cost $12,459.

Called the Miller-Meteor Landau Panoramic, this attractive coach featured Miller-Meteor's novel play on the landau theme with large wrap around rear corner windows and a full expanse of side windows. The slightly smaller landau panel featured a special size landau bow and an attractive landau shield. The landau Panoramic styling was optional at no extra cost on landau models. The Miller-Meteor landau in straight end-loading form sold for $10,759, while the Duplex combination car with either a reversible floor or a removable skeleton type casket table cost $11,033.

Called the Sentinel, this Miller-Meteor ambulance featured the styling characteristics of the Futura limousine models. This car with the limousine styling and a roof height of 42½ inches sold for $10,936, while the 48-inch headroom version carried a price of $11,309, and the First Aider with a headroom of 52 inches went for $16,323. A 42½ inch ambulance was also offered with landau styling and it sold for $11,157, while the landau ambulance with the landau panoramic roof line was called an Amblelandau and had a large chrome ambulance cross on the landau panel.

The attractive Miller-Meteor flower car carried the hefty price tag of $11,141. Notice the two-tone paint scheme and the slanted rear windows.

Imported to the United States briefly, this little Mercedes-Benz ambulance was distributed by Ambulance Imports of Warsaw, Ind. The car was built on a Mercedes 190 chassis with a body by Pollmann of Bremen, Germany. The car offered economy and prestige at the same time.

Just entering the professional car field was the Richard Brothers with their attractive Briarian combination on the Chrysler chassis. This car, mounted on a Chrysler station wagon, featured maximum versatility and would convert from funeral car to ambulance in minutes. This car was mounted on a Chrysler New Yorker chassis that had not been extended but did have a raised roof line.

1962

Eureka once again offered the industry's second most expensive line of professional cars. Although the styling was very slightly changed from last year, the Eureka landau was available with the Full View styling that included the wrap around rear corner windows. The landau was offered in six versions with the least expensive being the rear loading funeral car at $11,697. The three way car with a manaully operated casket table cost $12,526, while the three-way with an electrically assisted table went for $13,319. Combination versions of this car ranged in price from $11,691 to $11,867.

The Eureka limousine funeral car was a very distinguished looking car with elegant lines and a wide variety of model versions. The straight end-loading car with limousine styling went for $11,458, while the three-way manual car sold for $12,287 and the three-way with an electrically assisted casket table cost $13,083. Combination versions of the limousine ranged in price from $11,452 to $11,627. This car was also offered with the Full-Vision styling at no extra cost.

The standard Eureka ambulance sold for $11,996. An example of this attractive car is seen here with the optional Full-Vision styling. Note the mounting of the roof lights and the application of the two-tone paint job. Eureka also offered high headroom versions of this car at a significantly higher price.

Eureka's flower car had a style all its own and borrowed from no other maker. Selling for $11,998, this car was built in very limited numbers and featured a very low rear flower deck made of stainless steel.

This was a year in which the major manufacturers were content to continue with the styling from last year and to plan for next year. There were no new bodies or specific styles introduced in 1962 as far as funeral coaches and ambulances were concerned. The newest models introduced were limousines on extended chassis, not funeral cars or ambulances. Both Cotner-Bevington and Superior announced new seven-passenger limousines. Cotner-Bevington's offering was on an extended Oldsmobile chassis and featured a high, square look. Superior on the other hand, a larger and more style-conscious firm introduced an extended frame limousine on the Pontiac chassis. The offering was highlighted by the fact that the car followed the overall General Motors styling that was popular on passenger cars. The Superior Pontiac seven-passenger limousine followed that style even though lengthened, and was quite an attractive vehicle.

The way in which the rear door should open had for quite some time been a bone of contention, with the various makers all having differing views. Some coaches had rear doors that were hinged on the left while others hinged on the right. Flxible announced an innovation that would please the funeral director no matter what his preference in rear door hinging. Called the Flxidoor, this new rear door could be opened from either side with the only give-away being that cars equipped with this feature had two door handles, one on each side.

The rear bumpers on the 1959 and 1960 Cadillac commercial chassis were veed and the only maker that contoured the rear door to that of the bumper was S & S. This gave a more pleasing appearance and eliminated the dirt-catching space between the bumper bar and the body that was prominent with all of the other makes. It also had the more practical advantage of moving the loading edge of end loaders closer to the rear of the car, providing easier loading conditions.

Total U.S. production of hearses and ambulances this year was 3,764, of which 2,229 utilized Cadillac chassis.

Conrad Pollmann Norddeutsche Karosseriefabrik of Bremen, Germany, one of Europe's leading manufacturers of funeral service vehicles built this unusual Borgward chassis funeral coach this year. Note the large frosted glass panel in the rear compartment and the overall style of this attractive car.

1962

The dignified and distinguished Hess & Eisenhardt S & S Victoria was offered in six funeral car and combination coach versions and one ambulance version in this year. The Victoria prices began at $12,651 for the straight end-loading car and went to $13,226 for the end loading car with an extending casket table. The Victoria with a manually operated casket table sold for $13,072, while the three-way car with a power assisted table went for $14,381. The Victoria when outfitted as a combination with Duo Floor disappearing casket rollers sold for $13,072 and the Victoria Parkway ambulance went for $13,159.

The most elaborate and best equipped ambulance available was the S & S Eisenhardt Custom Professional Parkway that sold for $18,985. This car featured a significantly raised roof line with three built in tunnel-type roof warning lights and a long chrome tube-type pod on the roof sides that enclosed a warning lamp at the rear. The area just behind the driver's compartment on the roof was finished in blue plexiglass, and the cars were totally air conditioned.

The Flxible Buick ambulance on the long wheelbase Premier chassis was also offered in two ambulance versions. This car with a 42½ inch headroom was the standard form and sold for $9,408, while a special high headroom version went for $10,175. This car was delivered to the Goodwill-Potterstown Funeral Home, and was finished in a deep red.

The Flxible Flxette short wheelbase professional car line continued in production this year, with a total of 74 such units being built. The straight funeral car version of this little coach went for $7,185 in limousine form, while the same car in landau form sold for $7,414. Combination models were also offered at prices beginning at $7,278 in limousine form. The service car on the Flxette chassis sold for $7,183 and an ambulance version went for $7,466.

The Flxible Company of Loudonville, Ohio, continued to market their popular and successful Buick based professional car line through this year. The Flxible Premier landau shown was offered in straight landau funeral car form at $9,014, landau combination form for $9,216, and as a service car for $8,785. In this year Flxible built a total of 72 Premier models, all on the Buick Electra chassis that had been lengthened.

Built in small numbers, this attractive Buick eight-passenger limousine was mounted on the same lengthened Electra 225 chassis as the regular Flxible Premier funeral cars and ambulances. This version of this extremely rare car exists today in the hands of Roger Hannay of Westerlo, N.Y. It is fully air conditioned, has a rough finish roof, an electrically powered center partition, tinted glass all around, and standard Buick Electra ornamentation all around.

The Superior flower car was considered a part of the Royale series, as was the service car. This year Superior built a total of 34 flower cars at $10,785 per issue. Superior also built 127 side service coaches, 239 end-loading hearses, 427 combinations and 185 ambulances on the Cadillac commercial chassis in 1962. Total production of Cadillac chassis cars totaled 1,058, two units more than last year.

In a fitting setting is the 1962 Crown Royale Landaulet finished in black with egg shell color drapes. The Crown Royale and the Crown Sovereign models were the top models in their respective series. All Royale models featured the rear corner windows, while the Sovereign series cars had this area blanked in. The landaulet was offered in both series of the Crown and the least expensive of these was the end loading car. The Crown Royale landaulet went for $10,781, while the Crown Sovereign went for $10,969. The manually operated three-way car in landaulet styling sold for $12,051 in Crown Sovereign garb, and for $11,701 one could get the same car in Crown Royale landaulet trim. The three-way electrically operated casket table versions of the Crown landaulets sold for $12,615 (Crown Sovereign) and $12,427 (Crown Royale). Both series offered combinations and 42 inch ambulances in both landaulet and limousine styling.

Continuing in popularity and wearing the new famous Criterion styling, the Superior Pontiac full length professional car line was built in two body styles, limousine and landaulet, and in a wide variety of model choices. Superior built a total of 239 long wheelbase Pontiacs this year. Of these 27 were straight funeral coaches, 144 were combination cars, and 68 were ambulances. This is the Criterion styled Superior Pontiac landaulet funeral car complete with formal draperies.

The Superior Cadillac limousine body style was offered in this form in both the Royale and the Sovereign series. The limousine in straight end-loading form went for $10,194 as a Royale and $10,382 as a Sovereign. The Sovereign manually operated three-way car sold for $11,469, while the same car in Royale garb cost $11,281 and the three-way electrically assisted limousines went for $11,847 (Royale) and $12,035 (Sovereign). On the average the Sovereign models sold for $200 more than the comparable Royale series models. Other models in the Royale series consisted of the end loading landaulet at $10,361, the three-way landaulet $11,443, the power assisted three-way landaulet at $12,007 and the landaulet combination for $10,627.

One of the 185 Cadillac chassis ambulances built in this year by Superior was this Royale Super Rescue 54 inch ambulance. This car carried a price tag of $12,999, while other ambulance models with lower roof heights went for substantially lower prices. The Royale Rescue 48 inch went for $11,552 and the Royale Beau Monde limousine ambulance for $10,586 and featured a headroom of 42 inches.

The Superior Pontiac long wheelbase coach with Criterion styling was a handsome car when adorned with limousine styling. This limousine combination was one of 144 such vehicles built this year by Superior.

1962

Low in production was this attractive little Superior Pontiac Consort service car. Most funeral homes had found it increasingly less expensive to just purchase a station wagon and blank off the rear side windows instead of purchasing a custom made and more expensive service car. Note the distinctive Superior stylized chrome wreath and bars on the upper rear body sides.

The popularity of the Superior short wheelbase Consort line of professional cars soared, and a total of 171 Consorts were built in 1962. This is a Consort combination car, of which a total of 118 were built. Superior also constructed a total of 10 straight Consort funeral cars, and an additional 43 Consort ambulances were built this year. The Superior Pontiac line was built in their Kosciusko, Miss., plant, while the main plant in Lima, O., was kept busy producing the Cadillac chassis cars.

The new Miller-Meteor Olympic ambulance was offered in two headroom versions. The one shown here was the standard 42 inch version and sold for $10,967, while a 48 inch version went for $11,329. Note the sweeping panel that divides the rear corner windows from the rest of the side windows. The style of the corner roof warning lights was a Miller Meteor exclusive as was the two-tone paint scheme.

Called the Guardian, this Miller-Meteor ambulance style was appearing for the last time. This car featured a 52 inch headroom height and sold for $15,874. Note the massive roof overhang and the rear step built into the bumper. This car resembled a rolling Easter egg with lights.

The Futura limousine was offered in four versions of funeral cars and combinations plus two ambulance versions. Ths straight funeral coach in end loading form went for $10,251, while the three-way car with a manually operated table cost $11,341 and the power assisted three-way car went for $11,901. The Duplex combination with Futura limousine styling was offered at $10,518. A totally new series of Miller-Meteor cars called the Olympic was offered this year. These cars featured a special panel that divided the rear corner windows from those that flowed across the upper sides of the car. The Olympic in straight end-loading form sold for $10,574 and the three-way power assisted car went for $12,445. The other models in the Olympic line sold for about $320 more than the landau or Classic series models.

Miller-Meteor introduced a whole new line of professional cars in 1962 complete with new names to match. This funeral car called the Classic was offered in four versions of funeral car or combination plus three ambulance versions. The Classic end-loading hearse sold for $10,472, while the three-way car with a hand operated table sold for $11,560 and the three-way with a power assisted table went for $12,123. The Duplex combination on the Classic body cost $10,738.

1962

The distinguished craftsmanship of the Cotner-Bevington line was displayed with the 1962 Cotington Oldsmobile line of long wheelbase professional cars. The Cotington was offered in both landau and limousine body styles again this year. The landau funeral car went for $8,524 while the limousine version went for $8,391. The landau combination cost $8,657 and the limousine combination sold for $8,524. A service car version of the Cotington cost $8,391 and featured the same clean lines as the limousine funeral car with the notable exception of having blanked off upper rear side panels.

The Cotington ambulance was offered in both landau and limousine body styles in this year. The Cotington limousine ambulance carried a price tag of $8,221, while the landau ambulance commanded a full $8,354. A special 44-inch headroom version of the limousine ambulance was offered for $8,688.

To help the funeral director to match his fleet, many coach builders began to offer limousines with styling that matched their coaches. Cotner-Bevington offered this interesting Oldsmobile limousine that would seat 8 passengers in luxurious comfort. Powered by the large Olds 330 horsepower engine, these cars rode on the same 150-inch wheelbase that the Cotington professional coaches used.

Economical, practical, and always ready for action, the Automotive Conversion Corp. offered ambulance conversions of almost any make of station wagon, such as this Mercury Colony Park wagon wearing all of the Amblewagon equipment. One of the attractions of this type of ambulance was the low initial investment and the fact that the car could be converted back to passenger type station wagon duties when it was ready to be sold.

New in the Superior Pontiac model line-up for 1962 was this cleanly styled and attractive 8-passenger Pontiac limousine. This car was ideal for the funeral director that owned Pontiac funeral service vehicles because he could now have a matching limousine. These cars were to find a fair amount of public acceptance and were used by such organizations as hotels, airports, resorts and affluent private individuals.

Mounted on an Austin Princess chassis, this attractive Woodall-Nicholson hearse was the style popular in Great Britain in this era. Note the large glass plate area in the casket compartment and the roof mounted flower rack.

Built as a pilot model, this attractive landau hearse conversion was carried out by the W.S. Ballantyne Company of Windsor, Ontario. The car never went beyond the prototype stage due to the high prices that this station wagon conversion commanded. For a straight, though attractive, conversion, this car on a Plymouth chassis would have sold for around $8,000 and for this reason the entire project was dropped. This car still exists in Windsor, though in rather rough shape, and is being driven by a plumber.

1963

These fairly recent years in the history of the funeral coach and ambulance indicate that beginning with this year a new classic era began. It was to be the era of classic, crisp styling, when this form of coach really came into its own.

This was the year of the really big change at Miller-Meteor. Following record breaking sales in its 1962 model year, when the firm took the position as the world's largest manufacturer of funeral coaches and ambulances on the Cadillac chassis, Miller-Meteor announced a completely new line with totally new design and construction. Roof lines were lengthened and subtly squared, while the body lines were sculptured to blend with the new Cadillac hood fender and body contours. This resulted in a longer, lower, wider and more distinctive profile. The 1963 Miller-Meteor funeral coach models were available in five styles, all available as three-way, combination or rear loading vehicles. Four ambulance models were available in both 42 and 48 inch headroom styles. New on ambulances for this year were the exclusive Miller-Meteor Ful-Vu roof lights. These were the same type as were used on aircraft, fully visible for miles in any direction with 360 degree spread of the red emergency warning lighting.

With the optional Full-Vision styling, the Eureka limousine funeral car sold for $11,751 as a straight end-loading machine and for $12,610 as a three-way manually operated coach. The electrically assisted three-way casket table version went for $13,421, while combination versions of the limousine body style carried prices ranging from $11,744 to $11,962. The standard limousine ambulance sold for $12,300 and had a headroom of 42 inches. High headroom ambulances were built by Eureka to special order and carried prices that reflected this custom built type of vehicle.

Continuing to feature the styling introduced last year, the Eureka Company once again offered a full range of models in both landau and limousine styling. The landau was available with the Full-Vision styling that included the wrap around rear corner windows again this year. When seen as a straight end-loading vehicle, the landau sold for $11,996. The three-way manyally operated landau went for $12,847 while three-way manually operated landau went for $12,847, while $13,664. Combination versions of the Eureka landau ranged in price from $11,989 to $12,167.

A dramatically different hearse style was featured by Miller-Meteor in this model year. The Paramount Landau featured a new open look in a landau style. Combining all of the advantages of the extra visibility of the limousine with the distinctive privacy of the landau, the new Paramount had a shorter solid rear quarter with a newly designed landau bow that was now almost straight and accented the Cadillac fin. The glass expanse was increased by adding an additional window on each side. Limousine models were offered in two distinctive styles by Miller-Meteor in this year. The Classic limousine featured crisply sculptured lines with a Thunderbird type rear roof quarter, while the Futura limousine offered a wide expanse of deep windows outlined with bright aluminum trim and a thin rear roof quarter panel.

Superior unveiled a major styling change with the 1963 models. Each model in each of the four styling series were completely redesigned, making this the most significant styling change in the company's history. Major styling innovations were found in the Sovereign, Royale, Crown Sovereign and Crown Royale series coaches, each offering both limousine and landaulet models with special styling distinctions of their own. A completely restyled flower car was found within the Royale series.

The Sayer & Scovill, Hess & Eisenhardt line for 1963 also featured the crisp, classic type of styling that was reflected in models from other manufacturers. The S & S line was entirely redesigned for this year and featured a low, crisp roof line with a low broad look at the rear. Sharp, crisp, formal lines accented the new Cadillac styling seen on both the front and rear ensembles of these cars. The sales of the S & S Victoria had doubled in the last three years and the model was to again be the company's sales leader. Styling of this model had changed little in concept from its original in 1938, and the company gave special attention to styling of this model. It reflected the epitome of dignified, distinctive formality and carried a price tag in its standard form of $12,615.

The crisp angular styling was also prevalent in the models announced for this year by Eureka. Window lines were squared and blended perfectly with the low flowing roof line that featured a distinctive break over the rear door. Available as an option was the wrap around style of rear quarter windows and this added yet another variety of style within the Eureka line.

In its first model change since its introduction, the Superior Pontiac nine-passenger limousine incorporated 1963 Pontiac styling in a luxurious limousine with all of the quality of other Superior coaches and the appearance of a more expensive limousine. This car was a perfect styling companion to Superior's line of Pontiac chassis funeral cars in both long wheelbase and Consort models.

Total U.S. production of hearses and ambulances was 4,718, with 2,695 of these being on Cadillac chassis.

Seen in an appropriate setting, the standard Royale Landaulet featured subtle styling without the ostentatious chrome mouldings that were found on the Crown models. All Sovereign models cost about $200 more than the equivalent Royale versions. The Royale landaulet as a straight end-loading car sold for $10,483, while the manually operated three-way coach cost $11,565 and the power assisted three-way went for $12,130. Combination versions of the Royale landaulet sold for $10,749 and ambulance versions went for $10,940 with a 42 inch headroom.

With the Crown chrome band across the roof and the rear portion of the roof covered in vinyl, the Crown series limousine coaches were very attractive cars indeed. Crown Sovereign models were priced about $200 higher than Royale versions of the same cars. The Crown Royale limousine as a straight end-loading car sold for $10,736, while the three-way car with a manually operated table went for $11,823. The luxurious Crown Royale limousine with a power assisted three-way casket table sold for $12,389, while a combination version of the Crown Royale, as pictured here, went for $11,002. Ambulances were offered in both Crown series with 42 inch roof heights, and the Crown Royale limousine ambulance cost $11,199.

The Superior flower car was classified in the Royale series again this year. The flower car sold for $10,907, and Superior built 28 of them. Superior production of Cadillac chassis professional cars zoomed to 1,174 in 1963, with 131 of these being side service coaches, 235 being end-loading coaches, and 582 being combination cars.

Featuring new styling for 1963, the Superior Sovereign and Royale limousines were again called Beau Monde models. The Sovereign versions cost approximately $200 more than the Royale Beau Monde, which, as a straight end-loading car, went for $10,316. The three-way car with a manually operated table cost $11,403, the power assisted three-way car sold for $11,969 and combination versions cost $10,582. The main difference between the Royale models and the Sovereigns was the fact that Royale models featured a small wrap around rear corner window, while the Sovereign series had this area blanked in and looked somewhat more formal.

Always a styling pace setter, Superior offered the Crown Royale and Crown Sovereign series in landaulet styling again. This year the Crown models featured a heavy chrome quarter panel moulding over the roof chrome bands and special interior and exterior trim. The Crown Sovereign landaulet (shown) in a straight end-loading form sold for $11,013, while the Crown Royale version went for $10,903. Manually operated three-way landaulets sold for $11,985 as a Crown Royale and $12,205. Three-way cars with power assist on the casket table cost $12,594 as a Crown Royale and $12,788 as a Crown Sovereign landaulet. Combinations were offered with prices beginning at $11,169 and ambulances began at $11,360 in landaulet form.

Superior had always prided themselves as leaders in ambulance design and construction. In this year they built 198 such vehicles on the Cadillac chassis. The 50 inch Royale Rescue ambulance sold for $11,818. Other models in the ambulance range consisted of a Royale 48 inch at $11,227 and 42 inch ambulances in limousine styling for $10,780, and landaulet styling for $10,940. Note the illuminated nameplate on the side of the roof and the built in warning lights on the facing part of the high roof.

The Superior Pontiac long wheelbase cars were built on an extended Bonneville chassis and were powered by a 330-horse 389-cubic-inch engine. These cars rode on a wheelbase of 146 inches with an overall length of 241 7/8 inches. The short wheelbase Consort had an overall length of 218 inches and a wheelbase of 123 inches. This attractive full size Pontiac service car sold for $7,826, while a Consort version of the service car went for $6,360.

Superior Pontiac chassis professional cars continued to be popular. This year they built a total of 245 Pontiacs on the extended chassis and 205 Consort short wheelbase cars. The full size Superior Pontiac funeral car sold for $8,200 as a limousine and $8,453 as a landaulet. The combinations sold for $8,438 as a limousine, and $8,692 as a landaulet. The ambulance versions of the full size Pontiac went for $8,692 as a 40½ inch and $9,061 as a 46 inch headroom version.

Riding on the extended Superior Pontiac chassis with a wheelbase of 146 inches, the Superior limousine made an attractive addition to any funeral home's fleet. It matched the other Superior Pontiac professional cars, and this fact made it a popular item with funeral directors who already owned Pontiac equipment or could not afford the more expensive Cadillac limousines.

Cotner-Bevington continued to offer a strong line of coaches mounted on the Oldsmobile chassis. The long wheelbase Cotington models were offered as landaus, limousines, and as large 8-passenger limousines. The Cotington landau sold for $8,937 as a funeral car, $8,636 as a combination and $9,036 as an ambulance. A 46-inch headroom ambulance was offered for $10,501 and a service car sold for $8,551. The short wheelbase Seville models on the standard Oldsmobile 98 chassis ranged in price from $6,180 for the funeral car and service car to $6,393 for the ambulance.

Called the symbol of professional pride since 1876, the S & S line of professional cars built by Hess & Eisenhardt of Cincinnati were very expensive cars indeed. This S & S Superline Park Row limousine was offered as either a straight end-loading funeral car that sold for $12,131 or with the exclusive S & S rear extension table that moved back through the rear door, to facilitate loading, for $12,706. The Park Row as a manually operated three-way car sold for $13,188, while the power assisted three-way Park Row went for $13,860. Park Row combination cars sold for $12,550 with the customer's choice of casket roller types. The Superline Parkway ambulance was derived from the Park Row and utilized the same body while selling for $12,639. Note the Veed shape of the rear door. This was an exclusive S & S trademark.

The Hess & Eisenhardt S & S Victoria continued to excell in all of the areas that a high quality funeral car should, and carried a very exclusive price tag. The standard Victoria sold for $12,651 as a straight end-loading vehicle. This same car when equipped with an extension table for ease of rear loading sold for $13,226. The Victoria was again offered with the manually operated three-way casket table for $13,710, while the power assisted version of this luxurious car went for $14,381. Combination versions of the Victoria carried a price tag of $13,072 and were offered with either Duo-Floor or the skeleton type tube casket table that was removable. The Parkway Victoria ambulance cost $13,159, while the Custom Professional Parkway with a high headroom roof carried a price of $18,985.

1963

The attractive and appealing Flxible Premier limousine carried a price tag of $8,806 in straight funeral car form and $9,008 as a combination like the one seen here. Flxible also offered a service car in the Premier line, which sold for $8,806.

Modest touches of restyling made the 1963 Flxible Buick professional cars among the most attractive they ever built. Flxible continued to offer their professional cars in two distinct wheelbase sizes. The long wheelbase Premier landau seen here sold for $9,035 as a straight funeral car, while the Premier landau combination sold for $9,236. With every year since their re-entry into the field in 1959, Flxible production had risen. This year they built a total of 88 Premier coaches and a grand total of 199 Buick chassis professsional cars of all types.

As a service car the Flxette was a very attractive looking and versatile little vehicle. Although attractive, this short wheelbase coach carried a price tag of $7,211, and this accounts for the fact that very few of these service cars were built.

Selling for $9,428, the Flxible limousine ambulance was by far the most expensive car in the entire range. These cars were built on an extended Buick Electra 225 chassis and were powered by Electra's 401-cubic-inch 325 horse engine. The suspension and the frame were beefed up to handle the load of the ambluance or hearse body.

The Flxible Flxette short wheelbase line of professional cars were mounted on the standard Buick Electra chassis without an extension to the frame. Flxible built a total of 111 Flxettes in 1963, making this series the most popular in the maker's production program. The Flxette was offered in a straight ambulance, seen here, for $7,483; a combination for $7,304, and a straight limousine funeral car at $7,211. The landau funeral car went for $7,440, while the landau combination car sold for $7,533.

The Memphis based Pinner Coach Co. was now building cars on Pontiac, Chrysler, Ford and Chevrolet chassis exclusively. This interesting Pinner Chrysler ambulance features a high roof line and an unusual rear corner window treatment. The Chrysler Newport chassis has been extended and the complete car has an air of distinction. Pinner was also offering funeral cars with both limousine and landau styling. Very little is known about this maker, but it is believed that 1963 was the last year that Pinner built cars in any substantial numbers.

1963

The most outstanding new design to come into the field since 1959 was the totally new Miller-Meteor Paramount landau funeral car line. This car combined all of the good features of the landau with those of the limousine models, and featured an extra window on the rear body sides and a new, smaller, almost straight landau bow in the landau area. Also new was the crisp type of Miller-Meteor styling that took over from the more antiquated rounded lines of last year's cars. The new Miller-Meteor Paramount did not replace the traditional landau, only augmented it. In straight end-loading form this car sold for $10,472, while the three-way car with a manually operated table sold for $11,560 and the three-way car with power assisted casket table went for $12,123. The combination version of this car, as seen here, went for $10,738, while an ambulance with a 42 inch headroom sold for $10,858. The newly styled 48 inch ambulance cost $11,246.

The National Coach Company of Knightstown, Ind. continued to offer custom built coaches on any chassis the customer preferred. This National Minute Man ambulance is seen on a Chevrolet chassis that has been substantially lengthened. Note the unusual panel between the C pillar and the rear quarter window which carried a warning light. National also offered landau and limousine funeral cars, multi-door air port type limousines, and carry-all coaches. They also offered a selection of 9, 12 and 15-passenger limousines for use by funeral directors.

Also wearing new, crisp styling was the Miller-Meteor flower car. The car no longer had the swept back type of roof line, but now featured a more square coupe style. This car sold for $10,843, but very few of these attractive vehicles were built.

Called the Siebert Aristocrat 1045 ambulance, this version was mounted on a Ford chassis. Siebert specialized on the Ford and Mercury chassis and built some rather strange looking ambulances and funeral cars. Note the long wheelbase and the high roof line.

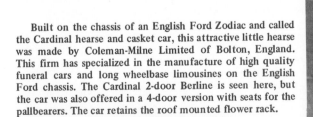

Built on the chassis of an English Ford Zodiac and called the Cardinal hearse and casket car, this attractive little hearse was made by Coleman-Milne Limited of Bolton, England. This firm has specialized in the manufacture of high quality funeral cars and long wheelbase limousines on the English Ford chassis. The Cardinal 2-door Berline is seen here, but the car was also offered in a 4-door version with seats for the pallbearers. The car retains the roof mounted flower rack.

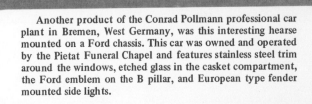

Another product of the Conrad Pollmann professional car plant in Bremen, West Germany, was this interesting hearse mounted on a Ford chassis. This car was owned and operated by the Pietat Funeral Chapel and features stainless steel trim around the windows, etched glass in the casket compartment, the Ford emblem on the B pillar, and European type fender mounted side lights.

1964

Superior offered many new designs and engineering advances on its four Cadillac commercial chassis series. Among the new features was a rear bumper that now included a license plate housing at the center, giving the appearance of being one piece. A distinctive crease line in the roof, just above the rear door, was duplicated by subtle creases in four directions at the center of the door. All Superior coaches featured tinted glass as standard equipment in the windshield of their models. The rear quarter of the Royale limousine models was accentuated with new ornamentation. A new chrome treatment was used on the sides and rear of the Crown Royale and Crown Sovereign series models.

The 1964 S & S line starred an impressive formal limousine in the Park Hill funeral coach. This coach featured an image of perfection with its vinyl covered padded top, crease line down the center of the rear door, and crisp styling.

Shortly after the announcement of the 1964 models, the Wayne works of Richmond, Ind., announced the purchase of the Cotner-Bevington Co. of Blytheville, Ark., makers of Oldsmobile funeral coaches and ambulances. The new acquisition would place Wayne in a better position to battle Superior for sales. By acquiring Cotner-Bevington, Wayne could offer a line of extended wheelbase coaches to market in the same price class as the Superior Pontiac line. The firm was to remain at its Arkansas location and retain its independent identity but would profit from the association through mechanical and technical progress and development.

Eureka launched its 1964 line that included a full line of coaches, ambulances and flower cars, but the fall preview was destined to be the firm's last. The company had for the past few years suffered from lagging sales and it was decided to cease production at the end of the model year. Thus, this company ended a 93-year history of professional car construction. Having been the first to design, patent and market the three-way table, an innovation that outlived its maker, this feature would remain as a reminder that Eureka had been an industry pioneer.

Total U.S. production was 3,647, of which 2,264 were on Cadillac chassis.

Built to special order was this attractive Eureka high headroom ambulance. The roof side trim was unique, as was the manner in which the roof lights were faired into the higher roof line. The air conditioning air intake is at the root of the rear fender fin.

Eureka flower cars were built to special order, and it is believed that there were at least three of these attractive cars built by Eureka in this, their last year of production. The low flower deck was finished in stainless steel.

In 1926 Eureka introduced the three-way casket table for funeral cars, and this innovation had a revolutionary impact on the future of these vehicles. The last Eureka limousine coach was offered this year and the three-way car with a manually operated table cost $12,750, while the power assisted version went for $13,560. The rear loading funeral car with limousine styling sold for $11,890, while combination versions of this car ranged in price from $11,885 to $12,065. Eureka's Full-Vision styling was optional at no extra cost, but air conditioning was an extra cost option.

After building high quality professional cars for 93 years, introducing the three-way casket table that revolutionized the industry, and carrying a rather exclusive air about them, the Eureka line of funeral cars and ambulances would cease to be offered after this model year. The 1964 landau, not surprisingly, appeared without any styling changes to the body. This attractive landau is seen with the optional Full-Vision styling, air conditioning and whitewall tires. The landau sold for $12,135 as a straight end-loading car, $12,985 as a manually operated three-way car, $13,805 as a three-way coach with an electrically assisted casket table, and combination versions from $12,130 to $12,305.

1964

The traditional and dignified Hess & Eisenhardt, S & S Victoria again continued without any major styling changes. This car carried a price tag that began at $13,060 for the straight end-loading car. The end loader with the exclusive S & S extension table sold for $13,655, while the three-way car with a manually operated table went for $14,152 and the electrically assisted version of the three-way car cost $14,840. Combination versions of the Victoria sold for $13,495, with the customer's choice of either a skeleton type removable casket rack or disappearing cakket rollers built into the floor.

The least expensive car in the Hess & Eisenhardt, S & S professional car range was the Park Row limousine funeral car with straight end-loading. This car sold for $12,525, while the same car with the exclusive S & S extension table went for $13,120. Three-way versions of the Park Row were offered as either manually operated for $13,615 or power assisted at $14,305. The Park Row combination sold for $12,955. The Park Row body was basically the same as that used for the S & S Parkway ambulance that sold for $13,050, the Kensington ambulance that went for $14,700, and the ultra-luxurious and high headroom S & S Custom Professional Parkway ambulance that carried a price tag of $19,825. Hess & Eisenhardt also offered a new line of limousine funeral cars in this year called the Park Hill. This car, although utilizing the same body as the Park Row, featured a padded, vinyl covered roof and a more luxurious interior than the Park Row. The Park Hill was priced about $450 higher than comparable Park Row models.

Continuing to offer custom station wagon conversions, the Automotive Conversion Corp. built this attractive Arlington funeral car from the body of a standard Chrysler station wagon. This car featured skeleton type casket rollers in the rear compartment, a rear door that opened from the side, and landau panels over the rear quarter windows. Automotive Conversion continued to market the highly successful Amblewagon ambulance conversions as well as this line of funeral cars.

Plowing through a rainy downpour, this Miller-Meteor 42 inch limousine ambulance has all lights ablaze. The 360-degree roof corner warning lamps were a Miller-Meteor exclusive design and were to be found on all ambulances by this prominent Piqua, Ohio, firm. In the standard 42-inch limousine form, this car sold for $10,845, while a landau version went for $11,058. Miller-Meteor was now a division of the Divco Wayne Company of Richmond, Ind.

Selling for $11,100, the newly styled Miller-Meteor flower car was a low production model in comparison with some of the other models.

Seen for the last time on a passenger car chassis was the line of professional cars from National of Knightstown, Ind. This example is mounted on an extended Pontiac chassis and featured limousine styling. Called the National Knighton limousine funeral car, this body was also offered on the Chevrolet chassis.

The large and attractive Flxible Buick Premier models were offered in six models this year, and a total of 88 were built during the model run. This Flxible Premier combination car featured removable panels over the rear quarter window that gave the car the look of a landau when in place and a limousine when they were removed. This coach sold for $9,195, while a straight landau version of this car went for $9,400. The Premier limousine funeral coach sold for $8,980 and the landau version cost $9,185.

Riding on a wheelbase of 126 inches with an overall length of 215 inches, the Flxible Flxette line of professional cars was offered in six versions. Flxible built a grand total of 68 of these attractive little coaches this year. This was the record low for this line, since their 1959 introduction. The Flxette landau funeral coach seen here sold for $7,465, while the landau combination car went for $7,330. Also offered in the Flxette line was an ambulance for $7,483 and Flxible's only service car at $7,211.

The Flxible Buick Premier long wheelbase ambulance cost $9,810 in this 40 inch headroom style and $10,510 as a 46 inch headroom version. These cars rode on a wheelbase of 149.5 inches and used the Electra 225 Wildcat engine with a 400 series transmission. The overall length of the Premier models was 242 inches. Note the interesting roof light design and the Electra 225 ornamentation that these cars carry.

Called a bearer's coach, this four door Daimler was built by the eminent firm of Woodall Nicholson of Halifax, England. Looking like a large Daimler station wagon, this car featured seats for the pallbearers to ride along side the casket which was again in full view. Note the flower rack on the roof.

The Cotner-Bevington short wheelbase vehicles were marketed under the name Seville, and they, too, were mounted on the Oldsmobile 98 chassis. The Seville in ambulance form as seen here sold for $6,800, while a combination version went for $6,685. The funeral car version of the Seville cost $6,575 with limousine styling and $6,715 with landau styling. The landau combination went for $6,830, while the same car as a combination coach cost $6,945. These Oldsmobile based coaches competed directly with the Flxible Flxette and the Superior Consort.

Cotner-Bevington continued to offer a complete line of long wheelbase professional cars on the Oldsmobile chassis. These cars competed directly with the Superior Pontiac line and those extended cars built on the Buick chassis by Flxible. The long wheelbase Cotington series offered a limousine body style in both funeral car and combination versions. The funeral car sold for $8,220, while the combination went for $8,335. Landau body styling was offered on the funeral car that sold for $8,410, while a landau combination cost $8,525. The Cotington series also featured two ambulances, one with a 41 inch headroom for $8,585 and one with a headroom of 46 inches that cost $9,160.

The magnificent Superior series of Crown models was continued through 1964 without any major styling changes. The symbol of the Crown models, a chrome over-the-roof moulding, remained and was supplemented with additional chrome trim and special Crown insignias on the rear door. This is the Crown Sovereign landaulet in three-way form. This year the Crown Sovereign three-way landaulet with manually operated casket table sold for $12,180, while the three-way power assisted version cost $12,745. The straight end-loading model went for $11,100 and the combination for $11,365.

Classified in the Royale series, the attractive and always distinctive Superior flower car cost $11,100 this year. Only 31 were built. This year saw a total of 1,223 Superior Cadillac chassis cars built. This was the most cars on this chassis that Superior had ever built during one model year. Of these, 129 were three-way cars, 247 were end-loading funeral cars, 630 were combination coaches, and 186 were ambulances.

Little changed from last year, the Superior Pontiac long wheelbase professional car line was built in a total of 239 units this year. Of these, 25 were of the end-loading funeral car style as seen here. This landaulet funeral car sold for $8,455, while a landaulet combination cost $8,690. The limousine version of the funeral car cost $8,200 and the combination went for $8,690. There were 147 Pontiac combinations built this year on this wheelbase. This car was a dark gray with a black vinyl roof and a deep red interior.

The Superior line of Royale and Sovereign limousines were very cleanly styled and quite brisk in the sales department. The Royale limousine in straight end-loading form sold for $10,510, while the three-way car with a manually operated casket table cost $11,600 and the same car with a power assisted casket table sold for $12,165. The combination version of the Royale limousine sold for $10,775, while a special service car was also available for $10,390. The same selection of models was offered in the Sovereign line at a slightly higher price.

Seen before delivery to the Kitchener-Waterloo Hospital of Kitchener, Ontario, this Superior Royale Rescue 50 inch ambulance carried a price of $12,013. This particular car was in daily use up until just recently. Other models in the Superior ambulance line for 1964 were the Royale 42 inch at $10,975, the Royale 50 inch for $11,422, and the Royale 42 inch landaulet at $11,135. Ambulances were also available in the Crown series, with the Crown Sovereign 42 inch limousine ambulance selling for $11,394 and the Crown Sovereign Landaulet going at $11,555 and featuring a 42 inch headroom.

The Superior Pontiac long wheelbase ambulance offered this year was built in two versions and 67 of these were constructed. The 40½ inch headroom version cost $8,690, while the 46 inch headroom version like the one seen here was priced at $9,060. The only other model in this line was a service car that went for $7,825.

The production of Superior Pontiac short wheelbase Consort models went up to 210 in this year to establish their all-time production high. Of these 21 were straight end-loading hearses, 141 were combinations like the one here illustrated and a further 47 were ambulances. The limousine funeral car sold for $6,400, while the limousine combination went for $6,520. The Consort was also available with landaulet styling and the landaulet funeral car went for $6,560, while the combination with landaulet styling sold for $6,685. The Consort limousine ambulance cost $6,550.

1965

With the end of the Eureka line, the industry was left with only three major manufacturers building coaches on the Cadillac commercial chassis. These were Superior Coach, Hess & Eisenhardt/S & S, and Miller-Meteor. Other makers actively engaged in the construction of funeral coaches and ambulances on extended chassis included the Flxible Company, National Custom Coaches, and the Wayne Corporation's Cotner-Bevington division. There were, of course, other makers involved in the conversion of station wagons into ambulances and hearses and some makers that built ambulances on truck chassis only, a type of utility ambulance. With the major manufacturers this is what was new for 1965.

Superior again offered four distinctive styling series for the Cadillac commerical chassis models. The Tiara styling option increased the variety of styles offered by Superior and was a new concept in limousine styling. Available in both Royale and Sovereign body styles, this option consisted of a slanted chrome "C" pillar that extended up and over the roof, similar to that seen earlier on the 1959 Crown Royale Laudaulet. All Superior coaches came with three warranties covering both the chassis and the body. The body styling of the 1965 Superior Pontiac line was again based on the Bonneville chassis with new variations created by the Superior designers. A Tiara type chrome roof bow extended across the top rear quarters of the limousine and ambulance models. The short wheelbase Consort line was revised with new interior materials and appointments that reflected a thoroughly professional design.

Crisp beveled styling continued to be featured on the Miller-Meteor line and all models were offered in 23 deep-luster paint colors that were chosen to harmonize with new color co-ordinated interiors. The new interiors of the Miller-Meteor coaches achieved greater width than on previous models with more headroom and increased length in the rear compartment through the use of curved windows and by following the general General Motors styling trend to more bulbous sided cars.

All S & S (Hess & Eisenhardt) coaches were both longer and wider. S & S body design emphasized the roof line with a unique halo roof treatment. The focal point of the new S & S interiors were panels of deep, selected woodgrains framed with stainless steel mouldings. These panels were incorporated into all doors on the coaches, the partition trim, and the sidewall trim.

Cotner-Bevington, now owned by the Wayne group of companies, was not marketed along with other Wayne products, but was offered in two series, both on the Oldsmobile 98 chassis. The Cotington series was available in both landau and limousine styles plus ambulances, and used the Oldsmobile chassis extended to a wheelbase of 150 inches, while the Seville series rode on a 126-inch wheelbase in the same variety of models.

Flxible announced its last group of models in this year. The 1965 Flxible coaches were available in two series, the Premier and the Flxette, all on the heavy duty Buick chassis. The Buick chassis was extended for the Premier coach line to a wheelbase of 149.5 inches, with the Flxette riding on a wheelbase of 126 inches. The company announced the end of funeral coach and ambulance production at the end of the model year and again concentrated on filling large orders of busses. With the Flxible decision to cease production of coaches, one of their salesmen established the Trintiy Coach Company in Duncanville, Texas, to construct Buick chassis funeral coaches and ambulances and built a few prototypes of these in this year.

U.S. production of funeral vehicles and ambulances hit an all-time record this year. The record, which still stands today, saw a total of 4,880 such vehicles roll out of the shops. Of this number, 2,961 utilized the special Cadillac professional chassis.

Flxible built this top-heavy car as a pilot model for 1965 production, and this car was the only one built by this Loudonville, Ohio firm. Flxible instead decided to drop the professional car manufacturing and turn all of their facilities over to the construction of passenger busses.

When Flxible decided to get out of the professional car field in December of 1964, they sold all of their professional car dies and some equipment to a new firm located in Duncanville, Texas. Called Trinity Coach, this infant organization was endeavoring to continue the flow of Buick chassis professional cars. Several prototype models were constructed this year on both the stretched and the regular wheelbase. When production was to finally get under way, the long wheelbase cars were dubbed Royal, while the passenger car wheelbase vehicles were called Triune models.

Cadillac set a new high with their attractively styled 1965 chassis. The commercial chassis was utilized for this magnificent Superior Crown Sovereign Landaulet. The Crown models were again offered in both Royale and Sovereign body styles and both featured a rich interior, chrome rear quarter panels and the traditional crown over-the-roof bow. The 1965 Crown Royale landaulet in standard end-loading form sold for $11,439, while the Crown Royale landaulet combination cost $11,705. Three-way versions of this car were offered in manual form at $12,521 and power assisted for $13,085. All 1965 Superior Cadillac coaches wore totally restyled bodies. This particular vehicle was owned by the Janisse Brothers Funeral Home in Windsor, Ontario.

Superior Cadillac limousines were offered with a novel option called the Tiara roof this year. The Tiara styling consisted of an over-the-roof chrome band at the D pillar and this area was vinyl covered and could be equipped with a small landau bow. This option was offered on both Royale and Sovereign limousines and is seen here on a Sovereign combination. In straight end-loading form the limousine sold for $11,019 in Royale garb. The Royale combination cost $11,118, while the manually operated three-way coach went for $11,939 and the power assisted three-way car cost $12,505.

All Superior Pontiac long wheelbase ambulances were built on a 146-inch wheelbase and had an overall length of 244 inches. They varied in roof heights, though. The 40-inch headroom version, shown here, sold for $9,044, while a 48-inch headroom version cost $9,413. These cars were powered by the famous Pontiac 389-cubic-inch 276-horsepower vee-eight engine and were built on a specially lengthened chassis. This chassis featured a heavy duty transmission, special heavy duty rear shock absorbers and front and rear springs. They featured a special commercial type center bearing and propeller shaft as well as a strengthened frame.

Superior built a total of 256 Cadillac chassis ambulances this year. This is the Royale Rescue ambulance with a headroom of 51 inches. This car sold for $12,354, while the Royale 42-inch ambulance cost $11,361 and the landaulet with a 42-inch headroom sold for $11,476. Note the way in which the car's designers made the high roof line look like part of the car and not an afterthought.

Completely restyled for 1965, the attractive and stylish Superior flower car cost $10,730. In 1965 Superior built 35 flower cars and total production of Cadillac chassis professional cars reached a new high with 1,324 units being built. Of these, 157 were side service coaches, 304 were end-loading funeral cars, 572 were combinations and 256 were ambulances.

All Superior Pontiac coaches were also wearing new styling for 1965. The Superior Pontiac long wheelbase funeral car line measured a full 244 inches from end to end and rode on a 146-inch wheelbase. In this year the Superior Pontiac landaulet funeral car cost $8,805, while a limousine version went for $8,552. Combinations went for $8,709 in limousine form and $9,044 as a landaulet. A total of 290 of these Pontiac coaches were built in this year and of these 38 were straight funeral cars, 127 were combinations and 123 were ambulances.

Sales for the Superior Pontiac Consort professional car line stabilized at 210 again for 1965. There were 17 funeral cars, 150 combinations and 43 ambulance versions of the Consort built this year. The Consort was built on the standard Pontiac chassis with a wheelbase of 123 inches and an overall length of 221 inches. The landaulet funeral car version of the Consort cost $7,064, while a limousine funeral coach went for $6,901. The landaulet combination sold for $7,186, while the limousine combination cost $7,024. Once again there was a service car offered in the Consort line and this sold for $6,861 in this year.

316 # 1965

Hess & Eisenhardt did not lack style on their cars and this fact is most easily demonstrated on the one-off flower cars that were built in this year. S & S did not list flower cars in their regular production schedule, nor were any prices listed for this type of vehicle. Prices and designs were available only upon specific customer requests. This is but one style built by this company in 1965. Another version featured a higher rear flower deck and a very high roof line on the cab. Note the vinyl covered roof and the low flower deck of this distinguished car.

A vinyl covered roof made the 1965 S & S Park Hill by Hess & Eisenhardt one of the true classic looking limousine funeral cars and gave the style unusual elegance. The Park Hill was a derivative of the Park Row limousine and was priced to fit between the Park Row and the Victoria. In straight end-loading form the Park Hill cost $13,397, but when equipped with the exclusive S & S rear extension table, the car sold for $14,006. Three-way versions of the Park Hill were also offered, with the manual version going for $14,527, and the power assisted version selling at $15,243. The Park Hill was offered as a combination with the customer's choice of either Duo-Floor or skeleton type removable casket rollers. The Park Hill combination went for $13,840. Other models in the professional car series consisted of the timeless S & S Victoria that sold for $13,485 to $15,332 in funeral car form. The Park Row offered coaches ranging in price from $12,935 to $14,774.

The short wheelbase Cotner-Bevington models were called Sevilles. These cars rode on a wheelbase of 126 inches and they were designed to compete directly with the Superior Consort. The Seville ambulance sold for $7,165, while the landau funeral car went for $7,093 and the limousine funeral car cost $6,921. Combination versions carried prices ranging from $7,065 for the limousine to $7,237 for the landau. A landau ambulance was also offered and sold for $7,337, while a service car went for $7,093.

Building some of the finest ambulances available, Hess & Eisenhardt offered this as the ultimate in professional emergency vehicles. Called the Professional High Body, this car featured a very high headroom and carried a price tag of $20,355. Other models in the large S & S ambulance range were the Superline Parkway that sold for $13,485 and the Superline Kensington that went for $15,391. The Parkway was a straight limousine ambulance and was derived from the Park Row body shell, while the Kensington offered a somewhat higher headroom roof.

The Seville Combination on Oldsmobile Ninety Eight Chassis

Cotner-Bevington aimed their Oldsmobile based cars at the Superior Pontiac price range and customer group. The ambulance version of the Cotington was offered in both 42 and 48-inch headroom versions. The 48-inch version shown here sold for $9,447, while the 42-inch model cost $8,910. The roof of this car was made of fiberglass to lessen the total weight of the vehicle and improve the handling characteristics.

Cotner-Bevington had become a part of the ever growing Wayne Co. professional car complex. Although they were not marketed through Wayne dealers this year, cooperation had begun with the new owners. The 1965 Cotner-Bevington landau sold for $8,702, while the limousine funeral car cost $8,472. The landau combination sold for $8,824, while the limousine version went for $8,587 and the service car sold for $8,702. Cotner-Bevington was now well entrenched in the field and had built up a strong line of vehicles based on the Oldsmobile chassis. The long wheelbase Cotington models had a wheelbase of 150 inches while being powered by the Olds. 360-horse Super Rocket engine.

1965

Miller-Meteor offered a range of distinctive ambulances in this year. The Paramount ambulance shown here was one of the industry's newest designs in emergency vehicles. With a 42-inch headroom as seen here, the Paramount ambulance sold for $11,406, while a special 48-inch headroom version with special heavy duty springs in the rear sold for $11,793. These cars were offered in either the Paramount or Landau Traditional body styles. Other ambulances in the 1965 Miller-Meteor line included Classic limousine ambulances in both 42 and 48-inch headroom versions. The 42 cost $11,191 and the 48-inch car sold for $11,553.

The distinctive and innovative Miller-Meteor Paramount landau blended the privacy of a landau with the wide visibility of a modern limousine coach. A special size landau bow was designed especially for this model. The Paramount landau was available as a straight end-loading car for the same price as the Landau Traditional, a manual three-way for $12,107, a power assisted three-way at $12,671, or as an ambulance with a large chrome cross in place of the landau bows.

The Freeman Funeral Home of Iron Mountain, Mich., took delivery of a 1965 Miller-Meteor Landau Traditional in this year. As an end-loading funeral car, the Landau Traditional sold for $10,984, while the same car as a combination carried a price tag of $11,276. Miller-Meteor also featured a full line of limousine coaches with prices beginning at $10,764. The limousine with manually operated three-way casket table sold for $11,889, while the same car with a power assisted three-way went for $12,448.

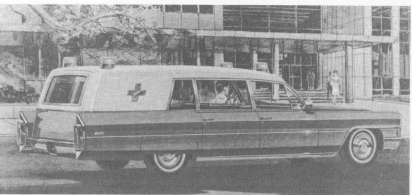

The Automotive Conversion Corp. of Birmingham, Mich., continued to offer outstanding ambulance and funeral car conversions on any station wagon chassis. This is an example of the landau panel funeral coach on a Ford station wagon chassis. These cars were equipped with casket rollers built into the tailgate and a removable skeleton type of casket rack within the car. The landau panel was removable as were the drapes. Amblewagons were also offered as ambulances with frosted glass in the rear quarter windows. This unit was called the DeLuxe Arlington funeral car and was very handy for use as a service car, an ambulance or an emergency coach. This was the type of unit that was killing the market for traditional service cars that were now very expensive.

Another of the interesting vehicles built by the Windsor, Ontario, based Ballantyne Co. was this attractive Chrysler ambulance that was used by the MacLean Ambulance Service of Windsor. All Ballantyne cars were built by craftsman Paul Renaud who was employed by them in this period. This car is seen in front of the offices of Chrysler Canada Limited in Windsor. Only one unit was built.

The most popular vehicle in the Soviet Union, and their most famous, is the Volga. This is an ambulance version of the Volga M22b. This car is based on the Volga station wagon, but has a metal partition between the driver's compartment and that of the patients. It is powered by a 4-cylinder engine with the customer's choice of 70, 75 or 85 horsepower versions. Note that this car features an illuminated ambulance cross sign on the roof and a spotlight on the fender. The car was finished in ivory with only a minimum of chrome trim. The rear side windows were of frosted glass.

1966

"Of course it cost more" was the headline of a Hess & Eisenhardt advertisement featuring the S & S Victoria. The Victoria was built to the most exacting standards to meet the demands of those who have the highest regard for their profession. This snobbish advertising attitude reflected the same thinking that Rolls-Royce traditionally exhibits in their ads. The 1966 Hess & Eisenhardt S & S Victoria continued with the successful styling formula that had been employed since 1938. To back up their claim to cost more, the 1966 Victoria sold for $13,883 as a straight end-loader and for $14,492 with the exclusive S & S extension table for end loading. The Victoria three-way car with a manually controlled table cost $15,015, while the three-way car with a power assisted table sold for $15,730. The Victoria was also offered as a combination with the customer's choice of either Duo-Floor disappearing casket rollers or the removable skeleton type. The combination sold for $14,104.

The Hess & Eisenhardt S & S Park Hill was the medium priced car in the range and the most expensive limousine funeral car offered by S & S. The Park Hill featured a luxury interior and a padded vinyl covered roof. In straight end-loading form the Park Hill cost $13,795, while the end loader with the S & S extension table sold for $14,404. The three-way car with a manually operated table sold for $14,925, while the power assisted version went for $15,641. A combination version of the Park Hill sold for $14,017. The S & S Park Row funeral car line shared a body shell version and exterior panels with the Park Row, but with a somewhat less luxurious interior and without the vinyl roof covering. On the average, the Park Row models sold for $450 less than comparable Park Hill models. S & S ambulances for the year consisted of the Superline Parkway for $13,674 and with a 42-inch headroom, the Superline Kensington with a 48-inch headroom sold for $15,538, and the Custom Professional Parkway ambulance with an even higher headroom went for $20,468.

New marks in quality, prestige styling and functio advancements were found with the introduction of 1966 Miller-Meteor models and, in particular, the lat Miller-Meteor innovation, the Citation. The new Citat was an impressive new coach and was declared the fin of all professional coaches. Featuring distinctive bri mouldings, deluxe landau shields, and new roof t styling, the coach presented a totally new look fo top-of-the-line coach. A gleaming moulding arched acr the top of the roof and swept to the rear of the t sheathing the crisp beveled edges and rear corners. front portion of the roof was finished in a mirror-l gloss which contrasted with the vinyl-covered textu finish of the rear portion of the roof. All of these uni styling items were features of the Citation and exclus on this model. The 1966 Miller-Meteor line was the m complete ever offered by this maker and was availabl a choice of 13 basic models. Interiors on the Mil Meteors were all new, too.

Superior continued to offer four completely differ styling series with all new interiors for the year. extended chassis Superior Pontiac coaches offered styl that reflected the very latest automotive body desi Lines were long and flowing and accentuated w chrome mouldings. The extra length of the Super Pontiac line further enhanced its styling. Every year si its introduction in 1961, the Consort short wheelb Superior Pontiac line increased in popularity. In 19 the Consort continued to ride the crest of that popular with clean modern styling with increased versatility utility.

The S & S models continued to stress a look of dign and refinement with less chrome and more pain surfaces. This approach achieved a rich appearance the 1966 models. The S & S models also feature completely new design in hearse table hardware. formal design, keeping with the crisp, formal look of S & S line of professional cars, the table now featu rollers with bearings made of "oilite" an oil impregna bronze material which provided lifetime lubrication a noiseless operation.

Cotner-Bevington, this year being marketed by Way along side its Miller-Meteor line, again offered its t basic series, the Cotington and the Seville. They c tinued to offer the limousine mounted on the Oldsmob 98 chassis. The Cotington limousine was a roomy vehi offering quiet luxury while accommodating six passeng on both the regular and the folding auxillary sea Superior was offering its Embassy limousine on Pontiac chassis as a competitor to this car and they w well accepted by the funeral directors. Both Superior a Cotner-Bevington could offer the trade matched fleets.

Production this year was still excellent by all st dards, but did not reach the record figures of 1965. S a total of 4,209 hearses and ambulances were built in U.S., with 2,221 of these being on Cadillac chassis.

1966

Seen from the rear and in the Miller-Meteor plant showroom in Piqua, Ohio, is the magnificent Miller-Meteor Citation landau. This car carried a price above that of the Miller-Meteor Landau Traditional. The straight landau sold for $10,044 in rear loading form, while the landau combination sold for $11,338. With a three-way casket table that was manually operated, the landau went for $12,175, while the three-way car with Tri-Matic power assisted casket table cost $12,743.

Miller-Meteor continued to offer the Classic limousine that was available in combination form with removable landau panels that snapped over the rear side windows. In the combination form the Miller-Meteor Classic sold for $11,116, while a straight end-loading funeral car cost $10,823. Three-way versions of the Classic were offered in manually operated form for $11,955 or power assisted versions for $12,518.

The emergency version of the Miller-Meteor ambulance featured a roof height of 48 inches and sold for $11,617 in limousine form. A landau version of this car sold for $11,859 and was offered in either the landau traditional styling or as a Paramount. Note the higher roof line of this car and the significant chrome cross on the D pillar. Another interesting feature is the way in which Miller-Meteor applied the roof side warning lights integrated with a distinctive moulding.

The standard Miller-Meteor limousine ambulance featured a headroom of 42 inches and sold for $11,253. This car was painted a deep Chicago gray with a black roof and the unique "Mars" light on the roof. Note the exclusive Miller-Meteor roof corner warning lights and the unusual fender mounted lights. Miller-Meteor also offered a landau version of the 42-inch ambulance and this car cost $11,469.

All of the Paramount models sold for the same prices as the Landau Traditional models. The distinctive and original Paramount landau featured a small rear side window in the landau area and smaller landau bows. This car was finished in metallic silver with a black vinyl roof and a deep maroon interior. The Paramount was one of the most impressive of the new styled coaches offered to the trade in many years.

Called the Embassy flower car, this attractive Miller-Meteor coach was offered in two versions, with a flower deck height of 22½ or 25½ inches. Both versions sold for $11,510. This is the version with the 22½ inch flower deck. The Miller-Meteor flower car was a beautiful way to lead a cortege. These long low cars appeared in all of their flower decked majesty at the head of a good many processions and forever left an impression of the beauty of the funeral director's service and were thus a large asset to a funeral home's fleet.

The attractive Crown Sovereign limousine also featured the attractive chrome over-the-roof bow that was first seen on the landaulets. This particular car features an additional chrome moulding and a two-tone paint finish. As a straight end-loader the Crown Royale limousine cost $11,503, while the combination version went for $11,776. Once again the three-way cars were offered with manual or power controls for the casket table's operation. The manual version cost $12,619, while the same car equipped with Lev-L-Matic sold for $12,518. With power assistance for the casket table, the car went for $13,198, while with power assistance and Lev-L-Matic the car cost $14,094.

The top of the Superior Cadillac professional car line was the distinctive Crown series offered in both Sovereign and Royale body styles. Here we see a Crown Royale landaulet with all of the chromium dress ups that made the Crown so appealing. The Crown Royale rear-loading landaulet sold for $11,675, while a combination version cost $11,947. Three-way versions were again offered in both manual ($12,784); manual with Lev-L-Matic ($13,682) for leveling the car during loading, unloading, or with a load in the car; power assisted ($13,360) and power assisted with Lev-L-Matic ($14,258).

The standard Royale and Sovereign limousine and landaulet models continued to be Superior's bread-and-butter money makers. The Sovereign models cost slightly more than the comparable Royale versions for which prices are given here. The Royale limousine with straight end-loading sold for $11,074, while the landaulet went for $11,246. The three-way cars with manually controlled casket table went for $12,190 in limousine form and $12,355 as a landaulet. The power assisted casket table three-way cars went for $12,768 as a limousine and $12,932 as a landaulet. All three-way coaches were offered with Lev-L-Matic for leveling the coach. This option cost $900 on the average for any three-way car.

Wearing the beautiful styling inherited last year, the Superior flower car was designated a Model 609 and sold for $11,683. In this year Superior built a total of 1,200 Cadillac professional cars. Of these, 30 were flower cars, 140 were side service hearses, 241 end-loading funeral cars, 541 combinations and 248 ambulances. These Cadillac chassis cars all rode on a wheelbase of 156 inches and an overall length of 249 inches. They held 20 gallons of fuel and were powered by a Cadillac engine that produced 340 horsepower from 429 cubic inches.

Superior Pontiac professional cars on the 146-inch wheelbase continued to wear the styling that they had acquired in 1965. In this year Superior built a total of 270 of these long wheelbase professional cars. The landau was available in both funeral car and combination versions. The straight funeral coach version of this car sold for $8,805, while the combination ambulance/hearse cost $9,044 and featured reversible casket rollers built into the floor.

The 1966 Superior Cadillac Royale Rescue 51-inch ambulance came with all of the emergency equipment that was needed to save lives. (The elephants were not part of the package.) This car sold for $12,611, while a landaulet version with a 51-inch headroom went for $12,007. A straight 42-inch ambulance in limousine form went for $11,550, while a landaulet with the same headroom cost $11,712. Note the illuminated nameplate on the side of the roof and the manner in which the lights are mounted.

Superior Pontiac limousines on the 146-inch wheelbase chassis featured an over-the-roof tiara-type chrome band running from the D pillars. The straight funeral car version of the Pontiac limousine cost $8,552, while the combination version went for $8,790. Other models in the limousine range consisted of a service car for $8,178, a 40-inch headroom ambulance for $9,044 and an ambulance with a 48-inch headroom that sold for $9,413. Of the 270 professional cars built on this chassis, 31 were straight funeral cars, 121 were combination cars, and 118 were ambulances.

Riding on the longest wheelbase of any Superior-built Pontiac, the Embassy 9-passenger limousine was a very dignified vehicle. This car, complete with a padded vinyl-covered roof, was rapidly convertible from a passenger car to an ambulance that would accommodate one stretcher patient and an attendant. Note the 1967 license plate on this car. The photo was most likely being prepared for retouching.

The Consort was also offered as a distinguished little service car. In this form the car carried a $7,288 price tag and was designated as a model 308. In this year Superior built a total of 150 Consorts. This was significantly lower than the previous two years. Of these 150 cars, 15 were straight hearses, 102 were combinations and a further 33 were ambulances.

Cotner-Bevington Oldsmobiles were now being sold along side the Miller-Meteor coaches through Wayne dealerships. The Cotner-Bevington Seville combination seen here sold for $7,363, while a landau version of this car went for $7,532. A straight funeral car version of the Seville in limousine styling cost $7,218, while a landau body style hearse sold for $7,389. Ambulances were offered in both limousine and landau styling, with the landau selling for $7,632 and the limousine costing $7,461. The Seville coaches rode on a 126-inch wheelbase. The long wheelbase Cotington models were mounted on a chassis with a wheelbase of 150 inches and an overall length of 247 inches, an overall height of 69 inches and a width of 79 inches. These cars weighed an average of 5240 pounds and had a fuel tank capacity of 25 gallons. The Cotington landau was offered as a straight funeral car for $8,745, a combination at $8,869, and an ambulance for $9,193. The limousine was available as a straight hearse for $8,512, a combination that sold for $8,628 and as an ambulance with a 42-inch headroom for $8,956 or with a 48-inch headroom for $9,495. Cotner-Bevington continued to offer the ungainly looking Oldsmobile 7-passenger limousine, and this model sold for $7,934.

In ambulance form, the short wheelbase Consort professional car carried a price tag of $7,483. This model was designated as a 66-306, while the combination car carried a model designation of 305 and the funeral coach carried a designation of 304.

A very attractive hearse, this Cadillac funeral car proves that Cadillac is truly the "standard of the world." Built in Germany by the Pollmann Karosseriefabrik of Bremen, this car was one of a very few such cars built on an exported Cadillac commercial chassis. From a styling point of view, it is easy to see that the North American professional car makers were setting the trends for the entire world. Note the over-the-roof chrome moulding and the kick-up over the rear fender. Note the small detail changes around the windshield and the vent, the higher roof line and the massive expanse of glass in the rear quarters.

1967

A wide variety of styling, both conservative and contemporary, were features of the offerings of the manufacturers for 1967. All makers using the Cadillac commercial chassis combined new engineering and styling features of their own with those of the Cadillac commercial chassis. Exterior details and curves were tastefully simple and pure in design and concept.

Again the highlight of the Miller-Meteor line was the Citation with its impressive corona roof trim. This chrome decor strip swept majestically across the top of the car and down along side the roof to the special Citation landau shield, continuing from the base of the shield to the rear corners of the car. Citation landau bows were of stainless steel and bore an embossed olympic medallion in the center. Wide stainless steel mouldings matching the rocker panel trim sheathed the bottoms of the rear fenders to complete the distinctive look. The Miller-Meteor "3-cars-in-one" combination coach came complete with landau decor panels that could be put in place or removed in seconds, converting the car from a landau to a limousine or an ambulance. When used as an ambulance, special ambulance window insignias snapped into the rear windows, and the floor panels could be easily reversed to a perfectly flat floor without casket rollers. A warning light was then attached to the roof. This was common with all combination cars for this era, and holds true to this day. The combination coach is a very versatile and convenient vehicle if only one coach can be afforded.

The styling of the 1967 Hess & Eisenhardt (S & S) line was described as formal but not harsh. The S & S cars of this year were again available in three funeral coach models and three basic ambulance models. Available on all makers' three-way coaches was an optional electric table control, but the S & S version of this option was called Compu-Trol. This table was programmed to eliminate mechanical failure or human error. Compu-Trol was a device that governed the stopping point of the table to such a degree that it assured perfect positioning for the easiest loading and unloading. This table had undergone two years of extensive testing before being announced on the 1967 S & S models as an option. During the testing the device was being continually operated both electrically and manually to the equivalent of 100,000 funeral services without any failures. The device contained a memory counter that always recalled the farthest point of extension and retraction.

Superior again offered its Cadillac line in four separate styling series. In each series there was a complete selection of rear service, side service and combination and ambulance models, as well as a choice of landau or limousine styling. Royale and Sovereign limousine models could be optionally styled as Tiara models with the addition of an across-the-roof moulding at the rear roof quarter. Superior designers had also created all new interior styling with a choice of appointments, convenience features, options and accessories for 1967. The Superior Pontiac line continued to be offered in two wheelbase series and with a full complement of models in each size. The nine-passenger limousine was again offered but for the last time. Called the Embassy limousine, the car had high formal styling that easily identified it as a custom-built car.

The Trinity Coach Company Inc. of Duncanville, Texas, was finally in full production of its line of Buick chassis professional cars. It offered coaches in two wheelbase series and in a full complement of models. Built on the Buick Wildcat chassis, the cars carried the full Buick warranty and the extension was reviewed and given the Buick Division's full okay. The company, founded to fill the gap left when Flxible ceased building Buick professional cars, was formed by G.M. Fulgham, a prominent Dallas businessman, and Joe W. Summers, who was well known in the funeral car and ambulance business and was owner of the Summers Funeral Car Co., one of the largest funeral coach and ambulance distributors in the Southwest. He had been a Flxible dealer and, with the death of the professional coach that Flxible built, had formed a partnership with Mr. Fulgham to build Buick coaches using the basic Flxible design.

Production reached a total of 4,254 ambulances and hearses, with 2,719 being on Cadillac professional chassis.

Timeless in its beauty and dignified styling, the Hess & Eisenhardt S & S Victoria seems to go on and on without any major styling changes, always looking modern and distinguished. These looks adorn the most expensive funeral car in the industry, a car put together with loving care and built in numbers of about 150 per year. In straight end-loading form this car sold for $14,165, while the same car equipped with the exclusive S & S extension table went for $14,783. The three-way coach was priced at $16,026. The power operated three-way cars offered by S & S featured a computer-controlled electrically operated table that was called Compu-Troll. This table was designed to eliminate human error and programmed to do away with mechanical failure, a large disadvantage to a power operated three-way car. Combination versions of the Victoria cost $14,589.

Hess & Eisenhardt flower cars were custom built to specific customer order only. Designs were as innovative as they were fresh and modern, with long, low lines. Access to the rear was through either the rear door or the rear side doors. The rear deck was electrically operated as well as being stainless steel. The entire interior of the rear compartment was lined with stainless steel and casket rollers were built into the floor.

Superior built a total of 300 Cadillac chassis ambulances in 1967 with this 42-inch headroom limousine version going for $11,901, while a landaulet version cost $12,063. The 51-inch headroom ambulance sold for $12,358 in limousine form and the exclusive Superior Rescue 51-inch ambulance cost $12,962. Note the large Federal siren mounted in the center of the roof and the air intake and outlet on the rear portion of the roof. This car also has three gumball type warning lights and dual windshield pillar mounted spotlamps.

The 1967 professional cars were introduced at the N.F.D.A. Convention held in Miami, Florida in October, 1966. Superior used the Miami skyline for the backdrop for many photos of the 1967 Cadillac professional cars. Seen here against the skyline is the 1967 Superior Cadillac Sovereign limousine. Both the Sovereign and the Royale series were offered in landau and limousine body styles, with the Sovereign featuring closed-in rear corners and selling for about $400 more than the comparable Royale coaches. The straight end-loading Royale funeral car in limousine styling sold for $11,689, while a landaulet went for $11,861. The manual side-service car in limousine styling cost $12,805, while the same car with landaulet styling sold for $12,907. The power three-way coaches sold for $13,384 as a limousine and $13,546 as a landaulet. The limousine combination sold for $11,937 and the landaulet combination went for $12,108. All three-way cars were offered with optional Lev-L-Matic for leveling the coach and this option cost $900.

The prestigious Superior Crown Royale limousine is pictured here among the Miami palms in all its splendor. The Crown Sovereign and Crown Royale limousine models were identical in exterior and interior trim with the notable exception of the fact that the Crown Sovereign cars had the rear corner windows blanked in in a brougham fashion, while the Royale coaches had these small windows. Sovereign models cost slightly more than the comparable Royale versions and the following prices are for the Crown Royale versions of the limousine. The ordinary end-load car sold for $13,704 and the three-way power assisted car with Lev-L-Matic cost $14,280. Combination versions of the Crown Royale limousine sold for $11,937 with reversible casket rollers, while the combination with a tubluar casket rack went for $11,937 also.

The symbol of affluence, the Superior flower car, cost $11,869 and in 1967 only 20 were built. The flower car carried a model designation S-608. This year was to see Superior build 1,100 Cadillac chassis professional cars. This figure was down 100 units from 1966. They also built 118 side service coaches, 256 end-loading funeral cars, and 406 combination ambulance/funeral car units.

Superior Consort sales also suffered a withdrawal in 1967 with only 108 of these vehicles being built. The Consort funeral coach cost $7,438, while the combination sold for $7,561 and the ambulance went for $7,588. The landaulet funeral car carried a price tag of $7,604 and the landaulet combination went for $7,727. A service car was also offered for $7,396. Of the 108 Consorts built in 1967, 14 were straight hearses, 62 were combinations and 32 were ambulances as shown here.

The 1967 Superior Pontiac line keynoted styling. These cars were totally restyled for 1967 with new bodies to match the newly styled Pontiac front ensemble. Sales of the full-length Superior Pontiac slipped down to 223 units in this year. The landaulet was offered as a straight hearse for $9,313 or as a combination car with reversible rollers for $9,554.

1967

The top of the Trinity Coach Company's offerings was this prestigious Trinity Royal Landau. Like all Trinity products, the Royal Landau was mounted on a stretched Buick Electra chassis. The Trinity Royal Landau funeral car sold for $9,415, while a combination landau went for $9,315. The Royal series was also offered in limousine body styles with the straight funeral coach selling for $9,250 and the limousine combination selling for $9,150. Trinity also offered ambulances on the Royale 150 inch wheelbase, with a 40-inch headroom version selling for $9,400 and a 42-inch headroom car going for $10,280. Royale models measured 243 inches from bumper to bumper.

Riding on a wheelbase of 126 inches with an overall length of 219 inches, the small Trinity coaches were called Triune. The Triune ambulance sold for $8,100, while a service car version went for $8,050. Both limousine and landau body styles were offered in funeral car and combination versions. The landau funeral car cost $8,115, while the combination version of this body style went for $8,015. The limousine funeral car cost $7,950, while the limousine combination went for $7,850. The fledgling Trinity Coach Co. was located in Ducanville, Texas, and built cars on Buick chassis only.

Riding on a wheelbase of 150 inches and with an overall length of 247 inches, the luxurious Cotner-Bevington 7-passenger limousine continued to offer everything but styling. Included among the standard features of this car were Oldsmobile air conditioning, AM-FM radio with signal-seeker tuner and rear speaker, power steering, deluxe steering wheel, power brakes, Heavy-Duty Double Action Shock Absorbers, Heavy Duty drive shaft and center bearing. All of this and more was included in the list price of $9,487. Folding jump seats could accommodate three passengers while the front seat was power adjusted.

The 1967 Cotner-Bevington models were offered by the Divco-Wayne Corporation of which Cotner-Bevington was a division. Once again these Oldsmobile based cars were offered in both limousine and landau body styles and on two wheelbases. The long wheelbase Cotington models sold for $9,277 as a landau funeral car, $9,296 as a landau combination car. The limousine version of the Cotington series sold for $9,038 as a straight hearse, $9,056 as a combination and $9,348 as a 42-inch headroom ambulance.

Screaming through the night with all lights ablaze, we see the Cotner-Bevington Oldsmobile 48" headroom ambulance. This car sold for $9,923, while a version with a headroom of 42" went for $9,384. Cotner-Bevington, a division of the Divco-Wayne Corporation of Richmond, Indiana, was located in Blytheville, Ark., and was the direct descendant of the old Memphis-based Comet Coach Company.

The small Cotner-Bevington models, called Seville, rode on a wheelbase of 126 inches with an overall length of 223 inches. The Seville series were also offered with either limousine or landau styling and in a full array of models. The landau funeral car cost $7,882, while the landau combination went for $7,959. The Seville limousine cost $7,711 as a hearse, $7,790 as a combination and $7,889 as an ambulance like the car shown here. The Seville models were also offered with a limousine combination with removable landau rear quarter window panels. These panels would quickly convert a limousine style car into one with the landau

The best-selling landau model in the Miller-Meteor stable was this Landau Traditional. In straight end-loading form this car sold for $11,199, while a Duplex or combination version cost $11,548 and featured either a reversible floor or a removable casket rack. The three-way car with a manually operated casket table cost $12,322, while the power assisted three-way coach went for $12,886. Miller-Meteor also offered the Embassy flower car with the customer's choice of flower deck heights for $11,661. These prices also applied to the stylish Miller-Meteor Paramount landau models.

The National Coach Co. of Knightwtown, Ind., offered a line of ambulances based on the chassis of light trucks. These vehicles were modified inside to come up to the expected ambulance comforts. Other ambulance add-ons included special rear side window glass, wiring in the roof for warning lights and special interior details. Here we see a 1967 National Chevrolet ambulance in front of the National Coach's plant in Knightstown.

The Miller-Meteor limousine models were dubbed Classic models. The Classic end-loading coach sold for $10,979, while the combination ambulance and funeral car, called a Duplex at Miller-Meteor, as shown here sold for $11,326. This car was also offered with a manually operated three-way casket table for $12,104 or with a power assisted table for $12,663. Note the squared off roof lines and the handsome lower body lines.

A whole range of ambulance models were offered by Miller-Meteor in varying body styles and roof heights. Here we see the Classic ambulance with a headroom of 48 inches. This vehicle sold for $11,827, while a 42-inch headroom version of the Classic sold for $11,463. A complete selection of ambulances were also offered with either Landau Traditional or Paramount styling. These were available with a 42-inch headroom for $11,679 or as a 48-inch headroom version for $12,069. All of the 48-inch vehicles had special heavy duty rear springs to help cope with the loads that the car was capable of handling.

The Universal Coach Corp. of Detroit converted any Dodge station wagon into an ambulance or a funeral coach in 1967. The standard Polara station wagon could be converted with custom work that included draperies, tailgate protection bar, cot holders and a floor plate. Also included in this attractive conversion package was the customer's choice of roof styles. Universal offered straight limousine styles, a regular landau style and this attractive and unusual roof style that was somewhere between a landau and a limousine. These cars were excellent for use as service cars or as emergency funeral cars and appealed to the small town or limited-finances funeral home. Universal also converted Coronet station wagons to professional car uses.

A popular vehicle in France for almost every purpose is the Citroen station wagon. Here we see one of these unusual looking vehicles adorned with extra chrome trim, a roof mounted flower rack and a central glass partition between compartments. Note the lack of drapes and the four side door styling that seems unusual for a European funeral car. This car was used by Roblot S.A. of Paris, France. Roblot is one of the largest funeral establishments and has branches throughout France.

1968

Style-conscious Superior models on the Cadillac commercial chassis for 1968 were new from the inside out. All models featured new interior designs while the exterior styling was quite similar to last year's. Superior rear service coaches incorporated a new extension table that extended several feet out of the rear door for easier loading and unloading. Combination models featured an accessory switch panel on the roof of the driver's compartment, with every switch at the driver's fingertips. Side service models were available with either Super-matic or Super-manual casket tables that extended out of either side with extra high side doors that provided space for a floral spray on the casket. The Super-Matic electric table could be operated either electrically or manually from either side of the coach. Once again, each of the four Superior styling series offered a full selection of models in a choice of limousine of landaulet styling. The Royale and Sovereign limousines could be styled as Tiara models with the addition of the across-the-roof chrome moulding at the rear roof quarters. The funeral director had a choice of rear side doors that were hinged at the third pillar and opened at the front or at the second pillar and opening at the back. This standard feature was available on all Superior coaches at no extra cost. In flower car models, the customer could choose either a stainless steel power-operated deck or a permanent stainless steel flower deck.

Miller-Meteor also upgraded the interiors, which no featured deep biscuted upholstery, simulated waln insert panels and other luxurious appointments. T exteriors were designed to stay new through the yea according to the press release, but were actually n much different from last year's styling. Miller-Meteor al offered an extending table for its rear loading mode called the ExtEND table that extended a full 33 inch beyond the rear doorway enabling bearers to grasp tl casket handles without crowding each other or reachir in the coach. Roller bearings seated in a precisic aluminum track made smooth operation of the tab effortless. A Miller-Meteor patented feature called Tr Matic was also available to electrically operate the caske table. Like the Superior device, the Tri-matic table woul extend through either side door or through the rear. complete set of controls (forward, reverse, stop) wer located at each front corner at the rear of the casket tabl mound and were level with the doorway. The Mille Meteor line of funeral coaches for 1967 featured th Citation landau, Landau Traditional, Paramount Landau Classic limousine, Classic Landau-Limousine with remov able landau panels, and three way landau and limousine coaches.

The Oldsmobile 98/Cotner-Bevington Super chassis was used for all of the Cotner-Bevington coaches of this year. Featuring Oldsmobile styling with Ninety Eight front and rear ensembles, the Cotner-Bevington coaches were again available in two wheelbases. The Cotington models had the long overall length of 247 inches, while the compact Seville line had an overall length of 233 inches. Both length series were available in landau, limousine and ambulance styles. Ten Cotner-Bevington Oldsmobiles were on 24-hour duty at Montreal's Expo '67 world's fair. Cotner-Bevington is a division of the Divco-Wayne Corporation, now part of the Indian Head group of companies.

Representing artistry in metal, the 1968 S & S models continued with the same overall styling seen last year with some improvements. Some were in the three-way table and included low friction swivel rollers and added ball bearing rollers in the rear extension feature. S & S always had the engineering philosophy of using only the finest materials available and coupling these with the most advanced manufacturing techniques.

Trinity, which had only begun to mass produce cars the previous year, now offered their line on the Buick Electra chassis instead of the Wildcat. Unfortunately, the company could not compete against the industry giants and the Buick professional car went out of production once and for all at the end of the model year. Trinity, with only one full year of production, had failed. Failed to be able to compete with the giants, failed to establish a nation wide sales chain, and failed to sell enough cars to survive.

Production again remained excellent, with 4,216 professional vehicles being built in the U.S. Of these, 2,651 were on Cadillac chassis.

The 1968 Miller-Meteor sales catalogue was entitled "Profiles of Prestige" and this was the general observation one would make with just a glance at their 1968 Cadillac professional car line. In this year the Classic limousine series contained six models with the least expensive being the end-loading funeral car that sold for $11,392. The Duplex combination car cost $11,677 and was offered with either reversible floor type casket rollers or with a removable casket rack. The three-way Classic limousine with manual casket table sold for $12,525, while the three-way with power assisted table went for $13,084. Ambulances were available with headroom heights of 42 inches for $11,814 and 48 inches for $12,178. This photo illustrates the front of a Miller-Meteor Ciatation three-way funeral coach.

1968

The prestigious Miller-Meteor Citation landau was the top of the extensive line of landaus offered by this innovative Piqua, Ohio firm. Other landau models included the Landau Traditional and the Paramount landau series. The following prices apply to the Landau Traditional series and the Paramount Landau versions. The straight end-loading funeral car sold for $11,981, while the combination Duplex carried a price tag of $12,279. The three-way car with a manually operated table went for $13,134 and the three-way electrically assisted version cost $13,713. Also offered in the landau series were ambulances. The landau ambulance with a 42-inch headroom cost $12,413, while the version with a 48-inch headroom sold for $12,805.

The Divco-Wayne Corp., in addition to offering the Miller-Meteor Cadillac coach line, offered the Cotner-Bevington professional car range. This range consisted of two wheelbase size cars. The Cotington rode on a wheelbase of 150 inches with an overall length of 247 inches, while the car's overall width was 79.7 inches and the overall height was 65 inches. Here we see the attractive Cotner-Bevington Cotington landau that sold for $9,810 as a straight funeral car, $9,798 as a combination. This car was powered by the Oldsmobile 98 power plant of 42 cubic inches.

Appearing for the last time was the Trinity Royal series from the fledgling Trintiy Coach Co. of Duncanville, Texas. Offered again with the same specifications as for 1967 and in the same body styles, the Trinity Buick line of professional cars found that they were not being readily accpeted in this crowded market. Trinity started building professional cars in late 1965 utilizing old Flxible dies, but never had enough money to really promote these fine cars. Here we see one of the last Trinity Royal limousine combinations with a padded vinyl roof on the Buick Electra chassis. The Royal was also offered as a landau.

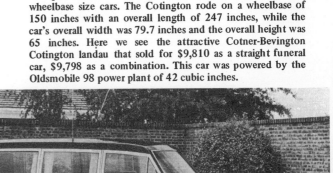

The Cotington limousine series featured the same specifications as those given earlier for the landau. The straight funeral car version of the limousine cost $9,569, while the combination went for $9,558. Ambulances were offered with a headroom of either 42 or 48 inches. The 42-inch version cost $9,886, while the 48-inch model carried a price tag of $10,425.

Once again the Trinity Triune was the maker's entry into the economy professional car field. This car was built on the standard 126-inch Buick passenger car chassis. This was the same chassis utilized for the Electra and was powered by a dependable 360 horse 430-cubic-inch power plant. The vinyl covered roof was optional at no extra cost and this model was the Triume combination ambulance funeral car. The roof mounted gumball type light snapped on or off in a few seconds. Drapes weren't fitted to this car when the photo was taken.

The Trinity Royal ambulance was truly a prestige vehicle designed very carefully to give practical service yet maintain a modern and economical style and price. Unfortunately, this was the last of the Trinity Royal ambulances and, as a matter of fact, the last of the custom built Buick chassis professional cars. The Royal rode on a wheelbase of 150 inches and measured 243 inches from bumper to bumper. These cars were powered by the Buick 360 Horsepower 430-cubic-inch eight with a four-barrel carb. They rode on Premium Firestone 500 tires.

The big news from Superior for 1968 came with revised interiors that were more luxurious and of much higher quaility. Little was done to change the exterior styling and the cars looked quite similar to last year's versions. The Crown Sovereign landaulet was offered in a wide range of versions, with the straight end-loading car selling for $12,135, while the landaulet combination carried a price sticker of $12,511. The Crown models were also offered as three-way cars with the manually operated three-way coach selling for $13,940 and the power assisted three-way version selling for $14,119. Superior continued to offer three-way cars with Lev-L-Matic for leveling the car during the loading or unloading process.

Superior's styling leadership is clearly seen with only a quick glance at the magnificent Crown Royale limousine. This car continued to feature the Crown over-the-roof chrome bow that signified this car's supremacy in the funeral car field. In straight end-loading form this car cost $11,969, while a combination version of this beautiful car went for $12,340. Three-way Crown limousines began with the manually operated Crown Royale limousine at $13,801, while the power assisted three-way limousine cost $13,664. Three-way cars equipped with Lev-L-Matic cost $13,960 with manually operated table and $14,540 with the power assisted table.

The perennial favorite, the Superior flower car, was as attractive for 1968 as it ever had been. This exclusive car was priced at $12,285 in this year and only 22 units were built. Superior production of Cadillac commercial chassis professional cars increased by 50 units in 1968 to a total of 1,150. Of these 1,150 cars, 127 were side-service or three-way cars, 365 were end-loading hearses, 299 were combination cars and an additional 337 were ambulances.

Superior built a total of 337 Cadillac chassis ambulances in 1968. This 51-inch headroom Rescue ambulance was the most elaborate model in the ambulance line and sold for $13,481. A 42-inch limousine version sold for $12,441, while the same headroom car with a landaulet body cost $12,604. The medium priced 51-inch headroom limousine ambulance carried a price tag of $12,877. This vehicle could accommodate four stretcher patients at a time plus one attendant riding in the rear.

The lower end of the Superior price scale was made up by the Royale and Sovereign limousines. Here we see a Royale limousine combination with a two-tone paint finish and optional windshield pillar mounted spotlights. This car sold for $11,549 as an end-loading hearse, $11,909 as a combination and $12,661 as a manually operated three-way coach. The three-way power assisted casket table version went for $13,245 The manual version with Lev-L-Matic cost $13,539, while the power assisted model went for $14,120 with Lev-L-Matic.

Superior again offered their volume sales Pontiac chassis professional cars in both the long wheelbase line of which 240 were built in 1968 and the short wheelbase Consort that saw a total of 105 units built in this year. Of the 240 long wheelbase cars built, 41 were funeral cars, 70 were combinations and 120 were ambulances. Of the 105 Consorts, 16 were hearses, 57 were combinations and a further 32 were ambulances. Here we see the long wheelbase Pontiac limousine combination that sold for $9,504, while the straight funeral coach version cost $9,129. The long wheelbase combination landaulet cost $9,764, while the straight landaulet funeral car sold for $9,308 and the service car went for $8,752. The ambulance with a 40-inch headroom roof cost $9,661, while the version with the 48-inch headroom sold for $10,039. The Consort landau funeral car carried a price tag of $7,728, while the landaulet combination went for $8,005. The Consort hearse sold for $7,579 with limousine styling, while the limousine combination went for $7,840. The Consort ambulance was a car with limousine styling that sold for $7,867, while the service car with blanked off upper rear sides cost $7,537.

1968

The medium priced Hess & Eisenhardt funeral car appeared with limousine styling and under the nameplate Park Hill. This car sold for $14,568 as a straight end-loading car and $15,207 with the exclusive S & S extension table for end-loading. As a manually operated three-way, the Park Hill cost $15,748, while as a three-way coach with the power assisted Compu-Trolled casket table the Park Hill went for $16,496. The Park Hill featured a combination car that sold for $15,018 and was available with the customer's choice of either a skeleton casket table that was removable or the Duo-Floor reversible casket rollers. Park Hill models had a vinyl-covered padded roof and a more luxurious interior than comparable Park Row models.

The elegant lines of the timeless Hess & Eisenhardt Victoria, refined to meet the test of time, continued to be offered through 1968. Lines that had been modified last year with the new Cadillac commercial chassis remained basically unchanged for this year. In straight end-loading form the magnificent Victoria sold for $14,661, while the end-loading car with the S & S extension table went for $15,300. With a manually operated three-way casket table the Victoria sold for $15,841, while the three-way car with the exclusive Compu-Troll power assisted casket table sold for $16,589. A Victoria combination car was offered for $15,111 with the customer's choice of either the removable skeleton type casket table or S & S Duo-Floor reversible casket rollers.

1968 S&S PARK ROW

The exclusive S & S Park Row was the least expensive car in the Hess & Eisenhardt range. This car sold for $14,568 as a straight end-loader, while the version with the exclusive S & S rear extension table went for $14,722. The Park Row with a manually operated three-way casket table cost $15,263, while the Compu-Trolled power assisted three-way version went for $16,011. The Park Row combination sold for $14,533, while ambulance versions of the limousine body style started with the Superline Parkway that sold for $14,648. The Superline Kensington ambulance with a slightly higher headroom roof cost $16,645, while the superb S & S Custom Professional Parkway ambulance that was the last word in emergency vehicles carried a price tag of $22,031.

Building some of the most beautiful flower cars available, Hess & Eisenhardt offered this style of coach in 1968. Flower cars were so expensive and in such a low demand that they weren't even listed in the S & S catalogue for 1968 and prices were available only upon request. Note the roof line and the extremely low flower deck. This photo was taken outside the Hess & Eisenhardt factory in Cincinnati, Ohio.

Built by the McClain Sales and Leasing Co. of Anderson, Ind., these custom crafted Cadillac flower cars were converted from Cadillac coupes. Note the well carried out rear deck of stainless steel and the rakish line of the roof. The car is shown here in front of the McClain showroom.

Troy, Michigan's, Automotive Conversion Corp. continued to turn out excellent conversions of station wagons for professional use. This example, mounted on a Ford chassis, could be used as an ambulance, a private passenger car or as a funeral car. Note the frosted glass in the rear quarter window, and the removable gumball light on the roof. Amblewagon would build these cars on any make station wagon.

1969

With Cadillac's all new chassis for 1969, the long look prevailed with most makers of professional coaches.

The Miller-Meteor line was all new for this year and were deliberately conceived and created to present a measure of excellence that surpassed anything ever before built. All new exterior lines and elegant new interiors highlighted the line's offerings for this year. The increased coach length blended well with new rear body lines which slanted forward a little more and were more rounded at the corners. A larger rear window increased rearward visibility and a larger rear door provided easier loading and unloading. All doors were extra wide and had checks to keep them open. The same model selection offered last year was again available with the same list of options and features.

The new 1969 S & S models were available in three funeral coach series, the Victoria, Park Hill and Park Row, like previous years; but the interiors were now highlighted with luxurious broadcloths, tapestries, and vinyls in a range of color co-ordinated deep tufted style Interior decoration included wood panels of Mountain walnut, striped teak, rosewood, and riced nacora. The rear servicing table on S & S models extended a full 4 inches out the doorway to facilitate loading and unloading. The 1969 S & S ambulances were designed to be literally hospitals on wheels. They included every bit of ambulance equipment recommended by the American College of Surgeons Committee on Trauma. The S & S ambulance was available in three basic models, the Kensington, Parkway and the ultimate ambulance, the Professional High Body.

Providing an all inclusive selection of styling and equipment to satisfy individual preferences and meet particular service requirements, the 1969 Superior Cadillac chassis line was introduced in the now famous four styling series line. The landau models featured a new flush landau bow in anticipation of federal safety requirements forbidding projections of exterior trim more than one inch beyond the body surface. Once again the Royale and Sovereign limousines could be embellished with the Tiara across the roof moulding at the rear quarter that identified them as Tiara models. The 1969 Superior Pontiac models took full advantage of the external and internal design advantages of the wide track Pontiac chassis. New to this line in this year were two full sized landaulets on the Pontiac extended chassis. They were available with or without any extension and provided versatility and convenience with professional performance and spaciousness wrapped up in a compact size. The Consort was also available in a full array of models.

Meanwhile, the Automotive Conversion Co. in Troy, Mich., continued to quietly convert station wagons to ambulances or hearses. Any standard model production station wagon could be completely converted to a funeral service vehicle or an ambulance by simply installing the Automotive Conversion Corporation's Amblewagon unit. It converted wagons to service cars, ambulances, or Arlington hearses through the use of removable landau panels, drapes and special rear flooring. The company had been doing this for quite a few years and continues to convert wagons and panel trucks for service as professional vehicles.

For the fifth year in a row, total U.S. production exceeded 4,000 ambulances and funeral vehicles. The total figure was 4,327, with 3,065 of these vehicles being built on Cadillac chassis.

The new Cotner-Bevington Oldsmobile series of Cotington professional cars rode on a wheelbase of 150 inches with an overall length of 248½ inches. The Cotington was offered in this year, as in past years, in both landau and limousine body styles. The landau funeral car sold for $10,079, while the combination with landau styling cost $10,108. As a limousine the funeral car version cost $9,836, while the combination went for $9,871.

The Cotner-Bevington Oldsmobile Seville series of professional cars rode on a wheelbase of 126 inches with an overall length of 244½ inches. Seville models were also offered with both limousine and landau styling and the limousine funeral car carried a price tag of $8,439, while the same body style as a combination sold for $8,639. Both series of Cotner-Bevington cars were powered by the Oldsmobile Rocket V-8 engine that produced 365 horsepower from 455 cubic inches.

The Opel nameplate is no stranger to most Americans but this large V-8 power Opel, called the Diplomat, just may be. This is an Opel Diplomat chassis adorned with a masterful Pullman hearse body with all of the German characteristics. Note the high roof line, rear window with a cross and a cycus leaf etched upon it and the unusual four side door styling. The Opel Diplomat used the Chevrolet small eight engine and was the GM car that began to give the luxury Mercedes-Benz a new run for the money.

The Hess & Eisenhardt S & S Victoria enjoyed the position of being the industry's most expensive funeral car line again for 1969. With a style as timeless as the funeral car itself, the Victoria reeked with understated elegance. As a straight end-loading car, the Victoria sold for $15,525, while the end-loader with the S & S extension table cost $16,070. As a three-way car with a manually operated casket table the Victoria commanded a $16,630 price tag, while the three-way version with a power assisted Compu-Trolled table went for $17,404. In combination form the Victoria cost $15,874 with the buyer's choice of either the removable skeleton type casket rack or the Duo-Floor disappearing casket rollers.

Occupying the middle of the Hess & Eisenhardt price scale was the dignified S & S Park Hill limousine funeral car line. The Park Hill series featured a more luxurious interior than the Park Row and a padded vinyl roof covering. As a straight end-loading car the Park Hill cost $15,220, while the end-loader with the exclusive S & S extension table went for $15,881. The Park Hill manually operated three-way funeral car sold for $16,441, while the same car with the power operated Compu-Trolled table sold for $17,215. Combination versions of the Park Hill went for $15,685 per issue again with the customer's choice of casket roller types.

At the top of the ambulance range offered to the buyer was the lavish S & S Custom Professional Parkway ambulance. This ultra emergency coach carried a price tag of $22,618 in this year. Some of this car's less lavish relatives were the high headroom S & S Superline Kensington that sold for $17,369 and the straight limousine, low headroom Superline Parkway ambulance that carried a price of $15,302. The Custom Professional shown here was the most expensive ambulance offered to the trade in this year on a Cadillac chassis.

The least expensive coach in the S & S Hess & Eisenhardt range was the Park Row series of limousine professional cars. The least expensive of these cars was the end-loading funeral car that carried a price tag of $14,718, while the end-loader with the S & S extnesion table cost $15,379. The three-way versions of the Park Row began with the manually operated casket table version that sold for $15,939, while the power assisted version with the Compu-Trolled table went for $16,713. For $15,183 the Park Row could be acquired as a combination car with a choice of casket table types. The Park Row, Park Hill series of professional cars utilized the same body shell and shared this with the ambulance line of coaches.

Caught in a funeral procession is this example of an Italian Fiat funeral coach. Note that the flowers are carried inside the casket compartment with the casket and the large expanse of plate glass in this area. The body maker is unknown as is the owner/operator.

A special economy flower car had been built for a good many years by the McClain Sales and Leasing Co. of Anderson, Ind. These cars, built from Cadillac two door coupes, feature specially crafted roof with a rakishly sloped rear window and a stainless steel flower deck mounted over the former passenger car's trunk and where the rear seat was. This flower deck is not powered and the car does not have facilities to carry a casket, but it does make a very attractive looking flower car at a fraction of the cost of one from the major manufacturers.

1969

The Classic Limousine series of Miller-Meteor professional cars featured prices that began at $12,024 for the end-loading car. The Duplex commanded a price of $12,308 with the customer's choice of either a reversible casket roller system or a removable casket rack. As a manually operated three-way coach the Classic sold for $13,156, while a power operated version of the three-way car sold for $13,715.

Miller-Meteor offered four ambulances in 1969 and these cars were available in two headroom versions. As a Classic limousine, the 42-inch ambulance carried a price tag of $12,690, while the 48-inch headroom version commanded a price of $13,055. These same headroom versions were offered with Landau Traditional or Paramount styling. With landau or Paramount styling the 42-inch ambulance cost $12,907, while the 48-inch version went for $13,297.

The Miller-Meteor landau series offered the Landau Traditional, the Paramount and the unique Citation again for 1969. The landau Traditional as a straight end-loader cost $12,310, while the landau Traditional as a combination or Duplex cost $12,530. Three-way versions of this car were offered with prices beginning at $13,376 for the manually operated version and with the power assisted coach costing $14,121.

Appearing for the last time in just this form, the Miller-Meteor Paramount landau would be dropped in both name and styling withh the 1970 models. The distinctive Paramount, introduced in 1963, combined the best features of both the limousine and the landau body styles. The Paramount landau sold for the same prices that the Landau Traditional commanded. Here is a Paramount landau as a combination and with the optional crinkle finish roof. This car was painted a deep brown on the lower half of the car with the roof done in a light sand.

The Wayne Corporation's Chevrolet Sentinel ambulance was designed for action and priced for economy. This vehicle had a headroom of 50 inches and was built on a wheelbase of 127 inches. The overall length of the Sentinel was 215½ inches with a huge 57 x 40 inch rear door opening. These cars were built by the Divco-Wayne Corp. in their Piqua, Ohio, plant. The Sentinel offered chrome bumpers, chrome hubcaps, ambulance insignia decal on the rear quarter window, beacon ray light, and spotlights as accessories at extra cost. The Sentinel was powered by the Chevrolet 307-cubic-inch 200 horsepower engine with the 350 and the 396-cubic-inch power plants being offered at extra cost.

Being phased out of production at Miller-Meteor because of high tooling costs and sales prices and low volume, the Miller-Meteor Embassy flower car was seen for almost the last time. In 1969 this attractive vehicle carried a price tag of 13,562 and featured the same styling that had been used in 1966. The entire rear deck was made of stainless steel and was electrically operated. Note the owner's nameplate on the lower side of the roof.

The Cadillac commercial chassis formed the basis for the Superior Cadillac series of professional cars for 1969. These cars rode on a wheelbase of 156 inches with an overall length of 250 inches. Here we see the magnificent Superior Crown Sovereign landaulet in an appropriate setting. The Crown Sovereign and the Crown Royale were identical with the exception of the fact that the Royale version featured small wrap around rear corner windows. The Crown Sovereign was priced somewhat above comparable Crown Royale coaches. As a straight end-loading funeral car, the Crown series landaulet sold for $12,800, while the combination version went for $13,167. Three-way cars were offered with the manually operated version selling for $13,940 and the power assisted coach going for $14,863. With Lev-L-Matic the manual three-way car cost $14,817 and the power assisted car went for $15,399.

The distinguished Superior limousines were always very attractive vehicles. The standard Royale and Sovereign series made up the bulk of Superior Cadillac chassis sales and these cars were offered in both landaulet and limousine body styles. The straight end-loading limousine cost $12,161, while the same car with a landaulet body commanded a price of $12,327. The limousine combination went for $12,521, while the combination with landaulet styling cost $12,693. The three-way coaches with a manually operated casket table cost $13,307 for the limousine and $13,467 for the landaulet. With Superior's Lev-L-Matic the three-way power assisted coaches cost $14,765 for the limousine and $14,924 for the landaulet. The manually operated three-way cars were also offered with Lev-L-Matic, while the power assisted coaches were offered without this leveling device, too.

Superior constructed 25 more units than they were able to build in 1968 and the total of Cadillac chassis coaches for 1969 was 1,175. Of these a full 24 were beautiful flower cars like the one shown here. This car commanded a price of $12,044 in 1969 and was complete with a stainless steel flower deck that was power operated. Note the nameplates on the front and back door of this particular car. Superior also built a total of 116 side-service coaches, 413 end-loading hearses, 265 combinations and 357 ambulances on the Cadillac commercial chassis in 1969.

Superior built a total of 247 Pontiac professional cars on the 147-cubic-inch wheelbase in this year. This was a full 7 units over 1968 model year's production of this series. Of these 38 were hearses, 75 were combinations and 137 were ambulances. In the landaulet styling shown above, the straight funeral car sold for $9,808, while the landaulet combination carried a price sticker of $10,181 and featured reversible casket rollers built into the floor.

The magnificent Superior Pontiac long wheelbase professional car line measured a full 248 inches from bumper to bumper. With limousine styling, this car, as a hearse, commanded a price of $9,553, while the combination version of the limousine cost $9,928 and also featured reversible casket rollers built into the floor. A standard 40" headroom version of the Superior Pontiac ambulance cost $10,085, while the special high headroom ambulance with a roof height of 48" sold for $10,890.

With a more modern trend sweeping Great Britain and a demand for lower more modern looking funeral cars, the Woodall-Nicholson Company of Halifax offered this attractive hearse on a Ford Zephyr chassis. Up until the spring of 1972, this was still being built. But in that year, Ford ceased production of the Zephyr series and announced a new line of large cars. While this car is longer and somewhat lower, it retains the traditional British funeral car styling. The flower rack is on the roof and the rear side window is long, narrow and devoid of any drapes. The rear door opens upward for loading and the casket deck floor is even with the base of the door opening.

1970

The modern landau evolved with the 1938 S & S Victoria and this car had been regarded as THE most exclusive landau by both industry and funeral directors alike. Carrying a price far above that of any other landau offered to the trade, with quality to match, the S & S Victoria was the car to own. The other makers realized this and without a doubt there was a market for high quality, higher priced, more exclusive landaus for the other makers. This year, 1970, was the year in which the other major manufacturers would announce and market new and exclusive top-of-the-line landau models.

The Miller-Meteor Division of the Wayne Corp., an Indian Head Company, announced the newest line of professional cars of the decade with the widest selection of models in the company's long history. In addition to the Citation, Classic Limousine and the traditional landau funeral coaches, Miller-Meteor announced two new surprise series hearses. The Olympian and Eterna bumped the model offerings for 1970 to 34 basic models. The new Olympian was the most luxurious car in the Miller-Meteor line and boasted a new decor arch moulding across the roof that wrapped around the rear at window sill level. This model also featured new future fashioned landau bows that accentuated the design. The bottoms of the rear fenders were sheathed with wide stainless steel mouldings that matched the rocker mouldings. An exclusive new landau shield further enhanced the luxury look. The interior of the Olympian set a new high in luxury for Miller-Meteor and was in a class by itself. Another newcomer to the Miller-Meteor line this year was the Eterna, that combined the best elements of both the limousine and the landau body styles as the earlier Paramount landau had done so well. But the Eterna featured a rear roof design that suggested a handsome landau, yet also presented a dramatic expanse of glass that resembled a limousine. Like the Olympian, the Eterna had a decor moulding that swept across the roof and wrapped around the rear at sill level. Miller-Meteor also introduced the most comprehensive line of ambulances in its history in this year. Four basic series were offered; the Guardian, Volunteer, Interceptor and Lifeliner were all offered in a choice of 42, 48, 50, and 54 inch headroom heights.

The Superior model line up for this year also offered new and exclusive landau models with one new one. The Crown Limited was introduced as a mid-year model and featured all of the standard equipment of the Crown series plus a new and distinctive landau bow and nameplates with old English lettering. The interiors of the Crown Limited were the most luxurious ever offered by Superior. Other models in the Superior Cadillac line for the year consisted of the Crown series models in both Sovereign and Royal styles and the Royale and Sovereign line of landaus and limousines. The Tiara option was again made available to those preferring this style. The changes to the Superior Pontiac line were modest, but many new features were revealed in the line. The Tiara decor bow was made available on the Pontiac line, and the coaches were available in either Royale of Severeign styling. The Royale series in either chassis always featured the small corner windows in the rear roof quarters while the Sovereign models were fully enclosed in this area.

The 1970 Hess and Eisenhardt (S & S) line of professional coaches were the longest offered to the trade in that year. The S & S cars were two inches longer than the competition because the S & S car was the only maker not using a shortened Cadillac commercial chassis. While the overall style of the S & S line had not changed for this year, they maintained that overall dignity and formality without severity of lines.

In the specialty car scene, the flower car had been losing sales with each succeeding year, and it was in this year that Miller-Meteor built their last flower cars. Hess & Eisenhardt would only build these coaches to special order and only then for prized customers and at an extremely high price. Superior was to be the only company to continue to build these unique and interesting cars, oblivious to the sales figures.

Total professional vehicle production dipped below the 4,000 mark for the first time since 1964 this year, with only 3,808 ambulances and hearses being built in the U.S. Of these, 2,503 were on Cadillac chassis.

Launching a completely new line of professional car lines with all new styling, Miller-Meteor introduced their new top-of-the-line Olympian. This new and exclusive landau was offered in four versions. The straight end-loading funeral car cost $13,714, while a combination Duplex sold for $14,004. The three-way manually operated version of the Olympian cost $14,850, while the three-way power assisted casket table version went for $15,415. Once again Miller-Meteor had started a revolution in the funeral car industry. Soon other makers would introduce new top-of-the-line exclusive landau models.

All of the Miller-Meteor ambulances carried new names for 1970 and new styling to boot. The special emergency squad type ambulance shown here featured a headroom of 54 inches and was called the Lifeliner. This car carried a price tag of $15,230, while the new Interceptor with a headroom of 50 inches cost $14,438. The Volunteer, a 48-inch headroom ambulance, cost $13,687, while the standard low headroom 42-inch Guardian ambulance cost $13,322. Miller-Meteor was still using the unique 360 degree warning lights on the roof and note that this car has one on the front fender. These lights were a Miller-Meteor exclusive and were designed especially for Miller-Meteor's use.

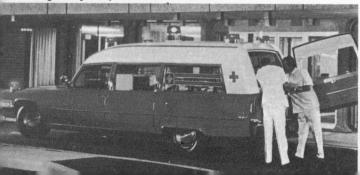

1970

Wearing its traditional styling, the 1970 Hess & Eisenhardt S & S Victoria continued oblivious to the other maker's styling fads and fancies. This car was a timelessly styled coach of enduring quality workmanship. As a straight end-loading funeral car, the 1970 Victoria sold for $16,613, while the same car equipped with the exclusive S & S extension table cost $17,188. As a three-way casket car, the Victoria sold for $17,716 in manual form and for $18,535 as a power assisted Compu-Trolled version. The Victoria combination car went for $16,919 with the buyer's choice of either a removable skeleton type casket rack or the Duo-Floor with reversible casket rollers.

The most expensive ambulance on the market was also one of the best emergency vehicles offered anywhere in the world. Built by Hess & Eisenhardt, this S & S custom Professional Parkway ambulance sold for $23,956 and could handle four patients on stretchers at one time. Other ambulances in the S & S line included the Superline Parkway at $16,263 and the Superline Kensington for $18,446.

For those addicted to the television series *Emergency*, this National Chevrolet ambulance should be no stranger. Built by National of Knightstown, Ind., this vehicle is based upon the Chevrolet truck chassis and features a high roof line made of glass fiber with built-in warning lights. This ambulance was competing in a very rough segment of the market where many body makers also entered their own versions of the truck type ambulance.

The Hess & Eisenhardt policy of not changing styles with any regularity and making a continuous line of consistently high quality products gained them the position of being the most exclusive maker in the industry. Even older model Victorias and Park Hills and Rows did not show their age when placed beside a newer model. This was the Park Hill for 1970. In straight end-loading form this car cost $15,709, while the end-loader with the S & S extension table sold for $16,989. With a three-way casket table, the Park Hill cost $17,517 and if this table was power assisted and Compu-Trolled, the car went for $18,337. Once again the combination versions were offered with the customer's choice of casket table types and sold for $16,719.

Built by the National Custom Coaches Co. of Knightstown, Ind., this ungainly looking vehicle was a first in the ambulance field. Called a Cardiac Ambulance, this vehicle was equipped with all of the known lifesaving equipment needed for the cardiac patient. A special suspension and interior fittings completed the package that was built on a Chevrolet Stepvan 20 chassis.

Entering a new field in professional car marketing, the Automotive Conversion Co. of Troy, Mich., offered this DeLuxe Arlington funeral car, or combination, on a Chevrolet light truck chassis. Amblewagon advertised that this funeral car had features comparable with any other coach at only a fraction of the cost. Amblewagon continued to build ambulance and funeral car conversions on any station wagon chassis as well as offering this interesting vehicle.

Always a styling leader, the Superior Crown Royale landaulet continued almost unchanged from last year. As an end-loading coach the Crown Royale sold for $13,404, while a combination version went for $13,795. The three-way manually operated casket table model was offered in two versions. The straight three-way manual car cost $15,140, while the same car with Lev-L-Matic cost $16,017. Also offered were power assisted three-way cars with or without Lev-L-Matic.

The Superior line of limousine professional cars stood out from the others because of styling that added distinction and elegance to these rather straight line vehicles. This is the 1970 Royale limousine funeral coach with a Tiara rear roof section. As a straight end-loading coach this car sold for $12,765, while a combination version carried a price of $13,140. As a three-way with a manually operated casket table the Royale limousine cost $13,925, while the manual car with Lev-L-Matic went for $14,803. The power assisted three-way cost $14,509, while this car with Lev-L-Matic sold for $15,384. Subtle styling changes distinguished the 1970 models from their predecessors and all models featured a richer type of interior appointments.

The magnificent Superior Rescue 54 ambulance had a headroom of 54 inches and cost $14,769, while the Rescue model with a headroom of 51 inches sold for $14,727. The standard 51-inch ambulance carried a price tag of $14,123, while the regular ambulance with a 42-inch headroom roof sold for $13,672. This car was finished in red and white and had illuminated owner's signs on the roof as well as a complete set of warning lights. Note the large circular warning light on the front fender.

New for 1970 from Superior was this interesting little Ford Econoline ambulance. Designed to add a new dimension to the ambulance field and to capture some of the squad type rescue business, the little Ford ambulances from Superior met with a fair amount of success. They were extremely easy to maneuver through city traffic and economical to operate and maintain.

The large Superior Pontiac professional cars were slightly restyled for 1970. The large cars rode on a wheelbase of 147 inches with an overall length of 248 inches. In 1970 production of these coaches totalled 250, a slight rise for the year. Of these, 43 were straight hearses and 66 were combination cars. The limousine as a straight funeral car sold for $9,945, while the same coach with landaulet styling cost $10,200. As a combination, the limousine went for $10,320, while the landaulet combination cost $10,580. Part of the combination package were reversible casket rollers built into the floor. These could be changed from casket rollers to flush chrome floor pieces in an instant.

Superior built only 23 of these beautiful flower cars in 1970 and those went for $13,588 each. Production of Cadillac chassis coaches at Superior was down slightly for this year to a total of 1,156 units. Of these, 135 were side-service coaches, 399 were end-loading cars, 232 were combinations and a further 367 were ambulances.

1970

The Cotner-Bevington Division of the Wayne Corp., an Indian Head Company, offered their Cotington Oldsmobile professional car line in both landau and limousine body styles. The Cotner-Bevington models were all sprayed with Ziebart rust-proofing to preserve the life of the coach and to make it look like new longer. The Cotington series rode on a wheelbase of 150 inches inches with an overall length of 248½ inches. These cars were powered by the famous Oldsmobile 365 horsepower 455-cubic-inch engine with a Quadrajet carburetor and a Turbo-Hydramatic 400 transmission. The limousine funeral carried a price tag of $10,397, while the limousine combination went for $10,432. The landau funeral car cost $10,641 and the landau combination went for $10,669.

Also offered in both limousine and landau body styles was the 126-inch wheelbase Seville series of Cotner-Bevington Oldsmobiles. These cars measured a full 224½ inches from bumper to bumper and were built on the standard Olds chassis. The limousine body style funeral car cost $9,033, while the limousine combination went for $9,091. The landau Seville funeral car sold for $9,197, while the combination with landau styling cost $9,254.

Abbott & Hast Company of Los Angeles builds these high quality station wagon conversions of any make car. The basic conversion cost $395 in 1970 and included full paneling and painting, chrome plated landau bows, interior side paneling, draperies and one way vision, a stainless steel rear compartment floor plate, and cot holders for two cots. The company also offered a full range of optional equipment such as a vinyl-padded roof, removable one-piece casket rack, nameplates and brackets, a rear door window with a cycus leaf pattern and frosted as to permit one-way vision. Abbott & Hast advertised these vehicles as being ideally suited for removals, flowers, utility, casket deliveries, graveside services, or for baby or child funerals. This unit is on a Ford chassis.

A popular unit with rescue squads and fire departments was the Divco-Wayne Sentinel ambulance mounted on a Chevrolet light truck chassis. These units were becoming increasingly popular with all types of ambulance services and rescue squads across North America. This unit featured economy of operation and maintenance with ease of maneuverability within congested city traffic. Note the raised roof line and the emergency warning lights that were optional.

Offering three ambulances on two wheelbases, Cotner-Bevington covered the field superbly. The 150-inch wheelbase car was offered in two headroom versions. The Cotington 42 sold for $10,722, while the 48-inch headroom version went for $11,262. The 126-inch wheelbase Seville ambulance was offered with only a 41-inch headroom roof and sold for $9,185. Here we see both the Cotington 42 (top) and the Seville 41 (bottom) and can clearly see the length differences as well as the styling similarities.

This is the type of ambulance now used by the Ontario Hospital Ambulance Services and the only type of vehicle that this government body will certify for use in the province. This unit built on a Ford Econoline truck chassis is operated by the Kitchener-Waterloo Hospital in Kitchener, Ontario. It is one of a fleet of about 10 such vehicles operating out of this facility.

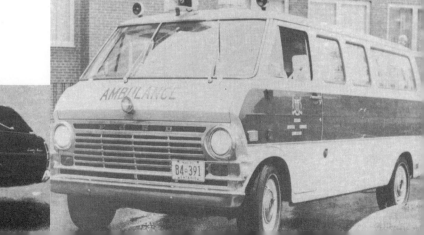

1971

Again for 1971 the S & S (Hess and Eisenhardt) line of professional funeral coaches and ambulances featured elegance and simplicity. They were dramatically new inside and out without looking new. Featuring the latest Cadillac commercial chassis, the 1971 version of the S & S Victoria, Park Hill and Park Row had fewer parts than any other professional car line, making them lighter stronger and more economical to operate and maintain. The styling carried that continuity of design that assured instant S & S identification. With a new, more graceful, slender roof line that blended with the natural body side curvature, and strong distinguishing character lines in the body sides running front to rear, the lowness of the car was emphasized. The rear view of the 1971 S & S was elegantly tailored so that from any angle this year's offering was among the most beautiful of the decade.

The Superior Coach Corporation introduced a new Chateau and Regency model line in its totally restyled 1971 model range. Only the Sovereign and Crown Sovereign names were retained to identify the broad line of limousines and landaulets. The entire line was completely new from the ground up, as was the Cadillac commercial chassis. The new Regency featured the appearance of a limousine with the dignity of a landaulet. A diminuitive landau bow was designed exclusively for the rear roof quarter panel to give an added accent to the sleek new Superior roof design. The new low roof line was a masterpiece of clean design and helped heighten the illusion of overall lowness. At the same time, the new coaches were designed to allow more interior space and larger door openings. The horizontal lines of the coach were highlighted through the use of discreet chrome mouldings at the belt line and above the windows. The new coaches eliminated the vent window in the front window area and thereby cleaned up the overall lines of the coach. The Crown Limited was again offered in limited numbers and available only on special order with special interior trim that was not offered on any other model in the line-up. The 1971 Superior Pontiac line reflected a complete restyling from the wheels up, both inside and out. The completely new vehicles were the creation of the Superior styling and design department. The same type of roof line was carried from the Cadillac models over to the Pontiac line, while the Pontiac chassis was lower overall. The absence of chrome trim on the side of the coach heightened the long, low appearance, and added to its overall dignity and grace of line.

The 1971 model year marked Miller-Meteor's 100th anniversary of funeral coach and ambulance craftsmanship and the company was celebrating with completely new cars. Reported to have been the most impressive and luxurious ever produced by the firm, the new coaches were restyled from the belt line down. The new Cadillac chassis made it necessary to redesign some aspects of last year's new bodies, but they blended in very well with the new chassis. The new models introduced last year were continued without major changes although the interiors were freshly restyled with an ever greater selection of colors and materials. For Wayne's other coach firm, Cotner-Bevington, this was the year of the big styling change. A more stately design with added elegance was bestowed to the stretched Oldsmobile chassis this year. All Cotner-Bevington roof lines were altered and new landau plates added. Landau bows were new almost completely straight with only an olympic crest in their centers. Window area was greatly increased and the cars were lower and longer looking than ever before. The models were still available in two series and two wheelbases in a full selection of body styles.

Total U.S. ambulance and hearse production returned to the 4,000 mark this year, with 4,153 such units being built. Of these, 2,478 were on Cadillac professional chassis.

Cadillac changed their styling radically for the first time since the 1969 models were announced and the bodies of many professional cars were altered to match the new chassis styling. But the traditional styling of the Hess & Eisenhardt S & S Victoria continued without any major changes. In 1971 the straight end-loading Victoria commanded a full $17,759, while the end-loader with the exclusive S & S extension table cost $18,374. The three-way Victoria with a manually operated casket table carried a price tag of $18,938, while the version with a power assisted Compu-Trolled casket table sold for $19,814 and the combination went for $18,087. This Victoria is seen with the new owner taking delivery in front of the factory in Cincinnati.

The S & S flower car was a beautiful vehicle that was built only to special order. Prices were available only upon request. This distinguished flower car would grace the head of any funeral procession with dignity and beauty. To the best of the author's knowledge, this is the only Hess & Eisenhardt S & S flower car built in 1971.

Superior production reached a total of 1,010 Cadillac chassis units in 1971. Of these, 68 were side-service or three-way cars, 450 were straight end-loaders, 15 were flower cars, 202 were combinations and a further 275 were ambulances. The standard 42-inch headroom ambulance, shown here, cost $14,560, while the 51-inch headroom version went for $15,015 and the 54-inch version cost $15,661. Note the retention of the squared-off roof line and the kicked-up body line near the back of the car.

The long wheelbase Superior Pontiac could also be ordered as an ambulance with the buyer's choice of three different headroom models. The 54-inch headroom version shown here cost $12,393, while the model with a headroom of 51 inches sold for $11,747 and the straight 42-inch headroom version cost $11,333. In this year Superior built a total of 153 long wheelbase professional cars and of these, 87 were ambulances, 29 were combination cars and 37 were funeral cars.

The Superior Pontiac Consort was hit with a drastic reduction of sales in 1971 and only 41 Consort professional cars were built. As a straight end-loading funeral car with limousine styling, the Consort cost $9,096, while a combination sold for $9,357. The Consort was offered with landaulet styling in either of these models for $213 additional. Chateau styling cost $160 and Regency styling went for $100 over the price of the basic limousine. In this year Superior built nine Consort funeral cars and 21 combinations.

The Consort ambulance saw 11 units built in 1971 and these sold for a basic $9,420 and had a headroom of 41½ inches. The Consort was built on the standard passenger car length Pontiac chassis and every component and feature of the car was carefully designed and engineered specifically for professional car service. See how the styling of the car follows that of the larger Pontiac and Cadillac chassis cars closely.

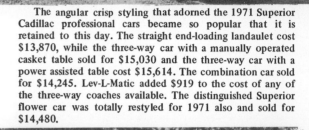

The angular crisp styling that adorned the 1971 Superior Cadillac professional cars became so popular that it is retained to this day. The straight end-loading landaulet cost $13,870, while the three-way car with a manually operated casket table sold for $15,030 and the three-way car with a power assisted table cost $15,614. The combination car sold for $14,245. Lev-L-Matic added $919 to the cost of any of the three-way coaches available. The distinguished Superior flower car was totally restyled for 1971 also and sold for $14,480.

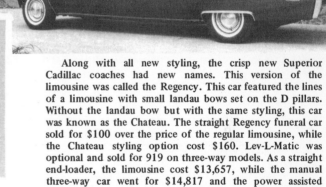

Along with all new styling, the crisp new Superior Cadillac coaches had new names. This version of the limousine was called the Regency. This car featured the lines of a limousine with small landau bows set on the D pillars. Without the landau bow but with the same styling, this car was known as the Chateau. The straight Regency funeral car sold for $100 over the price of the regular limousine, while the Chateau styling option cost $160. Lev-L-Matic was optional and sold for 919 on three-way models. As a straight end-loader, the limousine cost $13,657, while the manual three-way car went for $14,817 and the power assisted three-way cost $15,401. The combination version carried a price tag of $14,032.

Pontiac radically restyled their passenger cars and, thus, the Bonneville chassis utilized by Superior for their professional cars was also restyled. These cars featured the same crisp, angular lines that Superior featured on their Cadillac line. With limousine styling, the funeral car sold for $10,801 and the combination version sold for $11,176. Landaulet styling was available for an additional $213, while the Pontiac models could also be had with either Chateau or Regency styling. The Regency styling option cost $100, while the Chateau option went for $160 above the cost of the limousine.

1971

The 1971 Miller-Meteor Olympian added a new dimensi to professional leadership and the funeral director fortuna enough to acquire one was quickly established as a leader his community. The straight end-loading version of t magnificent Olympian cost $14,520, while the Duple combination sold for $15,616. The three-way manual operated casket table version went for $16,078 and t three-way car with a power assisted table went for $16,64 The Olympian was the leader of the Miller-Meteor range this, their centennial year.

Based upon the beginning of the old A.J. Miller Company, Miller-Meteor, now an Indian Head Company and a Division of the Wayne Corp., celebrated their 100th anniversary year in 1971. The innovative Miller-Meteor Eterna presented a rear roof design that suggested a handsome landau while featuring a dramatic expanse of glass that resembled a limousine. This sculpturing was enhanced by a gleaming over-the-roof moulding that wrapped around the rear of the car. As a straight end-loading hearse this car sold for $13,889, while a Duplex combination sold for $14,249. Three-way coaches were available in either manual or power assisted versions. The manually operated casket table car went for $15,095, while the power assisted version cost $15,660.

The 1971 version of the Miller-Meteor Citation was truly a masterful piece of sculptured metal. The Citation was renowned since its introduction as a regal professional car, though some thought it was a bit gaudy. Notice how the dual roof mouldings complement the slant of the roof sides and the crisp flowing lower body make the Citation look exceptionally long and low. The rear portion of the roof could be finished in either crinkle paint finish or with a deluxe padded vinyl roof insert to contrast the painted gloss of the front section.

The Miller-Meteor Landau Traditional carried on the distinctive landau styling without any of the modern styling accoutrements. This car sold for $13,679 as an end-loader and for $14,039 as a Duplex combination car. The manual three-way coach went for $14,885, while the three-way with a power assisted casket table commanded a full $15,540. The many individual landau styles offered by Miller-Meteor made it possible for the funeral director to clearly express his own tastes in professional car styles.

Based on the fabulous Citroen Safari 21 station wagon this interesting funeral car was built by an unidentified English coach builder. Note that the car has a long narrow rear side window, with a four side door style. Once again the flower rack is mounted on the roof and the back door opens upwards.

The currently popular Woodall-Nicholson Daimler funeral car features modern lines combined with traditional styling and ideas. Note that the casket deck floor is even with the base of the rear side window. The area under the casket deck could accommodate another casket or could be used for carrying chairs or other apparatus. This private compartment is also ideal for use during removals. Once again this English hearse has a roof mounted flower rack. Woodall-Nicholson Limited is located in Halifax, England, and is one of that country's oldest makers of high quality professional cars.

1971

Cotner-Bevington continued to offer Oldsmobile professional cars on two separate wheelbases. The large Cotington models rode on a wheelbase of 151 inches with an overall length of 250 inches. Again offered in either landau or limousine styling, the 1971 Cotner-Bevington models had prices beginning at $11,021 for the limousine style funeral car. The landau style funeral car sold for $11,286 and the landau combination went for $11,595. The combination with limousine styling cost $11,351 and featured reversible floor sections with casket rollers built into one side.

Cotner-Bevington's 127-inch wheelbase Seville series of compact professional cars had an overall length of 226 inches. The standard limousine funeral car cost $9,670, while a combination version with limousine styling sold for $9,890. With landau styling the Seville funeral car cost $9,863 and the combination went for $10,066. The Seville ambulance featured limousine styling and sold for $9,982. This car had a headroom of 41 inches and measured 99 inches from the inside of the rear door to the back of the front seats. The rear door opening of the Seville measured 37 inches high by 45¾ wide.

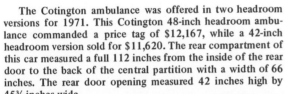

The Cotington ambulance was offered in two headroom versions for 1971. This Cotington 48-inch headroom ambulance commanded a price tag of $12,167, while a 42-inch headroom version sold for $11,620. The rear compartment of this car measured a full 112 inches from the inside of the rear door to the back of the central partition with a width of 66 inches. The rear door opening measured 42 inches high by 45¾ inches wide.

All new for 1971 from Wayne was this attractive little Vanguard ambulance built on a Chevrolet Chevyvan chassis. This car had a rear side door that slid back and was very easy to drive in city traffic.

This unique combination hearse/pallbearer's car is the product of Europe's largest professional car maker, Pollmann Karosseriefabrik of Bremen, Germany. Here this coach is mounted on a Mercedes-Benz chassis and has an illuminated black cross over the windshield to signify a funeral procession. The large rear window has been covered over and the casket would not be visible.

Wayne's popular Sentinel Chevrolet based ambulance continued in production through 1971 with a new and higher headroom roof over the patient's compartment. A quick glance at this car will show that this high fiberglass roof provided plenty of headroom for the attendants and for four stretcher patients. Note the placement of the spotlights on the sides of the roof and the fresh air intake mounted at the rear of the roof.

1972

In a year with minimal styling changes, Miller-Meteor was able to give its cars a new look without a major restyling. Through the development of a new design concept, the company worked on small details that gave the whole coach a new look. The landau bows were stylized through the use of a new olympic medallion motif and the restoration of some of the traditional curve to the old landau iron. The whole bow was stretched to cover the entire length of the landau area, thereby accenting the new extra wide belt mouldings at the top of the rear fenders. This same visual concept was carried through in the windows where a new cathedral valance added to the overall impression of distinction. In the Olympian, this styling concept was emphasized with deluxe landau shields forward of the bows and by carrying the extra wide belt moulding across the roof of the car. This new styling concept was carried on throughout the line and was further expressed with the contouring of the rear fender. The Eterna, Citation, Landau Traditional and the Classic limousine continued in this theme similar to last year's models. The line also included ambulances in 43, 48 and 54 inch headroom heights, all with styling similar to the rear of the Miller-Meteor line.

reputation. Attention was paid to maintaining a classic, ageless style with discreet touches of luxury. The lines of these cars were also enhanced with a totally new landau bow and emblem concept that emphasized the length and clean body lines of the Oldsmobile. Again, the funeral director could specify a choice of wheelbases without appreciable change in exterior styling. The Cotington was built as both a landau and a limousine with an overall length of 251½ inches and a wheelbase of 151 inches. The shorter Sevilles in both limousine and landau styles were on a 127-inch wheelbase with an overall length of 227¾ inches. A complete line of Cotner-Bevington ambulances were also offered in four models in two wheelbases.

Superior made significant engineering improvements along with distinctive styling refinements. A new model was added to the Superior line of funeral coaches on the Cadillac chassis to enhance the styling choices offered. The new model, the Baronet, was a limousine highlighted with a distinctive crest on the rear roof quarters. The Baronet joined the vast Superior styling selection with several other styling types offered in the limousine body style. Both the Chateau and the Regency limousines were carried into 1972 without significant styling revisions. Again heading the Superior line-up was the Crown Limited. To enhance the beauty of this already unique coach, an entirely new interior landau panel was designed, and there was a special emblem on the exterior landau bow. The Crown Sovereign Landaulet also bore the Superior hallmark and featured a gleaming chrome roofbow that ran over the top of this coach.

This complete 1972 Superior line of eight funeral coaches was complimented by a matching flower car with a power-operated stainless steel deck for display of floral pieces. The coupe-type roof of this unique flower car swept into the belt line and flowed through to the rear end of the coach. Superior remained the only manufacturer building these beautiful coaches and continues to retain that distinction to this day.

The latest achievement in the ambulance field emanated from the house of S & S in this year as the S & S Medic Mark I. A mobile emergency room, the Medic Mark I was equipped with all of the emergency equipment recommended by the American College of Surgeons Committee on Trauma. Although the Medic Mark I continued in the styling tradition of the Professional Highbody, the car created a new high headroom ambulance style that did not look top heavy. Other models in the S & S ambulance line included the S & S Parkway, which resembled the regular limousine funeral car.

The dignified and enduringly styled Hess & Eisenhardt S & S Victoria was at its unequalled best in 1972, and that without any major styling changes. As a straight end-loading funeral car the Victoria carried a price of $17,706, while the Victoria with the rear-loading S & S extension table went for $18,313. As a manually operated three-way coach the S & S Victoria commanded a full $19,041, while the three-way car with a power operated casket table carried a price tag of $19,881. As a combination with either skeleton type removable casket rollers or with a disappearing set of casket rollers built into the floor, the Victoria cost $18,029.

THE S&S PARK HILL

The superb engineering and construction built into Hess & Eisenhardt, S & S professional cars resulted in value that endured for the full life of the vehicle. This S & S Park Hill made up the medium priced entry and the most expensive S & S limousine offered. As a straight end-loader this car cost $17,370, while with the extension table this car went for $18,100. The Park Hill was also available as a three-way coach. The manually operated version cost $18,690, while the power assisted model with a Compu-Trolled casket table commanded a $19,529 price tag. The Park Hill combination with the buyer's choice of casket roller types sold for $17,817. The V-shaped rear door and rear door window conform to the general shape of the Cadillac rear bumper. A distinguishing feature of the Park Hill over the Park Row is the fact that the Park Hill had a roof covered with padded vinyl and more luxurious interior fittings.

1972

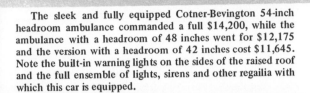

The Cotner-Bevington Seville models rode on the standard Oldsmobile Ninety-Eight chassis with a wheelbase of 127 inches. Once again the Seville series was offered with either landau or limousine styling. This Seville landau cost $9,905 as a straight hearse and $10,097 as a combination. The limousine body style sold for $9,728 as a hearse and $9,921 as a combination. The limousine ambulance had a headroom of 42 inches and commanded $10,024.

The sleek and fully equipped Cotner-Bevington 54-inch headroom ambulance commanded a full $14,200, while the ambulance with a headroom of 48 inches went for $12,175 and the version with a headroom of 42 inches cost $11,645. Note the built-in warning lights on the sides of the raised roof and the full ensemble of lights, sirens and other regailia with which this car is equipped.

The long wheelbase Cotington was also offered as a sleek limousine style professional car, as seen here. All of the 1972 Cotner-Bevington Oldsmobiles were powered by a 455-cubic-inch Olds Rocket V-8 coupled with a Turbo Hydramatic transmission. The Cotington models rode on a wheelbase of 151 inches and featured the "big car" styling that Oldsmobile had become famous for. The Cotington limousine cost $11,061 as a hearse and $11,369 as a combination car.

Set in a luxuriant field of daffodils is the 1972 Cotner-Bevington Cotington landau on the Oldsmobile chassis. This car as a straight funeral car sold for $11,196, while the landau body style when offered as a combination car cost $11,613. This car finished in a deep blue with a black crinkle type roof makes a distinguished looking coach. White side wall tires were now standard equipment.

Volga GAZ is the most popular passenger car in the Soviet Union. The Volga GAZ-22E is the ambulance version of this popular car. Riding on a wheelbase of 106 inches, this car is powered by a special ambulance roof light, spotlights and a red cross insignia on the doors.

The Wayne Corp., an Indian Head Co. of Richmond, Ind., offered a line of special ambulances based on Chevrolet truck chassis. The Sentinel was mounted on a Chevrolet 20 panel truck while the Vanguard, shown here, was based on the Chevyvan. The Vanguard 60 featured a headroom of 60 inches and had sliding rear side doors and a raised roof. These cars were very popular with big city ambulance squads and those that operated out of fire halls and rescue squads.

Offering a large selection of ambulances in 1972, Superior built a total of 281 such Cadillac units in this year. This 54-inch ambulance topped the Superior ambulance offerings for this year and sold for $15,408. These cars rode on a wheelbase of 157.5 inches with an overall length of 254.22 inches. Other ambulances offered in the Cadillac line were the 51-inch headrrom model that went for $14,755 and the straight limousine ambulance with a 42-inch headroom that sold for $14,328.

Superior continued to offer the distinguished Crown Limited landaulet in 1972. This rather exclusive coach was offered in three versions in this year and these cars were aimed at the top of the customer segment. As an end-loader the Crown Limited sold for $13,877, while the three-way car with a manually operated casket table cost $15,012 and the automatic, power assisted three-way was priced at $15,112. The attractive Crown Limited has curvaceous landau bows that were finished in black, exclusive interior appointments and special lettering for nameplates.

Designated a Model 609, the Superior flower car continued to carry styling introduced on it in 1971. This car sold for $14,224 and in this year only 26 were built. These flower car units represented a notable production rise for this model over previous years, and was the highest number of such units built by Superior since 1966. On the whole, Superior production was very good for 1972 with a total of 1,130 Cadillac chassis units being built. Of these, 88 were three-way cars, 534 were end-loaders, and 175 were combinations.

Sitting in front of a Lima, Ohio, fire hall is the 54-inch headroom version of the 152 inch wheelbase Pontiac ambulance. This car, the top ambulance offering in the Pontiac line, sold for $12,111, while a verison with a headroom of 51 inches went for $11,461 and the standard 42-inch version cost $11,068. Note the styling similarity to the high headroom Cadillac chassis ambulances.

The Superior Crown Sovereign limousine offered new styling touches such as the vinyl-covered rear portion of the roof and full formal draperies. The Crown limousine cost $13,877 as an end-loading car, $15,012 as a three-way manual car, $15,579 as a three-way power assisted coach and $14,252 as a combination car. Superior, always a style setter, also offered optional trim packages for the Sovereign limousines. The Chateau trim package cost $167, while the Regency styling package went for $104. This was above the price of the standard limousine. Another innovative styling trim package offered in mid-1972 was the Superior Baronette. This package added to the straight limousine featured a large chrome crest mounted on the "D" pillar that included the owner's initials.

The Superior landaulet, long one of this maker's most popular styles, was again offered in a full range of versions. The straight end-loading landaulet cost $13,621, while the combination version sold for $13,985. Three-way versions of the high-selling landaulet began with the manual version that sold for $14,325, while the three-way car with a power assisted casket table cost $15,323. This car was finished in silver metallic with a black crinkle type roof and dark blue draperies and interior appointments. Note the styling in comparison to that of the Crown Limited.

1972

Production of the 152-inch wheelbase Superior Pontiac professional cars reached a high in 1972 of 259 units. This was the most long wheelbase Pontiac cars Superior had built since 1965. Once again these cars were offered in either limousine or landaulet styling, and here we see a limousine combination with the removable landau panels in the rear quarter window. As a limousine, the straight hearse cost $10,549, while the combination with limousine styling went for $10,914 and featured reversible casket rollers built into the floor. Chateau styling was an optional trim package that sold for $167. The Regency trim went for $104. The landaulet versions of these coaches cost $208 over the limousine style. Of the 259 cars built in 1972, 40 were hearses, 58 were combinations.

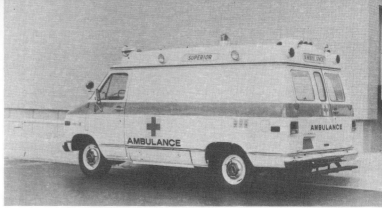

Superior's 61 ambulance was mounted on a Chevrolet G-30 chassis that was designed to meet or exceed ambulance design criteria and recommended specifications. This car was notably wider than the standard Chevyvan and measured a full 93 inches in overall width with a rear door opening of 68.4 inches. This car measured 206 inches in overall length and rode on a wheelbase of 125 inches. It has a headroom of 61 inches and an overall height of 91.2 inches. Superior Coach is a division of the Sheller-Globe Corp.

The distinctive Miller-Meteor Eterna, a car that combined the luxury look of a landau with the light airy look of a limousine, continued through the 1972 model year without any major change. For 1972 the straight end-loading version of the Eterna sold for $13,899, while the Duplex went for $14,257 and the three-way car with a manually operated casket table cost $15,095. The most expensive Eterna was the three-way car with a power assisted casket table and this car sold for $15,653. Other landau models in the 1972 Miller-Meteor range included the distinctive Citation, the top-of-the-line Olympian and the previously shown Landau Traditional.

Miller-Meteor's top ambulance offering for 1972 was the Lifeliner with a headroom height of 54 inches. This car was fully equipped and sold for $16,347 and could accommodate four patients on stretchers plus two attendants. The 48-inch headroom ambulance was called the Volunteer and carried a price tag of $14,573. The regular low headroom ambulance shared the same body shell as the Classic limousine with a headroom of 42 inches. This car cost $14,203.

Owned by the Futher Brothers Funeral Home of Wellesley, Ontario, and leased to the Gruhn Funeral Home of Kitchener, Ontario, is an attractive 1972 Miller-Meteor Landau Traditional three-way funeral car. As a straight end-loader this car went for $13,690, while the Duplex combination with this styling went for $14,047. As a manually operated three-way car, the Landau Traditional sold for $14,887, while the power assisted verison carried a price tag of $15,443. This car, built originally for a fleet, was brought to Canada and is equipped with a three-way manual table, air conditioning, and a full formal interior. It is but one of three almost identical such cars from the same fleet that operate in the Kitchener-Waterloo area.

The Miller-Meteor limousine offering for 1972 was once again dubbed the Classic Series. This car, with smooth lines and airline type draperies, was finished in a metallic blue with a dark blue interior. As an end-loading car the Classic limousine cost $13,487, while the Duplex or combination version of this attractive limousine was offered with reversible floor mounted casket rollers or with a removable skeleton type casket rack and sold for $13,827. The three-way car with a manually operated casket table carried a price tag of $14,747, while the power assisted or Tri-Matic version of the three-way commanded a price of $15,218.

1973

The current crop of funeral coaches and ambulances are the heirs to a heritage of over 100 years of hand-crafted coachwork and remain the sole contributors of hand-built, high quality carriage work to the automotive industry. Thus far we have traced this heritage for 73 years and shall wrap up this story with a look at this year's offerings.

Superior Coach Corporation, celebrating its 50th anniversary in the coach building field, offers a line of beautifully distinctive coaches for 1973. The vehicles on the Cadillac commercial chassis display long, low exterior lines with elegantly appointed interiors. The Crown Sovereign Landaulet, the finest coach in the Superior line, is a study in understated beauty. This vehicle is crowned with a rich chrome roofbow, and the interior sets a styling pace with appointments worthy of this unique coach. Its hallmark as the top of the Superior line is its landaulet bow with exclusive crown emblem integrally worked into the design. The 1973 Sovereign landaulet completes the line with gleaming chrome textures and colors and a well-appointed interior. The leading limousine style is seen in the Crown Sovereign limousine, which also wears the unique crown roof bow. This model is followed by the Sovereign limousine, Chateau, Regency, and the Baronet, a coach style introduced last year. Complementing and completing the comprehensive Superior Cadillac line is the 1973 Superior flower car, which matches the other models in the Superior line.

The Superior Pontiac line of funeral coaches, offered on the Pontiac chassis in two wheelbases, offers the funeral director a versatile selection of models designed with sleek profiles and rich interiors. Continuing to be the most popular vehicle in this line, the full-length Sovereign landaulet features luxurious interior appointments that are unique in this price class. The Consort, the short wheelbase model, continues to be available in both landaulet and limousine form while being built on the standard Pontiac chassis without any extension. The Consort landaulet features an unusually efficient design with professional styling.

The 1973 Miller-Meteor and Cotner-Bevington funeral coaches and ambulances place heavy emphasis on quality of design and craftsmanship. The Wayne Corporation, the only manufacturer producing coaches on both the Cadillac and the Oldsmobile chassis, introduced a broad selection of funeral cars and ambulances for 1973. The Miller-Meteor models feature totally new interiors with woodgrain inserts in the doors and partition. An exceptionally handsome "V" motif and deeply embossed upholstery details distinguish the interiors of the new Miller-Meteor line of funeral coaches. Landau models are offered in four distinct body styles, the deluxe Olympian with a wide over-the-roof chrome moulding, the unique Eterna with a combination of landau and limousine styling, the Landau Traditional with that elegant simplicity of design, and the beautiful Citation with paired parallel mouldings that enhance the roof. The limousine design of the Classic is ideally suited to the modern treatment of the Miller-Meteor interiors and gives full view to the casket and floral tributes. Cotner-Bevington models are available in four body styles in the funeral coach line and in two separate wheelbases. The big car styling of the line, featuring the Oldsmobile 98 chassis, is further enhanced by greater overall length that accentuates the clean-flowing lines of these coaches. In addition, all Cotner-Bevington models feature the famous Oldsmobile G ride system and are powered by the 455-cubic-inch Olds Rocket V-8 engine coupled with Turbo-Hydramatic transmission.

Representing a fresh approach to professional coach perfection, the 1973 S & S line features Hess & Eisenhardt's advanced styling concepts while making even more engineering progress. Each car is given meticulous handcrafted care, as the company will not automate fully. Again in 1973, the S & S line offers the most usable length of any coach built on the same commercial chassis. The side body metal sweeps its full length with the roof blending gracefully with the body side curvature. One of the unique features found only on S & S coaches is the distinctive bent V metal and glass shape of the rear door that blends with the overall concept of the Cadillac chassis and design.

The S&S Park Hill

The fresh approach to perfection is seen in the Hess & Eisenhardt S & S Park Hill limousine professional car. The Park Hill represents the medium price range for this manufacturer. All Park Hill models feature a padded vinyl roof covering and luxurious interior appointments.

The least expensive coach in the Hess & Eisenhardt S & S range continues to be the Park Row limousine funeral car. Every detail of S & S engineering reflects nearly 100 years of experience in accommodating the funeral director's needs. Every detail of the construction is meticulously executed and inspected to assure the owner of years of virtually maintenance-free service. The inset photo shows the exclusive Compu-Trolled three-way casket table utilized by S & S. This table functions precisely at the touch of a lever. Note the finish of the table and the interior appointments.

The S&S Park Row

1973

Miller-Meteor reached new horizons in leadership with their 1973 line of professional cars. The 1973 Miller-Meteor Olympian is the most dignified and luxurious funeral car ever constructed by this Piqua, Ohio, firm. The gleaming chrome roof band and the impressive landau shield quickly distinguish this as one of the most attractive professional cars on the market today.

The Hess & Eisenhardt Medic Mark I is virtually an emergency room on wheels. This one car is equipped with every device that can possibly be accommodated to save lives under all circumstances. The high headroom and the functional styling make this the most expensive ambulance on the Cadillac chassis built by any maker in North America.

Special smaller landau bows have become an integral part of the styling of the unique Miller-Meteor Eterna professional car. Today this car remains the best combination of the best qualities of both the landau and the limousine body styles. The Eterna fills the medium price gap in the line-up of Miller-Meteor landaus.

Like a circle, the limousine body style may continue on forever. These cars never look out of place and they maintain a special quality that cannot be captured by a landau. This Miller-Meteor Classic limousine captures all of the dreams that the original designers of this body style had in mind in the early 1920's — grace, beauty and dignity.

Continuing in all of its regal splendor, the Miller-Meteor Citation is crowned with dual chrome roof mouldings and a special roof treatment that is distinctly Miller-Meteor. The rear portion of the roof is covered with vinyl or finished in crinkle type paint. This distinguished car, first introduced in 1966, may well live on to be regarded as a classic in its own time. The 1973 Miller-Meteor Paramount line of professional cars may also be considered classics one day.

The Wayne Corp. also offered a complete range of van type ambulances mounted on either the Chevrolet or the Dodge chassis. This is the Wayne Medicruiser powered by Dodge and custom built for ambulance work. The Wayne Vanguard seen in earlier chapters is Chevrolet powered and is based on the Chevyvan 20.

The extensive range of Miller-Meteor ambulances includes the special rescue, high headroom Volunteer with a 48-inch headroom. Other cars in this line of high quality ambulances include the Guardian with a 43-inch roof height and the Lifeliner with a 54-inch headroom. These cars ride on a wheelbase of 157.5 inches with a wide stance of 65 inches.

Celebrating their 50th anniversary with the most attractive line of professional cars ever built, Superior offers a complete line of cars again for 1973. Always ahead with styling, this 1973 Superior Crown Sovereign limousine is sculptured in the tradition of 50 years of professional car heritage and finished with the unique beauty of traditional Superior styling. The Crown Sovereign is also offered as a landaulet. These models add the crowning touches to 50 years of professional car craftsmanship.

The distinctive and beautiful Superior flower car holds the esteemed position of being the only production flower car offered by any maker. Its unique Superior profile imparts a new aura of tradition to any fleet of professional cars. The styling is distinctly Superior and the car carries the traditions of Superior flower cars since 1948.

The expert blend of tradition and practicality are seen in the 1973 Superior Sovereign limousine. This car features all of the design thinking that has epitomized Superior styling over the past 50 years. The sleek, low richness of the limousine body combine to make this one of the truly sought-after professional car beauties of today.

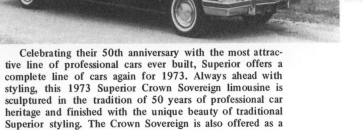

The Superior Pontiac Sovereign limousine is a coach that combines efficiency and performance with beauty and grace. Sculptured in the tradition of Superior, this car is also offered with a distinctive landaulet bow that adds a unique touch to the limousine styling. This styling option is called the Regency. Another trim option consists of a silvery coronet that runs over the roof and highlights the rear portion of the roof. This option is dubbed the Chateau.

The most popular Superior Cadillac body style has the matchless grace of Superior styling and meticulous attention to all of the engineering and construction details that make any coach a value over the years. The landau shield and the area over the side windows has imitation woodgrain inset to add to the distinctive styling. This car is finished in black with a white vinyl covered roof and off white draperies.

Unsurpassed for value and dollar-for-dollar service, the 1973 Superior 54-inch high headroom ambulance on the Cadillac commercial chassis represents yet another progressive step in Superior's 50 years of progress. Superior also offers a line of ambulances with 51 and 42-inch headrooms on the Cadillac chassis. The Superior ambulances offer a selection of seven interior arrangements.

With headrooms of either 51 or 54 inches, the Pontiac ambulances on the 152-inch wheelbase chassis offer a complete selection of interior floor plans from which to select. This 54-inch ambulance features an illuminated ambulance roof sign and built-in tunnel lights. Note the rear bumper step and roof-mounted air extracter.

Continuing to be the most popular vehicle in the Superior Pontiac model range is the long wheelbase landaulet. This car features Superior performance enhanced by exquisite styling. In the medium price range this car represents a true summation of dignity and decor. Once again the landau panel is finished in imitation woodgrain.

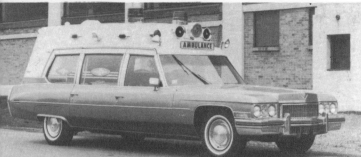

1973

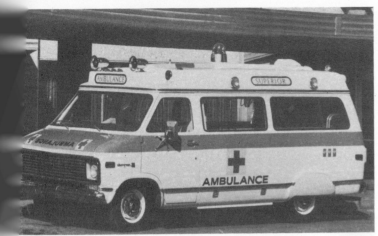

The versatile Superior Consort limousine is offered as a straight funeral car, a combination and as an ambulance, as seen here. This car was designed to meet the needs of the modern funeral director and is placed in the medium price range.

Wearing fender skirts and a full green house of windows, the 1973 Superior 61 Ambulance is built on a Chevyvan G-30 chassis. This car combines maximum headroom and elbow room with the widest back door opening on a van type ambulance. This car features the Superior designed steel-supported fiberglass roof that offers maximum headroom and protection with minimum weight.

Dignity at a low price has always been the goal of the Cotner-Bevington line of prestige professional cars on the Oldsmobile chassis. This distinguished Cotner-Bevington Cotington limousine rides on a 151-inch wheelbase and has all of the luxury qualities that every Olds Ninety-Eight features. This version is the combination.

The Consort by Superior is a fully professional vehicle on the standard Pontiac wheelbase chassis. This rich looking landaulet carries the traditional Superior styling and has assured performance. Note the almost straight landau bows and the rich warmth of the imitation woodgrain inlaid landau shield.

The Cotner-Bevington Seville series of short wheelbase professional cars makes up the least expensive segment of the Olds professional car range. These cars ride on a wheelbase of 127 inches and are again powered by the Olds 455 V-8. This limousine Seville is offered as a hearse, a combination and an ambulance with a 42-inch headroom.

The distinguished economy coach, the 1973 Cotner-Bevington is a direct descendant of the Comet Coach Company that was established in Memphis in 1955. The name was changed to Cotner-Bevington in 1960 and in 1965 the firm was purchased by Wayne. As a division of Wayne, they, too, are considered an Indian Head company.

The Cotington landau also rides on the 151-inch wheelbase and offers the Cotner-Bevington buyer a more dignified type of styling on a low priced chassis. These cars are powered by the Olds 455-cubic-inch V-8 and feature Oldsmobiles famous G-ride system.

NORTH AMERICAN HEARSE & AMBULANCE MANUFACTURERS, 1900-1973

The following roll call of U.S. and Canadian funeral car and ambulance manufacturers includes all firms known to have produced at least one funeral car or ambulance, whether in its entirety or as a conversion of some other vehicle. An asterisk (*) denotes firms still engaged in this activity in 1973.

*Abbott & Hast, Los Angeles, Calif. (station wagon conversions)
A.B. Greer & Son, London, Ontario, Canada
Able Coach Co., Buffalo, N.Y. (station wagon conversions)
Acme Motor Co., Sterling, Ill. 1950-1955 (Pontiac chassis only)
A. Geissel & Sons, Builders, Philadelphia, Pa.
A.J. Diefenderfer Corp., New York, N.Y.
A.J. Miller Co., Bellefontaine, Ohio, Est. 1853 (became Miller-Meteor, 1957)
Albert E. Lattimore Co., San Francisco, Calif. 1897-
Alden-Sampson Motor Truck Co., Division of U.S. Motor Co., Detroit, Mich.
Alliance Manufacturing Co., Streator, Ill.
Ambulance Imports, Inc., Warsaw, Ind. (imported Mercedes-Benz ambulances)
Anchor Body & Top Works, San Francisco, Calif.
Auburn Automobile Co., Auburn, Ind. (built hearses, ambulances 1935-1936)
Autocar Co., Ardmore, Pa.
Auto Hearse Manufacturing Co., Bridgeton, N.J. (J. Paul Bateman)
*Automotive Conversion Corp., Troy, Birmingham, Mich., Port Credit, Ont.
August Schubert Wagon Co., Oneida, N.Y.

Banks Motor Corp., Louisville, Ky.
Bateman Body Co., Vineland, N.J.
Bender Body Corp., Cleveland & Elyria, Ohio, 1938-40 Studebaker, Cadillac
Bernard Arntzen, Inc., Chicago, Ill. 1938 sedan hearse conversions
Biehl's Wagon and Auto Body Works, Reading Pa.
Blue Ribbon Auto & Carriage Co., Bridgeport, Conn.
Bradley Motor Co., Chicago, Ill.
Brantford Coach & Body Ltd., Brantford, Ont. Canada (former Canada Carriage)
Bridgeton – Bridgeton, N.J.
Brownell-Burt, Inc., Taunton, Mass.

Canada Carriage & Body Co., Brantford, Ont. Canada (became Brantford Coach)
Cadillac Motor Car Division, General Motors Corp. (built hearses 1926-1931)
Carmichael Motors Ltd., Tillsonburg, Ont. Canada (conversions, 1947-1951)
Champion Wagon Co., Owego, N.Y.
Cole Motor Co., Indianapolis, Ind.
Cole-Stratton Co., New York, N.Y.
Columbia Body Co., Detroit, Mich.
Comet Coach Co., Blytheville, Ark. 1955-1959 (became Cotner/Bevington)
Continental Memorial Coach Sales Corp., Cleveland, Ohio, 1963
Copple Auto Works, Los Angeles, Calif.
*Cotner-Bevington Division, Miller-Meteor, Blytheville, Ark. (Oldsmobile)
Crane & Breed Manufacturing Co., Cincinnati, O. 1850-1924 Winton chassis
Crow-Elkhart Motor Co., 1911-1923
Cynthiana Carriage Co., Cynthiana & Covington, Ky.

Davis Funeral Car Co., Raleigh, N.C.
Derham Body Co., Rosemont, Pa.
Des Moines Casket Co., Des Moines, Iowa
Dietrich, Inc., Detroit, Michigan 1925-1933
Dominion Manufacturers Ltd., Toronto, Ontario, Canada 1921-
Dorris Motor Car Co., St. Louis, Mo.

Economy Coach Co., Memphis, Tenn. 1950-1955
E.M. Miller Co., Quincy, Ill. 1856- (built Miller-Quincy vehicles)
Eureka Co., The, Rock Falls, Ill. 1887-1964 (formerly Eureka Body Co.)

Fred Groff Co., Lanchester, Pa.
F.S. Wood Co., New York, (built electric ambulance, 1900)
Fitz Gibbon & Crisp, Trenton, N.J. (built funeral omnibusses)
Flxible Co., The, Loudonville, O. (built hearses, ambulances, 1925-1952, 1959-65)

Gardner Motor Car Co., St. Louis, Mo. (St. Louis Coffin Co.)
G.A. Schnable & Sons, Pittsburgh, Pa. 1914-
Gate City Body Co., Gate City, N.C.
G. Dessecker Co., New York, N.Y.
General Vehicle Co., New York, N.Y.
Glesenkamp & Sons Co., Pittsburg, Pennsylvania, (built funeral omni 1912)
Godeau & Co., San Francisco, Calif. 1886-
Grabowsky Motor Truck Co., Detroit, Mich.
Great Eagle, Columbus, Ohio (U.S. Carriage Co.)
Guy Barnette & Co., Inc., Memphis, Tenn. 1949-1955 (Chevrolet & Ponti.

Harley-Davidson Co., (built some motorcycle ambulances)
Harry W. Smith Co., Minneapolis, Minn.
Hayes-Diefenderfer Co., New York, N.Y.
H.H. Babcock Co., Watertown, N.Y.
Henney Motor Co., Freeport, Ill. 1868-1954 (former John W. Henney Co
*Hess & Eisenhardt Co., Cincinnati, Ohio (successor to Sayers & Scovill, S & in 1942)
Holcker Manufacturing Co., Kansas City, Mo.
Hornthal Co., New York, N.Y.
Horton Co., Columbus, Ohio.
Hoover Body Co., York, Pa. (also known as York-Hoover)
Houghton Motor Car Co., Marion, Ohio

Ingersoll Body Corp., Ingersoll, Ont. Canada (Chevrolet, Pontiac 1948-1950
Isenhoff Auto Rebuilding Co., Grand Rapids, Mich.

James Cunningham, Son & Co., Rochester, N.Y. 1838-1936
J.C. Brill Co. (railway equipment, built funeral omnibusses)
J.I. Case TM Co., Racine, Wisc.
J.M. Kairwisch Wagon Works –
John J.C. Little Garage, Ingersoll, Ontario, Canada 1940-1958
John V. Morris, Wilkes-Barre, Pa, (Converto-Coupe flower car conversions)
J. Paul Bateman Co., Bridgeton, N.J.
Juckem Co.

Kelley Manufacturing Co., Bucyrus, Ohio
Keystone Vehicle Co., Columbus, Ohio
Kimble Bros., Galion, Ohio
Kissel Motor Car Co., Hortford, Wis. (KisselKar, National-Kissel Funeral Cars)
Knightstown Body Co., Knightstown, Indiana (Silver-Knightstown Funeral Cars)
Knox Automobile Co., Springfield, Mass.
Kunkel Carriage Works, Galion, Ohio, Est. 1875

Lagerquist Auto Co., Des Moines, Iowa
Leo Gillig Automobile Works, San Francisco, Calif.
Lippard-Stewart Co., Buffalo, N.Y.
L & M Co.–
Luverne Automobile Co., Luverne, Minn.

McCabe-Powers Carriage Co., St. Louis, Mo.
McClain Sales & Leasing, Inc., Anderson, Ind. (flower car conversions)
Memphis Coach Co., Menphis, Tenn.
Memphis Coffin Co., Memphis, Tenn.
Meteor Motor Car Co., Piqua, Ohio (also Mort, became Miller-Meteor, 1957)
Michigan Hearse & Carriage Co., Grand Rapids, Mich.
*Miller-Meteor Division, Wayne Corp., Piqua, Ohio 1956-
Millspaugh & Irish Body Co., Indianapolis, Ind.
Mitchell Hearse Co. Ltd., Ingersoll, Ont. Canada (formerly O.J. Mitchell Co.) 1860-1938

*Modular Ambulance Corp., Dallas, Texas
Montpelier Manufacturing Co., Montpelier, Ohio
Mort Motor Car Co., Piqua, Ohio (division of Meteor Motor Car Co.)
Motor Hearse Corp. of America Richmond, Ind. (Lorraine)

National Automobile & Electric Co., Indianapolis, Ind.
National Body Manufacturing Co., Knightstown, Ind.
National Casket Co., Boston, Mass. (marketed National-Kessel, Henney-Reo)
National Custom Coaches, Inc., Knightstown, Indiana
National Hearse & Ambulance Co., Toledo, Ohio (Shop of Siebert)
Northwestern Casket Co., Los Angeles, Calif.
New York & Brooklyn Casket Co., New York, N.Y.

O.J. Mitchell Co., Ingersoll, Ontario, Canada (preceded Mitchell Hearse Co.)
Oltman-O'Niell Co., Centerline, Mich.
Owen Bros, Lima, Ohio, Est. 1867-1935

Paul DeMers Conversions, Beloeil, Que. Canada
Peerless Motor Car Co., Cleveland, Ohio
Peter Kief Co., New York, N.Y. (successor to Diefenderfer)
Pfeiffer Auto & Carriage Works, Omaha, Nebraska
Pilot Motor Car Co., Richmond, Ind. (Lorraine)
Pinner Coach Co., Memphis, Tenn., Olive Branch, Miss.
Premier Motor Manufacturing Co., Indianapolis, Indiana, 1914
Proctor-Keefe Body Co., Detroit, Mich.
Progress Funeral Car Co., New York, N.Y.

Reo Atlanta Co., Atlanta, Ga.
Reilly Sales, Belmar, N.J. (flower car conversions, 1950-1952)
Richard Bros., Division of Allied Products Corp., Eaton Rapids, Mich.
Riddle Coach & Hearse Co., Ravenna, Ohio 1831-1926
Rhode Island Vehicle Co., Providence, R.I.
Robert Thompson Co., Los Angeles, Calif.
Rock Falls Manufacturing Co., Rock Falls, & Sterling, Illinois, 1877-1925

Sayers & Scovill Co., Cincinnati, O. (now Hess & Eisenhardt Co., S & S) est. 1876
S.C. Pease & Sons, Merrimas, Mass.
Sievers & Erdman Co., Detroit
Seaman Body Co., Milwaukee, Wis.
Shop of Siebert Associates, Waterville & Toledo O. (Nat'l. Hearse & Amb. Co.)
Smith Motor Body Works, Toronto, Ontario, Canada
Sowney Bros., Philadelphia, Pa.
St. Catharines Auto Bodies Ltd., St. Catharines, Ontario, Canada (conversions)
St. Louis Casket Co., St. Louis, Mo. (Gardner funeral cars)
Staver Carriage Co., Chicago, Ill.
Stratton-Bliss Co., New York, N.Y.
Streator Motor Hearse Co., Streator, Ill.
Studebaker Corporation of America, South Bend, Indiana
*Superior Coach Corp., Lima, Ohio 1923-Also plant at Kosciudko, Miss.

Trinity Coach Co., Duncanville, Texas, 1965-1968 Buick chassis

U.S. Carriage Co., Columbus, Ohio
Universal Coach Corp., Detroit, Mich.

Velie Vehicle Co., Moline, Ill.

Weller Bros., Inc. Memphis, Tenn.
Welles Corp., Windsor, Ontario, Canada
White Motor Car Co., Cleveland, Ohio
William A. Carroll Co., New Bedford, Mass.
Williams Carriage, Hearse & Auto Co.
William Erby & Sons Co., Chicago, Ill., Est. 1866
William Wagon Works, Macon, Ga.
Willman Lumber Co., Hartford City, Indiana

About the author

Thomas A. McPherson has had a fascination for motorized funeral vehicles and ambulances since the age of 10. This fascination has manifest itself to the point where representatives of this industry now refer to Mr. McPherson as "the official automotive historian for the funeral profession."

This rather unique interest started innocently enough when young Tom received a large stack of funeral business magazines from a funeral home that was cleaning house. Already endowed with a distinct interest for anything automotive, Tom began clipping the hearse ads from the magazines. By the time he had worked through the entire stack, he was hooked on the subject.

From that point on, Mr. McPherson became a confirmed collector of material on these interesting but virtually unknown vehicles. At this point he is credited with having the largest collection of hearse and ambulance material, photographs, literature and advertising to exist anywhere in the Western Hemisphere, and possibly in the world. The contents of this book, though covering in depth the major items and events in the history of funeral vehicles, barely scratches the surface of Mr. McPherson's collection.

On the professional side, Mr. McPherson is a full-time automotive journalist, and not surprisingly, also has written a multitude of articles on historic and current funeral vehicles for the major funeral trade journals in both North America and England. In addition, he is an automotive writer for various newspapers in Canada, and writes, directs, and produces his own monthly automotive program on TV in which he tests, evaluates and comments on various makes and models of cars.

The planning for AMERICAN FUNERAL CARS & AMBULANCES SINCE 1900 began several years ago when he and Walter McCall, another hearse buff, decided to compile the scattered history of the funeral car business into one concise reference. At first the idea was to just consolidate the material for personal reference, but over the course of time it appeared that a worthwhile history book could be derived from these efforts. At this point, Crestline Publishing Co. was contacted, and after many discussions, it appeared that the material had a definite place in Crestline's growing automotive series.

Mr. McPherson lives in Kitchener, Ontario, with his wife Cynthia. Currently he is the proud owner of a 1971 Citroen DS 21 Pallas and a 1970 Envoy Epic GT.

Among his interests he enjoys photography, history and travel — all of which will stand him in good stead as he continues to work on other books for the growing Crestline Automotive Series.